T0310118

A Practical Introduction to Human-in-the-Loop Cyber-Physical Systems

A Practical Introduction to Human-in-the-Loop Cyber-Physical Systems

David Nunes
University of Coimbra

Jorge Sá Silva
University of Coimbra

Fernando Boavida
University of Coimbra

This edition first published 2018

Registered Office(s)
John Wiley & Sons, Inc., 111 River Street, Hoboken, NJ 07030, USA
John Wiley & Sons Ltd, The Atrium, Southern Gate, Chichester, West Sussex, PO19 8SQ, UK

Editorial Office
The Atrium, Southern Gate, Chichester, West Sussex, PO19 8SQ, UK

For details of our global editorial offices, customer services, and more information about Wiley products visit us at www.wiley.com.

Wiley also publishes its books in a variety of electronic formats and by print-on-demand. Some content that appears in standard print versions of this book may not be available in other formats.

Library of Congress Cataloging-in-Publication Data

Names: Nunes, David, 1987- author. | Silva, Jorge Sá, author. | Boavida, Fernando, 1959- author.
Title: A practical introduction to human-in-the-loop cyber-physical systems / David Nunes, Jorge Sá Silva, Fernando Boavida.
Description: First edition. | Hoboken, NJ : John Wiley & Sons, 2018. | Includes bibliographical references and index. |
Identifiers: LCCN 2017025006 (print) | LCCN 2017042126 (ebook) | ISBN 9781119377801 (pdf) | ISBN 9781119377788 (epub) | ISBN 9781119377771 (cloth)
Subjects: LCSH: Cooperating objects (Computer systems) | Human-computer interaction.
Classification: LCC TJ213 (ebook) | LCC TJ213 .N86 2017 (print) | DDC 621.39–dc23
LC record available at https://lccn.loc.gov/2017025006

Cover Design: Wiley
Cover Image: © ipopba/Gettyimages

Set in 10/12pt Warnock by SPi Global, Chennai, India
Printed and bound in Malaysia by Vivar Printing Sdn Bhd

10 9 8 7 6 5 4 3 2 1

To my parents, Jorge and Eulália, and to my brother, Telmo.

David Nunes

To Fátima, Catarina, Pedro, Jojó, and my parents

Jorge Sá Silva

To Maria João and our three daughters – Susana, Inês, and Catarina

Fernando Boavida

Contents

List of Figures

List of Tables

Foreword

Our world keeps being an increasingly technological one. As first put forward by the renowned computer scientist Mark Weiser, we continue to see that, as devices get smaller in size, more mobile, powerful, and efficient, they begin to "disappear". Technology is now so intrinsic to our everyday lives that it has become an inherent part of our existence. This is the premise behind concepts such as the Internet of things and cyber-physical systems, in which distributed technology is used to monitor and control the environment. However, our current technological advancement still falls short of Weiser's ideas. Each time we have to hurdle through unintuitive configuration menus, errors, and software incompatibilities we become stressed by our computers and appliances. Weiser argued that the ultimate form of computers was an extension of our subconscious. To him, the ideal computer would be capable of truly understanding people's unconscious actions and desires. Instead of humans adapting to technology and learning how to use it, it would be technology that would adapt to the disposition and uniqueness of each human being.

In fact, systems that consider the human context are becoming increasingly more important, and there are strong indications that most future technologies will most likely be much more human-aware. This book focuses on the realm of human-in-the-loop cyber-physical systems (HiTLCPSs), that is cyber-physical systems that take human response into consideration. HiTLCPSs infer the user's, intents, psychological states, emotions, and actions through sensors, using this information to determine the system's actions. This involves using a large variety of sensors and mobile devices to monitor and evaluate human nature. Therefore, this technology has strong ties with wireless sensor networks, robotics, machine learning, and the Internet of things.

This book is useful to BSc and MSc students, as well as to PhD students, researchers, and professors addressing the areas of ubiquitous computing, Internet of things, cyber-physical systems, and human–computer interaction. It can also be useful to professional developers that intend to introduce HiTL concepts into their mobile apps and/or Internet of things/cyber-physical system applications.

Throughout its pages, the book will guide the reader through a journey into this novel and exciting area of research and technological development. As such, it is intended to be used as a primer on HiTLCPSs, providing some insights into the research being done on this topic, current challenges, and requirements. One of the book's objectives is to introduce the reader to the practical usage of HiTL paradigms within software development. Therefore, we included a comprehensive hands-on tutorial where the major theoretical concepts behind HiTLCPSs are applied to a sample mobile application and explained from a practical perspective. This tutorial requires some knowledge of Android and the Java programming language, as well as some notions about databases and RESTful web services. It is accompanied by a base source code repository and several code snippets which the reader can extensively modify.[1] It is not our intention to provide in-depth knowledge about the programming languages, and/or the machine learning techniques, necessary to create complex HiTL systems. Instead, the tutorial aims at illustrating and consolidating some of the book's theoretical ideas.

Finally, we would like to thank you, the reader, for your interest. We would also like to ask you to contact us and tell us about your experience with our book. Your feedback is a very valuable resource towards improving the book. Send your email to dsnunes@dei.uc.pt, sasilva@dei.uc.pt or boavida@uc.pt.

1 The source code repositories are located at: https://git.dei.uc.pt/dsnunes/happywalk.git https://git.dei.uc.pt/dsnunes/happywalkserver.git

Preface

The Internet has changed our whole life and it will have further impact on how we live and how we work. Most of the **cyber-physical systems** (CPSs) make use of the Internet and even define parts of it. Let me cite Wikipedia in this preface, even though it is not very scientific so to do. Understanding the CPS as *"a mechanism controlled or monitored by computer-based algorithms, tightly integrated with the internet and its users"* means that users, humans, are essential for any CPS. The National Institute of Standards and Technology of the US Department of Commerce (NIST) goes even further, stating that *"these systems will provide the foundation of our critical infrastructure, form the basis of emerging and future smart services, and improve our quality of life in many areas"*. Looking at the examples mentioned in Wikipedia, *"smart grid, autonomous automobile systems, medical monitoring, process control systems, robotics systems, and automatic pilot avionics"*, human are always involved.

Humans are not only involved; humans are the essential part of CPSs; CPSs have to serve us! With the basic idea, to incorporate humans as being in the system, we encounter **human-in-the-loop** (HiTL). It comprises a model, an adequate representation of the human behavior in order to treat it as an integral part of the whole system. Just as one example, let me cite Carsten Binning *et.al.* at his preface of the Proceedings of the first Workshop on Human-In-the-Loop Data Analytics HILDA of June 26th, 2016, in San Francisco, California: *"A major bottleneck in data analytics today is to efficiently leverage the human capabilities to formulate questions and understand answers of data analytics systems … Recent technology trends (such as touchscreens, motion detection, and voice recognition) are widening the possibilities for users to interact with data, and data-driven industries are shifting to personalized processing to better target their services to users' needs"*.

Hence it seems somewhat natural to look at both topics together in a kind of textbook and survey. In my six years as editor-in-chief of the journal *ACM Transactions on Multimedia Computing, Communications, and Applications (ACM TOMM),* I have, unfortunately, not come across a comprehensive high-quality survey paper of CPS HiTL; it has been even more serious: nobody even tried to cover with a survey this essential area on multimedia computing, communications, and its applications. No one did so far!

At the present time, writing this preface, I was only able to read parts of this book; I am looking forward to reading it all together–the whole book.

The authors of this book, David Nunes, Jorge Sá Silva, and Fernando Boavida from the University of Coimbra provide an in-depth view to HiTLCPS evolution, theory, technologies, and applications. Moreover, they illustrate how to apply HiTLCPS concepts to a sample smartphone application, through a hands-on approach that guides the reader from the development environment to the final product, including data acquisition, state inference, and actuation. With (1) their profound technical knowledge of many areas in computing and communications, as well as with (2) their expertise and experience as authors of other textbooks, the authors are certainly key for this book being a long-term successful scientific book in this area. Congratulations!

DR. RALF STEINMETZ
Fellow of the IEEE and Fellow of the ACM
Director, Multimedia Communications Laboratory, Technische Universität Darmstadt
Chairman of the Board, Hessian Telemedia Technology CompetenceCenter, Germany

Darmstadt, March 2017

Acknowledgments

A book such as this would not have been possible without the help and support of many people and institutions.

First of all, we would like to thank our base institutions—the Department of Informatics Engineering, and the Center for Informatics and Systems, both from the University of Coimbra—in the scope of which we carry out our teaching and research activities, for the provided facilities and research environment. With their effort and contributions, enthusiasm, discussions, and suggestions during several years of joint research activities and human-in-the-loop social interaction, our students and our colleagues were instrumental in making this book a reality.

We also thank IMDEA Networks Institute, in Madrid, for the support provided during Fernando Boavida's sabbatical in 2015/2016, and especially to its leading computer scientist, Arturo Azcorra, for his support; to Antonio Fernández Anta, Miguel Péon, Jeanet Birkkjaer; and Rosa Gómez for their encouragement; and to all its researchers and staff in general.

Some of the research that formed the basis for this book was carried out in the scope of financed research projects and initiatives and, thus, it is also right to thank the entities that made the referred research possible, namely the Portuguese Foundation for Science and Technology (FCT), FCT's POPH/FSE program, and the SOCIALITE Project (PTDC/EEI-SCR/2072/2014), supported by COMPETE 2020, Portugal 2020, Operational Program for Competitiveness and Internationalization (POCI), and the European Union's ERDF (European Regional Development Fund).

We would also like to thank David Hutchison, from Lancaster University, for believing in us and putting us in contact with the excellent editorial team at John Wiley & Sons.

Finally, we would like to thank our families, for their unconditional love and support.

List of Abbreviations

AI	Artificial Intelligence
ANN	Artificial Neural Network
API	Application Programming Interface
AS	Android Studio
AV	Autonomous Vehicle
BCC	Body-Coupled Communication
BCI	Behavior Change Interventions
CHIL	Computers in the Human Interaction Loop
CoAP	Constrained Application Protocol
cOre	Constrained RESTful environments
CPS(s)	Cyber-Physical System(s)
CPU	Central Processing Unit
DAO	Data Access Object
ECG	Electrocardiography
EEG	Electroencephalography
ESM	Experience Sampling Method
FCT	Fast Cosine Transform
FFT	Fast Fourier Transformation
GPRS	General Packet Radio Service
GPS	Global Positioning System
GSM	Global System for Mobile Communications
HiTL	Human-in-the-Loop
HiTLCPS(s)	Human-in-the-Loop Cyber-Physical System(s)
HTML	HyperText Markup Language
HTTP	Hypertext Transfer Protocol
HVAC	Heating, Ventilation, and Cooling
ID	Identification
IFR	International Federation of Robotics
IoA	Internet of All
IoT	Internet of Things
IP	Internet Protocol
IDE	Integrated Development Environment
IEEE	Institute of Electrical and Electronics Engineers
IETF	Internet Engineering Task Force
ISM band	Industrial, Scientific, and Medical radio bands

Java EE	Java Enterprise Edition
Java SE	Java Standard Edition
JDK	Java Development Kit
JSON	JavaScript Object Notation
LTE	Long-Term Evolution
M2M	Machine-to-Machine
MPTCP	MultiPath Transmission Control Protocol
NAT	Network Address Translation
NSF	National Science Foundation
OSI	Open Systems Interconnection
OS	Operating System
P2P	Peer-to-Peer
POI(s)	Point(s) of Interest
RAM	Random-Access Memory
REST	Representational state transfer
RF	Radio Frequency
RFID	Radio-Frequency Identification
RSSI	Received Signal Strength Indication
SCTP	Stream Control Transmission Protocol
SDK	Software Development Kit
sMAP	Simple Monitoring and Action Profile
SMS	Short Message Service
SOAP	Simple Object Access Protocol
SQL	Structured Query Language
TCP	Transmission Control Protocol
UDP	User Datagram Protocol
URI	Uniform Resource Identifier
URL	Uniform Resource Locator
UUID	Universally Unique Identifier
VoIP	Voice Over Internet Protocol
WSDL	Web Service Description Language
WSN(s)	Wireless Sensor Network(s)
XML	Extensible Markup Language

About the Companion Website

Don't forget to visit the companion website for this book:

www.wiley.com/go/nunesloop

There you will find valuable material designed to enhance your learning, including:

- Source codes

Scan this QR code to visit the companion website.

1

Introduction

Humans are a remarkable species. For most of our history, we have used our intellectual ability to create and develop many different tools and processes to assist us and ease our lives. Since the days our ancestors discovered how to control fire, around 300,000 years ago, we have achieved an exponential technological progress. From the invention of wheeled vehicles, around 6,000 years ago, to the transistor, invented just 70 years ago, many were the technological advances that have drastically changed the way we experience and perceive our reality.

The last few decades have seen an unprecedented surge of technological advancement, particularly in the area of computer science, resulting in some of the most revolutionary human inventions yet: we have developed personal desktop and portable computers, as well as a global network that interconnects all kinds of computerized devices, aptly called the Internet. Despite the fact that they have been in existence for an extremely short time, these technologies have transformed, and will continue to transform, the way our world and society work, at a very fundamental level and at an incredibly fast pace.

1.1 The Rise of Cyber-Physical Systems

Interestingly, once the Internet was in place, we quickly achieved the power to extend it to our traditional tools and appliances, which then became "interconnected". One of the first "tools" ever connected to the Internet was the Carnegie Mellon University Computer Science Department's Coke Machine, in the early 1980s [19], which was able to report its stock and label it as "cold" or not, depending on how much time it had been inside the machine. An idea began to spread: a vision of an interconnected world where information on most everyday objects was accessible.

Since then, scientists and engineers have developed this idea into a concept that is known as the "Internet of Things" (IoT). The idea started small, considering scenarios where radio-frequency identification allowed the "tagging" and managing of objects by computers. Each object would carry a radio-frequency identification (RFID) tag, a small, traceable chip which could be wirelessly scanned by a nearby RFID reader. The RFID tag enabled the automatic identification of the object and allowed it to be traced/managed through the Internet.

The continued advances in miniaturization allowed us to go beyond the simple tagging and identification of everyday objects. As predicted by Gordon Moore, back in

A Practical Introduction to Human-in-the-Loop Cyber-Physical Systems, First Edition.
David Nunes, Jorge Sá Silva and Fernando Boavida.

Companion Website URL: www.wiley.com/go/nunesloop

1965, the amount of computing power in integrated circuits has been doubling every 18 months for the last 50 years [20]. The remarkable work of computer industry engineers and scientists has led to many new technologies. The continuous integration of computational resources into all kinds of objects made our tools "intelligent". Everything from light bulbs to refrigerators, microwaves, and coffee machines will soon be connected to the Internet. In fact, some studies estimate that we will have an IoT with 26 billion connected devices by 2020 [21].

We can see evidence of this trend all around us. The Internet now interconnects a large number of highly heterogeneous devices, from traditional desktop PCs to laptops, tablets, and smartphones.

For example, the area of sensing technologies and wireless sensor networks (WSNs) is becoming increasingly prominent. WSNs are composed of dozens or even hundreds of autonomous "sensor nodes", small computerized devices that are capable of collecting physical world data and forwarding it by means of wireless communication. They can be used to monitor environmental luminosity, temperature, pressure, sound, and many other parameters, and can be spatially distributed in an ad hoc fashion. These technologies have been receiving a great deal of attention from the research community due to their potential in almost every application area. In fact, WSN deployments can now be found in many industrial, medical, and domestic environments. Recent studies in WSNs have brought great advancements in this area, namely in terms of energy efficiency and integration capabilities, with sensors being provided as services [22, 23], accessible through the Internet [24]. Sensors are now indispensable devices, for they allow us to collect data from real-world phenomena, handle this data in digital form, and ultimately extend the Internet to the physical world.

In fact, the number of sensors that nowadays can be deployed on humans can turn them into walking sensor networks. Humans can use smart-shirts; carry a smartphone with several sensors and networking capabilities (e.g. global system for mobile communications (GSM), *Bluetooth*, long-term evolution (LTE)); and use Google glasses, iPods, smart watches, and shoes with sensors. In terms of sensing applied to individual users, Bosch Sensory Swarms and the Qualcomm Swarm Lab at UC Berkeley estimate that the number of sensors in personal devices can add up to 1000 wireless sensors per person, to be deployed over the next 10 to 15 years [25], resulting in large amounts of data being available for processing, and allowing a wide range of sensing applications to be deployed. This reality depends, of course, on the drastic reduction of sensor production costs, which are expected to come down to negligible values over time, as with most silicon-based hardware [26].

As for automated actuation, the world has seen a gradual increase in the number of installed robots per year. The 2015 World Robot Statistics study, issued by the International Federation of Robotics (IFR) [27], indicates that the total number of professional service robots sold in 2014 rose by 11.5% compared to 2013, from 21,712 to 24,207 units. IFR expects that, for the 2015–2018 period, sales of service robots for professional use will increase to about 152,375 units, while sales of robots for personal use will reach about 35 million units, with a total estimated value of about $40 billion. Global sales of industrial robots, on the other hand, will experience a yearly growth of 15% until 2018, while the number of sold units will double to around 400,000.

Interwoven with the concept of IoT is the concept of cyber-physical systems (CPSs), which consist in the sensing and control of physical phenomena through networks of devices that work together to achieve common goals. These CPSs represent a confluence of robotics, wireless sensor networks, mobile computing, and the IoT, to achieve highly monitored, easily controlled, and adaptable environments.

The IoT and CPS concepts have been pushed by two distinct communities. IoT was initially developed using a computer science perspective, mostly supported by the European Commission. The goal was to develop a network of smart objects with self-configuration capabilities on top of the current Internet. This development effort included hardware, software, standards, and interoperable communication protocols and languages that describe these intelligent devices [28]. IoT builds on several requirements, namely the development of intelligence in devices, interfaces and services; the assurance of security and privacy; systems integration; communication interoperability; and data "semantization" and management [29].

On the other hand, the concept of CPSs was initially supported by the US National Science Foundation (NSF). CPSs stem from an engineering perspective and concern the control and monitoring of physical environments and phenomena through sensing and actuation systems consisting of several distributed computing devices [30]. These systems are mostly interdisciplinary, requiring expertise and skills in mathematical abstractions (algorithms, processes) that model physical phenomena, smart devices and services, effective actuation, security and privacy, systems integration, communication, and data processing [31].

Thus, IoT tended to focus more on openness and the networking of intelligent devices, while CPSs were more concerned with applicability, modeling of physical processes, and problem solving, often through closed-looped systems. While their core philosophy and focus were initially different, their many similarities, such as intensive information processing, comprehensive intelligent services, and efficient interconnection and data exchange, have led to both terms being used interchangeably [32] without clearly identified borders [30].

CPSs combine elements from robotics, wireless sensor networks, and mobile computing, among others, to achieve specific goals. From industrial applications that monitor and actuate on several industrial processes, to social applications that aggregate data from various users in order to achieve goals, such as reducing pollution and traffic in metropolitan areas, CPSs can encompass a multitude of domains. For example, improvement of personal health can be achieved through body networks that integrate the user's vital signs and activity levels with environmental information on pollutants or noise to suggest healthier and more pleasant walking routes, restaurants, and leisure activities. CPSs can also be used in transportation, as many modern vehicles feature cruise control systems that maintain the automobile's speed or perform parking maneuvers, not to mention autonomous driving. All these systems combine sensors, actuators, and the computational capabilities of the devices to achieve the desired results. In fact, these sensors and actuators can be used not only in individual objects but also in structures and buildings in order to monitor, for example, their structural health.

While IoT and CPS technologies do exist, current systems are still designed with a specific scientific, industrial, or engineering application in mind. They are, for example, typically responsible for collecting data from sensors and analyzing it for a certain task. This objective-driven approach results in academic or industrial systems that may be effective for their targeted scenarios but are very constrained in applicability and, therefore, narrow in their usability.

Nevertheless, we know from previous experience that this happens during the beginning of any new paradigm-changing technology, as was the case with most information technologies. The most striking example is the Internet itself, which initially only connected the University of California at Los Angeles, the Augmentation Research Center at Stanford Research Institute, the University of California at Santa Barbara, and the University of Utah's Computer Science Department. This and other initial computer networks continued to grow and merge, giving birth to the Internet as we know it today.

Much like the Internet, it is very likely that existing disconnected and restricted CPS deployments, whose primary beneficiaries are privileged users who already benefit from and explore their capabilities, are just the initial steps towards a future where the vast majority of intelligent devices are interconnected in massive, non-centralized networks. In fact, some researchers argue that future CPSs will become ubiquitous and distributed, with many data streams overlaying the network, provided by large amounts of sensors. They also argue that these streams should be open for use, without centrally controlled authorization, through self-advertising and discovery by nearby users. Thus, data acquisition, processing, and visualization should be focused on users, not administrators or scientists [33].

1.2 Humans as Elements of Cyber-Physical Systems

The reduction of production costs of silicon-based hardware [34] continues to fuel this increasingly pervasive technological world, endowing people with the ability to access extremely rich and dynamic pools of data pertaining to their surrounding environment. The epitome of these ideas was first put forward by the renowned computer scientist Mark Weiser, in his famous 1991 article "The computer for the twenty-first century" [35]. Weiser maintained that, as devices became smaller in size, more powerful, and efficient, they would eventually disappear. Technology would become so intrinsic to everyday life that we would no longer perceive it as an isolated concept but as an inherent part of our existence. This idea came to be known as "ubiquitous computing", and the concept of "calm" technology arose. This concept is a direct antitheses of the stressful use of technology, which is still prevalent. Each time we have to navigate menus, errors, bugs, or unintuitive setups, we become stressed by our computers and appliances. On the other hand, Weiser suggested that the true purpose of computers was to help us in a way similar to intuition. He propounded the view that the ideal computer would be something invisible that could truly understand human nature and interpret people's unconscious actions and desires. Instead of humans adapting to technology and learning how to use it, it would be technology that would adapt to the disposition of human beings.

Weiser also predicted that these "calm" interactions would be informative but not intrusive, not demanding the user's attention, and would make use of human intuitive clues. He was right, since we can see his vision materializing with every passing day. Computers no longer require people to sit in front of them; machines now enter the human environment embedded in all kinds of objects, integrating computing in the course of everyday human activities. In fact, current technology is quickly evolving towards these principles predicted 25 years ago: current mobile devices replaced traditional buttons with much more intuitive touchscreens, and software developers give an ever-increasing importance to usability and non-intrusiveness. The number of portable mobile devices has also grown exponentially, and the number of communication interfaces used by them has also grown. It is not hard to imagine a near future where we get up in the morning and our home also "wakes up" and automatically launches and executes many of the routines corresponding to that particular day of the week, under the control of several computing devices. In fact, as computation evolves, humans will most likely stop "using" computer devices, that is human–computer interaction will no longer require direct user attention and will become intuitive, as if it is second nature. In the words of Weiser, "The more you can do by intuition, the smarter you are; the computer should extend your unconscious."

It is not sufficient for interconnected and intelligent tools to communicate with each other without any human involvement. Human technology is made by humans, for humans. In order to promote the creation of systems that are useful to the average person, it is necessary to consider efficient and intuitive operation. Therefore, in addition to providing basic functionality, openness, heterogeneity, and integration capabilities, it is equally important to discern how systems or tools can be used within a certain context.

These ideas have been previously explored as *context awareness*, or *context adaptation*, for mobile and wireless networking [36] and IoT [37]. Actually, increasing context awareness is a cross-cutting challenge for the design of highly optimized networking systems that support distributed autonomic decision making and reconfigurable aspects [36]. However, current trends on context awareness research encompass a broad definition of context. "Context" can be defined as any information that can be used to characterize an *entity*, that is a person, place, or object [38]. Thus, several works in the area attempt to assess context [39] and use this information to achieve several goals, such as mobility management [40] or energy efficiency in ubiquitous environments [41]. There are also remarkable proposals for frameworks that manage and distribute this contextual data [42, 37].

Yet, outside of the area of e-health, whose primary objective is the monitoring of patients [43], there is still scarce knowledge on the actual effects of this human "context" on the CPS control loop. Indeed, one important element often left out of current CPS research is the human user [33]. Most current CPSs that involve control loops still keep humans as external elements to the control system. This is apparent if we think on the technology that we currently find around us. For example, aircraft pilots decide for themselves when to engage the autopilot or when to take manual control of the plane, and cruise control systems for automobiles simply maintain the desired speed without taking the driver's behavior into consideration.

Systems that consider the human context will become increasingly more important, and most future technologies will be human-aware. Future CPSs will most likely bolster a much stronger tie between humans and control loops. This involves using a large variety of sensors and mobile devices to monitor and evaluate human nature, making humans an integral part of the CPS.

We are now in the realm of human-in-the-loop cyber-physical systems (HiTLCPSs), that is cyber-physical systems that take human response into consideration. Human presence and behavior are no longer seen as external and unknown factors but have become a key part of the system instead. HiTLCPSs infer the user's intentions, psychological states, emotions, and actions through sensors, integrating this information into the control loop as feedback to determine the actions of the CPS. By considering humans an integral part of the system, the control loop's performance and accuracy can be vastly improved. For example, cruise control HiTLCPS systems will be able take the driver's psychological state into consideration (e.g. fatigue, attention-levels, etc.) in order to generate alarms and suggest the activation of cruise control [13]. In fact, previous research has already proposed image-based processing of facial expressions to detect irritation in drivers, and use this information to improve driving safety [44]. HiTLCPSs also include brain–computer interfaces, controlled assistive robots, intelligent prostheses, and monitoring systems, among others [16]. In order to improve the accuracy and timeliness of the system by considering the human element, it is essential to develop and integrate reliable and accurate human behavior modeling techniques that attempt to learn and predict human behavior. Since human behavior is quite hard to predict, making humans part of CPSs is a colossal challenge, as it requires modeling of complex behavioral, psychological, and physiological aspects of human nature. Nevertheless, a multitude of variables regarding a person's state may be measured, including movement, vital signs, and attention level, among many other things, which may be crucial to control the task at hand.

The maturing of HiTLCPS's design has yet to be achieved. For the most part, we have not reached a consensus or even a general understanding regarding the underlying requirements, principles, and theory. This drives us to ask questions, such as why do current IoT solutions still leave behind the human factor? Why have we yet to integrate the human component into CPSs? What challenges do we still face in order to achieve true HiTLCPS deployments? How can we take advantage of new ubiquitous sensing platforms such as smartphones and personal devices used massively by people throughout the day?

In the paper written by Stankovic *et al.* [45] three main challenges were identified for HiTLCPSs.

- First, there is a need for a comprehensive understanding of the spectrum of HiTL applications, which requires a study of existing, emerging, and potential solutions so that common underlying principles, requirements, and models may be found. As HiTLCPSs have a wide spectrum of applicability, it is necessary to amass examples of HiTL solutions from multiple domains before such an understanding may be achieved [13].
- Second, it is necessary to improve the techniques that derive models of human psychological states, emotions, physiological responses, and actions. In other words, we need reliable mechanisms for modeling, detecting, and possibly predicting human behavior, such as advanced mathematical models or machine learning techniques. Current state-of-the-art techniques are either very coarse and general or too application-specific. The development of dynamic human behavior models that are both accurate and general enough remains an enormous challenge.
- Finally, human behavior models need to be incorporated into the formal methodology of feedback control, either outside or inside the loop, within the system model or at any other hierarchical control level.

As a consequence of our research activity, we have come to believe that current research is close to providing answers to many of these obstacles, and that HiTL concepts will become increasingly more common. Despite being in its infancy, we have found promising research that indicates we may be reaching a tipping point in our technological evolution. More than having intelligent IoT and CPS systems that autonomously control our environment, these systems will, more importantly, adapt to the human context and needs. In fact, with HiTLCPSs we may be on the verge of achieving an unprecedented control over our environment, one that we could only conceive of in our wildest dreams.

1.3 Objectives and Structure

The next few years will most likely converge into not an Internet of things, but an "Internet of all" (IoA): an Internet that includes the emotions, psychological states, actions, and drives of the ordinary user, the human user, as part of larger-scale systems.

In this context, this book has two main objectives. First, it is intended to be a primer on HiTLCPSs, providing some insights into the research being done on this topic, current challenges, and requirements. As such, throughout the book we will lead the reader on a journey through this new and exciting area of research and technological development. The book's second objective is to initiate the reader in the development of HiTLCPS systems, using a hands-on approach. We will guide the reader through a comprehensive tutorial where the major theoretical concepts behind HiTLCPS are applied to a sample application and explained from a practical perspective. To cope with these objectives, this book is divided into three major parts.

Part I provides an overview of HiTLCPSs, encompassing their evolution, theory, technologies, and some applications. Chapter 2 presents the evolution of these systems, beginning with the scope of simple "things". From there, we evolve to whole environments and, finally, we consider the monitoring of human beings. HiTLCPSs have a wide spectrum of applicability. As such, in an attempt to cover as many solutions and domains as possible, Chapter 3 presents a general taxonomy of human roles within HiTLCPS. Concluding Part I, Chapter 4 presents several pieces of research work and several technologies that can be or have been used in HiTLCPS deployments. The purpose of this chapter is to provide the reader with an idea of current, real-world HiTL implementations.

Part II addresses our second objective, that is to provide a hands-on tutorial that will consolidate the previously presented theoretical concepts. To do so, we guide the reader through the creation of a simple, smartphone-based HiTLCPS. In Chapter 5, we present a sample Android application that we will endow with HiTL control. The objective is to begin with a bare-bones map application to a system capable of rough estimations of the user's current mood, providing suggestions to improve their physical and mental well-being. After explaining how to set up the necessary development environment in Chapter 6, we delve into actual HiTL development, from Chapters 7 through 9.

Part III discusses topics that will affect future and emerging HiTLCPS applications, providing the reader with pointers on several aspects that must be taken into consideration when implementing HiTLCPSs. As such, we first discuss existing requirements and challenges for HiTL applications in Chapter 10. Subsequently, we conclude our book in Chapter 11, with some remarks and conclusions on the covered material, identifying the main technical and ethical limitations that may be expected in future endeavors.

Part I

Evolution and Theory

In this first part of the book, we will introduce the concept of human-in-the-loop cyber-physical systems (HiTLCPSs), through an overview of the evolution of technology and research that has led us to this new paradigm. We will begin with the sensing of "things" and "places", converging on the more recent phenomenon of sensing of humans and their lives through smartphones in a social networking context. Afterwards, we will try to provide the reader with an overview of the possible types of HiTLCPSs and human roles in them, through a taxonomic analysis. At the end of this part, the reader will have a sense of the extent and challenges of bringing technology closer to understanding humans.

A Practical Introduction to Human-in-the-Loop Cyber-Physical Systems, First Edition.
David Nunes, Jorge Sá Silva and Fernando Boavida.

Companion Website URL: www.wiley.com/go/nunesloop

2

Evolution of HiTL Technologies

As we discussed in the introductory chapter of this book, the technological progress of the last few decades has been particularly remarkable. Still, how did all of this come to be? First, in Section 2.1, we will see how researchers began to address and understand the scope of simple "things", and we will subsequently consider work that targeted more complex and larger environments. Finally, in Section 2.2, we will discuss more recent examples of monitoring of human beings.

2.1 "Things", Sensors, and the Real World

The linkage of physical objects and sensors to the Internet and their integration with web and enterprise applications has long been considered in the literature. As previously mentioned, early works began by proposing the use of physical tokens (such as barcodes or electronic tags) to relate objects to the web. An example of such an approach was described in [1], as Figure 2.1 illustrates. Since these early times, researchers were concerned with the heterogeneity and complexity of the available network protocols. In the world of pervasive computing, "client devices" like PDAs were used to access services on "server devices" like printers, light switches, or smart home appliances. Since these regular server devices have limited computation, memory, and power capabilities, they require extremely small and inexpensive servers that consist of low-performance micro controllers with only a few kilobytes of memory. However, they are expected to be able to support complex ad hoc networking protocols, and bê capable of handling computationally heavy communication tasks like parsing and generating XML messages. In an attempt to solve this problem, Shaman [2], a Java-based service gateway, was proposed. As shown in Figure 2.2, it worked as a network proxy that supported various standards for ad hoc networking, allowing for the integration of small, limited and low-power server devices (known as *Lite*Servers) into heterogeneous networking communities.

Another interesting advancement in this area was achieved by the Cooltown project [3], which provided an infrastructure for "nomadic computing", a term used by the authors to describe the human interaction with mobile and ubiquitous devices. In the Cooltown project, the authors intended to push web technology into common digital appliances such as printers, radios, and automobiles, and also to non-electronic things like CDs, books, and/or paintings, connecting each "thing" to a web "presence"

A Practical Introduction to Human-in-the-Loop Cyber-Physical Systems, First Edition.
David Nunes, Jorge Sá Silva and Fernando Boavida.

Companion Website URL: www.wiley.com/go/nunesloop

Figure 2.1 In [1], books and other common objects were augmented with RFID tags and associated with virtual documents by PDAs.

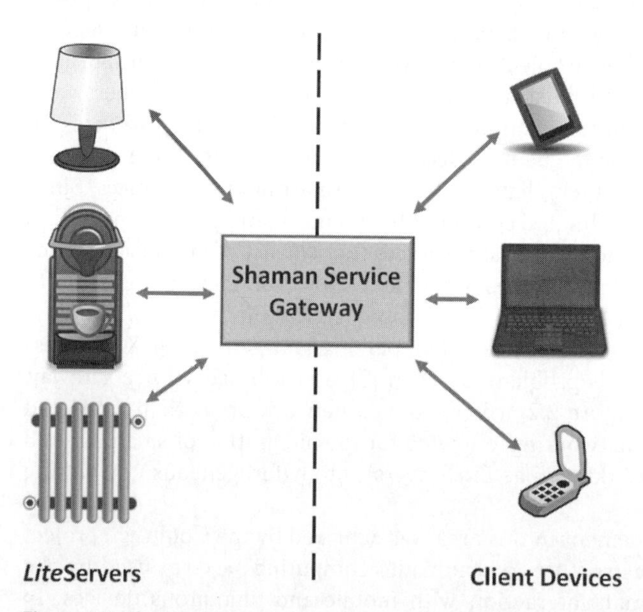

Figure 2.2 Shaman [2] acted as a representative for the connected *Lite*Servers, offering Java and HTML interfaces.

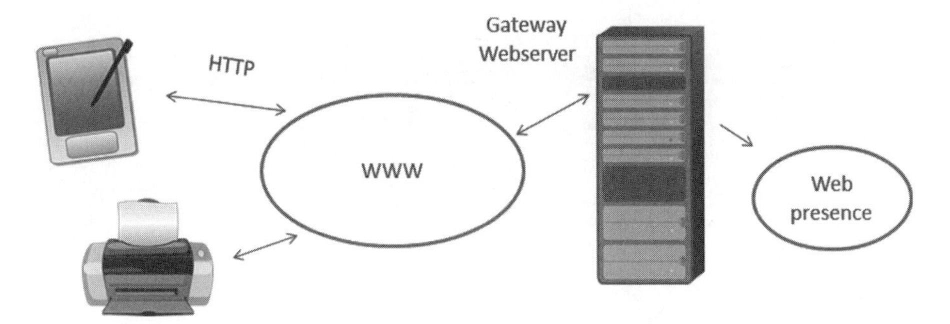

Figure 2.3 Device web presence in Cooltown [3]. *Source:* Adapted from Kindberg *et al.* 2002.

(see Figure 2.3). The web presence extended the concept of a web page to every physical entity, basically a page with information and services for every entity of the physical world. The authors designed the "Cooltown Museum" test environment, where they implemented two different methods of web presence recognition. One method consisted of the use of infrared beacons that supplied PDAs with the URL of the corresponding point of web presence, where the other consisted of the use of tag identifiers which were sensed by the PDAs and sent to a service that maintained a collection of bindings from identifiers to URLs and which then returned the corresponding URL.

The success of Web 2.0 and the advent of web services and associated technologies like SOAP (simple object access protocol) and WSDL (web service definition language) brought brand-new approaches to the integration of devices and sensor networks. In this vein, the "web of things" vision is concerned with providing a concrete architecture where actuators and embedded devices expose their data and functionality as an integral part of the web.

The open-source Project JXTA tried to specify a standard set of protocols for ad hoc, pervasive, peer-to-peer (P2) computing as a foundation for the web of things. It standardized a thin and generic network protocol layer to build a wide variety of P2P applications, where each peer benefits from being connected to a high number of other peers, and where information is shared and maintained among the community of embedded devices. The network protocol itself is independent from software and hardware platforms, and defines a virtual network on top of the existing physical network infrastructure in order to hide all the complexity of the underlying physical protocols (see Figure 2.4). Thus, as long as the JXTA protocols are implemented in a given platform, all peers in the network would be able to communicate with the members of that platform. The members of the JXTA initiative have already ported the protocol to a few different platforms; namely, the JXTA-C project delivers a small and efficient implementation of the JXTA protocols that can be used for small memory embedded devices without requiring any proxy servers [4].

Another approach was Tiny Web Services [46], a system that deployed web services technology on resource-constrained sensor nodes, allowing their functionality and data to be directly accessed by multiple applications. Data could be carried through SOAP-formatted packets, while web service bindings were defined through XML and the WSDL.

However, some members of the scientific community felt that these types of approach were not ideal, arguing that both the interface definition (WSDL) and the messages

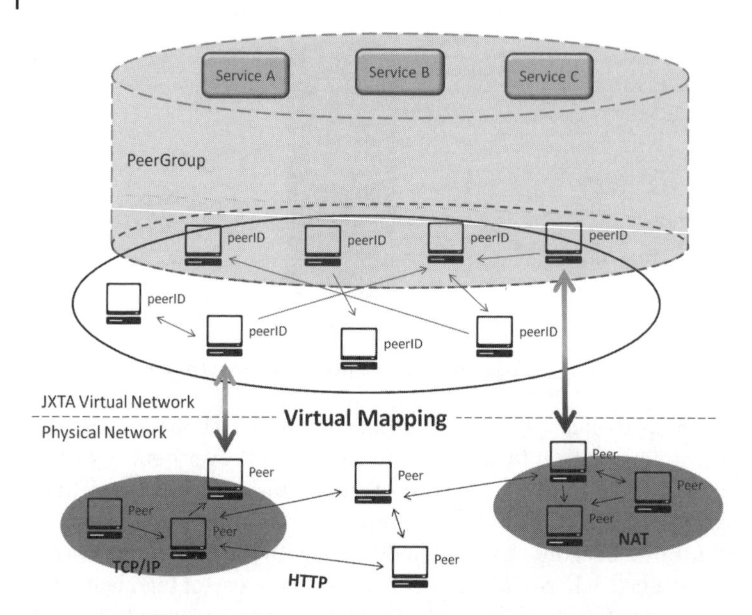

Figure 2.4 JXTA [4] peers created virtual ad hoc networks which served to abstract the real ones.

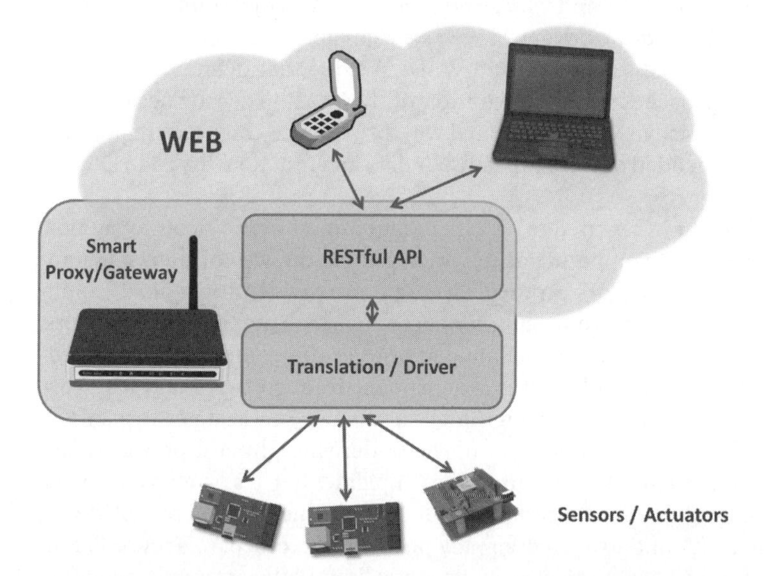

Figure 2.5 Works such as [5] and [6] used proxies to offer embedded devices' capabilities through RESTful web services.

(SOAP) were too complex for devices with limited capabilities and also that the overall system was not truly loosely coupled [47]. Instead of relying on proprietary and tightly coupled systems, REST principles were then applied to embedded devices, allowing the use of web languages like HTML, JavaScript, and PHP to create novel web applications (see Figure 2.5). With this approach, the interaction with a sensor node becomes as easy as typing a URI into a Web browser, and it allows for traditional web mechanisms to be applied to embedded devices such as browsing, searching, and bookmarking [5].

Several other projects focused on REST protocol principles for sensor integration with the web. One of these projects was the sMAP project [6], which presented an architecture, specifications, and implementations of a simple monitoring and action profile (sMAP) that promoted data interoperability between sensor and actuators in building environments and the Internet. The sMAP architecture allowed clients to communicate with embedded devices in buildings through the Internet. In order to support resource-constrained devices, this communication was dependent on several proxies that compressed and decompressed data between IP end points. The proposed architecture was built on HTTP/REST protocols and used JSON as the object interchange format. The authors applied the architecture to several resource monitors and actuators inside a commercial building, including mote-based wireless sensors. Overall, the authors believed that the approach based on REST was widely implementable and efficient, while the communication API definitions were expressive and concise. They also believed that the use of proxies is ideally suited for resource-constrained embedded devices.

More recently, the the Internet Engineering Task Force (IETF) Constrained RESTful environments (CoRE) Working Group has standardized the Constrained Application Protocol (CoAP). CoAP is a web transfer protocol for use with constrained nodes with very few resources. The protocol is designed for machine-to-machine (M2M) applications such as smart energy and building automation. According to its standards track, "CoAP provides a request/response interaction model between application endpoints, supports built-in discovery of services and resources, and includes key concepts of the Web, such as URIs and Internet media types. CoAP is designed to easily interface with HTTP for integration with the Web, while meeting specialized requirements such as multicast support, very low overhead, and simplicity for constrained environments" [48].

These advances in the integration of sensing led to the use of IoT and wireless sensor nodes, soon evolving beyond "things", and also became a doorway between virtual environments and real-world information. This sensing data has tremendous potential, particularly when we consider the power of crowdsourcing. This is evidenced through the number of organizations that freely open and share data with their users. MTA [49], for example, provides open-source transit-related data for the development of applications. Also, the OpenDataBCN [50] open data portal shares the city of Barcelona's data regarding geography, demography, economy, city services and utilities, and administration. The cities of Toronto [51], Edmonton [52], Ottawa [53], and Vancouver [54] have also joined forces to collaborate on an "open data framework" initiative that openly offers city-related data sets to users.

Several initiatives are dedicated to applying all of this sensory information from real-world locations to virtual representations of those locations.

One example is SenseWeb [7], a scalable infrastructure for sharing sensing resources among sensor owners and exploring geocentric sensor data streams. This infrastructure offered a web-based front-end, called SensorMap, which enabled users to visualize the sensor data streams on an interactive map. Instead of depending on closed and monolithic solutions, sensor deployments shared their data to make their resources re-usable by other systems or concurrently used by multiple entities. The environment map was a virtual representation of real-world locations, allowing users to analyze environmental phenomena through the combination of multiple sensor streams dispersed in space.

The SenseWeb proposal tackled two main problems. First, it succeeded in combining information from groups of heterogeneous sensors that differed in resource, mobility, and network connectivity. The solution used an open and extensible architecture that resorted to remote sensor gateways to host the different sensor data streams, exposed through uniform interfaces. These gateways communicated with a coordinator, which served as a common point of access for the various sensor contributors and for applications to gain access to the available data (see Figure 2.6). The second problem tackled by SenseWeb was the management of scalability when dealing with large amounts of geographical data. To this end, the authors provided several techniques for caching computationally expensive visualizations derived from sensor data and for efficiently reusing them to serve user queries. The various components of the SenseWeb system exposed their functionality by using a set of web service API interfaces that allowed applications developed on different platforms to access SenseWeb data.

The work in [55] introduces the concept of "reality mining", the data mining of sensor streams that monitor specific environments. The manipulation of massive amounts of sensory data can be used in detection and actuation systems, allowing users to use sensor data in valuable ways. The authors designed a prototype of a sensor information system that used geographic information systems software, mission planning/terrain visualization systems, and sensor networks in conjunction with a photo-realistic, 3D visualization of the prototype's environment. They used the prototype to propose several systems where the use of sensors and virtual representations would be useful. These propositions included a fire-detection system which used sensors in order to help

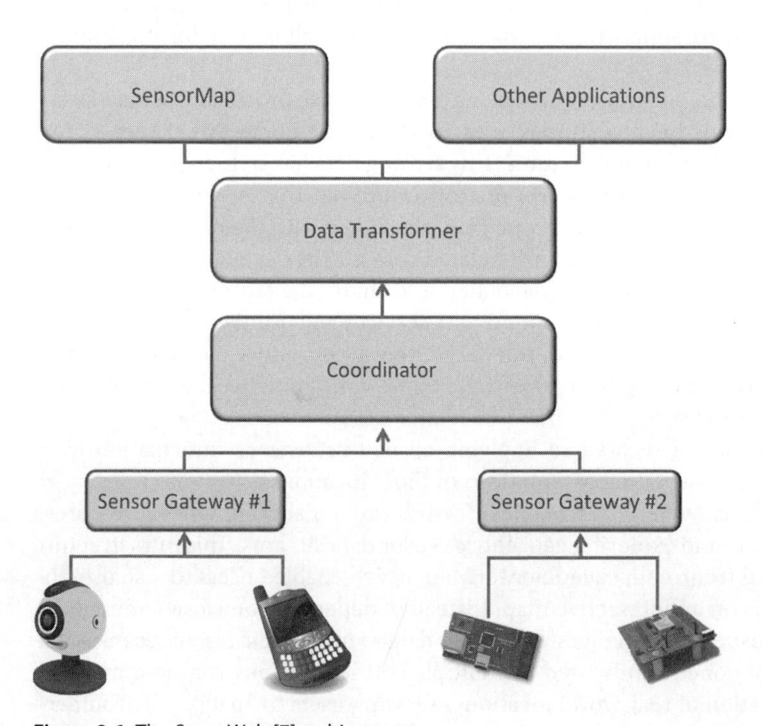

Figure 2.6 The SenseWeb [7] architecture.

anticipate the initial spread of the fire, virtual tourism, and a live view of stock price changes as clouds over a 3D map.

Other research projects and companies focus on the use of sensors and intelligent devices in an urban context. For example, the Urban Sensing research project [56] sought to develop cultural and technological approaches using mobile and embedded sensing to enrich civic life. The idea was that many ubiquitous sensors for urban sensing are already deployed and mobile phones can provide sounds and imagery from these sensors. Thus, users will have access to a great diversity of sensors in future urban settings that will allow them to know more about their homes, neighborhoods, and communities. By sharing and cross-referencing sensed data with publicly available data from private and municipal monitoring systems, a user can have access to information about the city, such as traffic, weather, air quality, and pedestrian flow.

Sense Networks, Inc. was a company that indexed real-world data using real-time and historical location data for predictive analytics across multiple industries (it has since been acquired by YP Marketing Solutions [57]). Sense Networks developed machine learning technology that indexes and ranks real-world places, based on movement data between these places at different times. This movement and location data was collected in real-time from devices with GPS or WiFi positioning technology, such as mobile phones and automobiles. This information was used to create applications that can create profiles of various locations within a city and use them to better understand visitors and anticipate their needs. Earlier products were CitySense, a local nightlife discovery and social navigation platform, and CabSense, an application that helped users find available taxicabs near them.

WikiCity [8] was a project with the objective of developing a platform that allowed an entire city to become a real-time control system. In order to achieve this, the platform required sensors able to acquire information about several aspects of the city, intelligent mechanisms that evaluated the performance of the system, and physical actuators that performed actions on the system (see Figure 2.7). One interesting aspect of this project was the fact that it considered the city's own inhabitants actuators. The platform was capable of storing and exchanging data with the users through mobile devices and web interfaces, and it enabled people to "become distributed intelligent actuators, which pursue their individual interests in cooperation and competition with others, and thus become prime actors themselves in improving the efficiency of urban systems".

2.2 Human Sensing and Virtual Communities

More recently, some researchers began to focus on the human side of sensing. Many of these works also presented a strong social networking component.

One example is the work of Lifton, J. *et al.* at the Massachusetts Institute of Technology media laboratory, who coined the term "dual reality" [58, 59] to indicate the ability to merge the real and virtual realities by using sensor networks. They designed several prototypes where they performed experiences in merging a real-world location, the MIT Media Lab's third floor, and virtual worlds, in this case Second Life®. One of these prototypes is described in [60] in which the authors present the ShadowLab, a Second Life® map of the Media Lab's third floor animated by data collected from a network of several

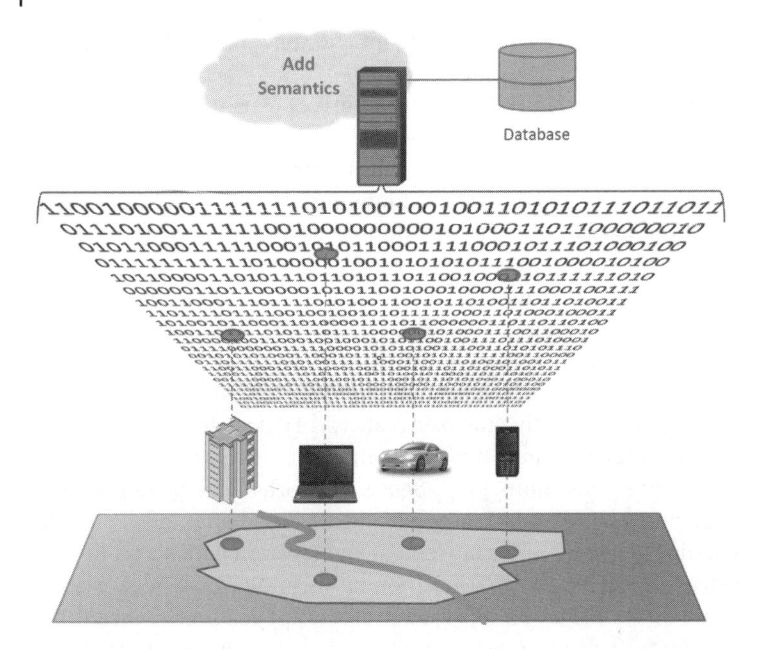

Figure 2.7 WikiCity [8] interfaced between virtual data and the physical world through a semantically defined format for data exchange.

sensor/actuator nodes. The sensor nodes used in ShadowLab could sense light, vibration, sound, motion, temperature, and measure the amount of alternating current drawn from each outlet. They also hosted a low-power radio for wireless communication. The data provided by these sensors was animated in virtual environments in an engaging way that naturally suggested the sensed stimuli. This was accomplished by resort to virtual objects called DataPonds, that changed appearance depending on the activity level in their respective location. The authors also began to experiment with virtual/real-world interaction by allowing Second Life® avatars to interact with virtual objects that would play audio clips through speakers, and by designing physical versions of DataPonds that would be stimulated by avatar motion in a particular region of ShadowLab. They also used ShadowLab to experiment with user avatar transformations based on real-world data, as avatars could "metamorphose" according to the activity level outside of the avatar's corresponding user's office.

Another dual reality implementation described in [60] was the Ubiquitous Sensor Portal, a device designed for two-way cross-reality experience and communication. These portals streamed information in both directions, from the user's environment to ShadowLab, and from ShadowLab to the real world. The portals hosted several environmental sensors that measured motion, light, sound level, vibration, temperature, and humidity. They could also communicate with a family of badges designed to identify individuals facing them and capture audio and video. Because the portals could stream private data, an important requirement was to manage privacy. A system of user badges was implemented, where each badge beaconed a unique ID, to wirelessly mediate privacy settings. Portals knew which badges were potentially in sensor capture range and controlled data access according to the badge user's preferences.

Sensor nodes have also been used as a means of transmitting mobility of people into virtual worlds. In [61] a framework is proposed which maps a sensor node to an object in Second Life®. The location of the sensor mode is calculated by the framework and reflected on an avatar in Second Life® which moves according to the real-world movement of the node. The location of the node is calculated from the received signal strength indication (RSSI) values from three or more fixed reference nodes, thus requiring a carefully designed WSN architecture.

Despite the importance of all of these research initiatives, one particular invention has undeniably changed not only the landscape and prospects of human sensing but also our very own society and daily life. This revolutionary invention surged in the form of a small rectangular device that tends to accompany us everywhere we go: the mobile phone. While traditional phones have been with us for the last 140 years, mobile phones have drastically changed the paradigm of long-distance communication, owing to the mobility they provide. In fact, the mobile phone has nearly become a basic need and an important part of our lives; people often claim their day is "ruined" when they forget theirs at home.

Mobile phones are so important that have been disseminated even in places that lack much-needed basic infrastructure. The International Telecommunications Union found that, by the end of 2011, the number of mobile phone subscriptions reached 5.9 billion, representing a penetration of 87% of the entire world and 79% of all developing countries [62]. In fact, the development of mobile phone networks surpasses other infrastructures, such as paved roads and electricity, in many low- and middle-income countries, diminishing the need for fixed Internet deployment [63, 64].

Thus, mobile phones are extremely common, increasingly cheap, and provide mobile Internet connectivity almost everywhere, even in less favored environments. These characteristics make them excellent candidates for gateways for new types of large-scale HiTLCPSs aimed at solving real-world social problems. Some research suggested the use of these extended mobile networks to help low-income patients in under-developed countries to manage chronic diseases, such as diabetes. This idea was tested on patients with diabetes from a clinic in a semi-rural area of Honduras, through a system that delivered automated phone calls that helped manage the disease [65]. The lack of technological infrastructures also prompts the mobile phone to serve as a very resourceful device, sometimes even more so than in developed countries. One interesting example of this is mobile banking: Kenya's mobile network Safaricom introduced a service called M-Pesa, which allows users to store money on their mobiles. Users can then pay utility bills or send money to friends through a simple SMS and the recipient converts it into cash at their local M-Pesa office. This allows millions of Africans to have cheap, mobile, and easy access to a bank account [66].

Perhaps because of their usefulness and dissemination, the evolution of mobile phones has been extremely rapid and the market remains very volatile. In fact, mobile phones are quickly being replaced by "smartphones", devices possessing a computing power that matches a desktop computer, and a size compatible with our pockets. And even in the latest iteration, the evolution continues to be astonishing.

Consider the Nokia 6101 (Figure 2.8), a very popular smartphone at the time of its release; current smartphones such as the iPhone 6s and Nexus 5X make this mobile look almost archaic, yet it was released in 2005, a mere 12 years ago. Smartphones

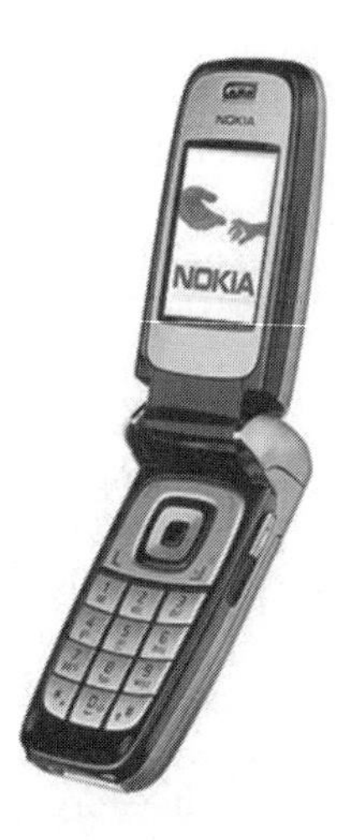

Figure 2.8 Nokia 6101 vs iPhone 6s/LG Nexus 5X.

and tablets have become personal portable computers, representing a versatile computational resource; nowadays, even the most basic and cheap smartphones are capable of processing considerable amounts of information through basic programming platforms. Modern smartphones are actually more powerful than desktop computers from a decade ago. For example, an iPad 2 tablet, introduced in 2011, has a peak calculation speed equivalent to that of the Cray-2 supercomputer, introduced in 1985 [67]. However, tablets and smartphones also possess advanced sensors such as gyroscopes, accelerometers, and digital compasses, feature quad-core processors and up to 2 gigabyes of RAM. In a very real sense, these devices have brought us pocket-size, supercomputer-like computational power in a matter of a few years. They have also brought us incredible mobile connectivity, providing Internet access almost anywhere.

Even when seen in perspective, it is difficult to grasp how fast the mobile market has been evolving. As it is the case with most silicon-based technology, mobile phones also tend to become cheaper over time. This means that they are more easily adopted by the general population, namely in developing countries. In fact, smartphone sales are globally outpacing those of regular phones [68].

The possibilities of such advanced mobile platforms are already apparent in the diversity of existing applications made available for them. However, these are only primordial examples. Smartphones are not evolving alone, having grown together with the Internet boom and closely accompanying the evolution of the World Wide Web and social networking. Almost all newer smartphone models offer native support for the integration with several social networking services (such as Twitter[1] and Facebook[2]) also offering advanced Internet browsers that function almost as well as their personal computer counterparts.

Not surprisingly, considering the social beings we humans are, as our Internet-connected devices evolved so did the means we use to communicate

1 https://twitter.com/
2 https://www.facebook.com

and interact with the people we deem close. In the past, people's interactions were mostly face-to-face amongst their peer groups, with occasional long-distance relationships through letters or telephone calls. In today's world, we see a social revolution where people use their smartphones to share, in real time, funny stories, thoughts, feelings, photographs, and other pieces of their lives with their family and friends, some of which they have not physically been in contact with for a long time, and in some cases not even ever seen in "real life".

There has also been a considerable evolution over earlier iterations of social networking when it comes to the sharing of personal information: whereas users used to simply fill their personal pages with static personal information (such as their hobbies or self-descriptions), we are now seeing mobile social networks that use collaborative feedback to acquire real-world information in order to provide more useful services. We are also seeing an enormous increase in the sharing of social activity, with users posting more multimedia items about their lives and social interests. According to research by Pixable [69, 70], the changing of Facebook profile pictures seems to increase every year; in fact, the number of profile photos per user per year tripled from 2006 to 2011, independently of the user's age, since older users upload as much as younger users do. The research indicated that, on average, a Facebook profile picture had two comments and three likes and the average person had 26 profile pictures. From a social networking point of view, the representation of the current status of people in virtual environments is a very interesting concept: access to social networking nowadays is becoming increasingly more mobile, and it is not uncommon to see people use their smartphones to share and discuss daily experiences shortly after their occurrence, updating thoughts, and responding to feedback from their friends as the situation develops and the user's life continues. Current users can announce social events to their group of friends, share experiences through photos and comments, and show their opinions and hobbies through "likes" and their own "private wall".

Thus, social networking is a phenomenon that bloomed and continues to connect an astonishing number of users, becoming the fastest-growing active social media behavior online. The sheer scale at which these changes are happening is astonishing: a 2014 statistical analysis by Browser Media, Socialnomics, and MacWorld suggested that Facebook, one of the largest social networks, had around 1.4 billion users worldwide, and that 98% of 18 to 24-year-olds already used social media websites [71]. This social networking tendency continued as the number of Facebook users increased 12% from 2014 to 2015 [72]. In fact, in 2016, another study claimed that 31% of the global population used Facebook [73].

Since social networks are becoming so important in the interconnections between humans, it is expected that they will play a prominent role in HiTLCPSs. As mobile technology develops, social networking websites become increasingly more pervasive. This is evident when we take into consideration the ubiquity of social networking smartphone applications. In 2016, of the 1,721,000,000 monthly active Facebook users around 1,104,000,000 were mobile ones [72] who spent 68% of their total Facebook time on a mobile device [73].

Despite these advancements and the general public's interest in these social services, their current functionality does not yet reflect the true dynamics of people's relationships and personal lives. Instead of being pre-determined and unique events in time, social group activities can, in fact, happen very frequently and, most of the

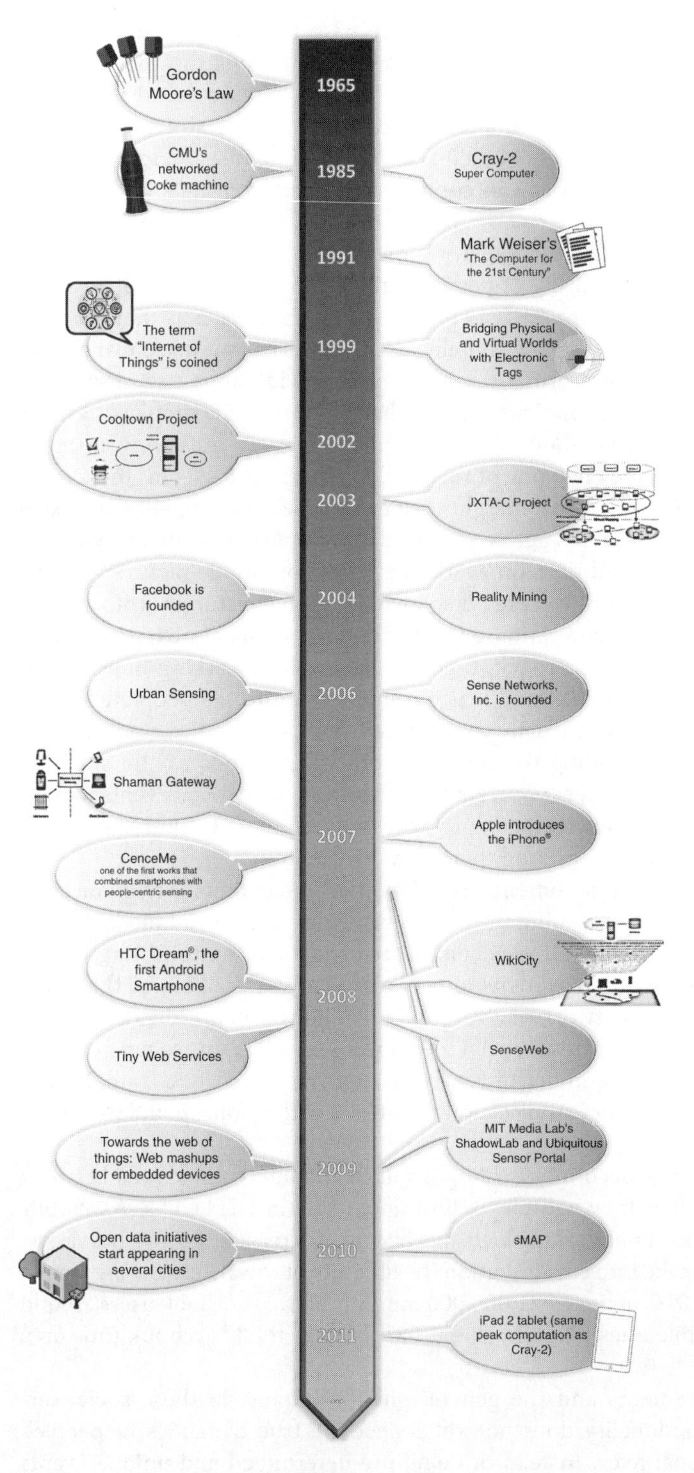

Figure 2.9 HiTL technologies evolution timeline.

time, spontaneously. Current systems are not capable of providing this "real-time" component to social networking, which diminishes its true potential. In a sense, we can classify current social networks as still very "static" when compared to a more complete system capable of closely following the extent of human social interactions. While the use of collaborative contributions is still an important part of social applications and can provide meaningful and useful data, sensing systems can provide a more reliable and responsive feedback that is crucial in achieving "real-time and human-aware" social-networking. In fact, an HiTL approach to social networking may well prove to be a technological leap over current social networking of the same magnitude as the one provided by mobile phones over traditional telephones.

2.3 In Summary...

This chapter showed us a certain evolution in terms of IoT and CPSs. Figure 2.9 shows a timeline containing some of the relevant events and works that were referenced.

Research that began with a focus on "things" has then evolved to the monitoring of entire environments and, more recently, of human users. This makes sense, since real-world objects are more controllable: they are made by humans and we understand the full extent of their uses and states. Thus, they became the initial targets for extending CPSs into the web. Later development of WSNs enabled CPSs to monitor wide geographical locations. However, only more recently did we achieve the necessary advancements in miniaturization, computational power, sensing, information linkage, and machine learning that allow us to focus on the most complex aspects of our reality, including ourselves. These possibilities have been brought forward by the tremendous advances in mobile devices, such as smartphones, and social-networking.

All of these advancements and ideas make it difficult for us to understand the range, limits, and possibilities of these new "human-aware" paradigms. In an attempt to organize ideas, we will now focus our attention on organizing these HiTL concepts into taxonomic concepts.

3

Theory of HiTLCPSs

So, what exactly is a HiTLCPS? Understanding the principles and theory behind these systems is the main objective of the current chapter. First, in Section 3.1, we will establish a taxonomy that will be used throughout the book. Then, in Sections 3.2 through 3.4, we will address the human role in the three basic processes of HiTLCPSs, namely data acquisition, state inference, and actuation, respectively.

3.1 Taxonomies for HiTLCPSs

As HiTLCPSs have a wide spectrum of applicability, it is difficult to cover all possible examples of HiTL solutions for a multitude of domains. Thus, we will begin by presenting the most basic, common processes in HiTL control, shown in Figure 3.1. The first phase is known as "data acquisition". Data related to the human individual is gathered from the available sensors. This data is then processed in the "state inference" stage with the objective of inferring the human's physical and/or psychological state. Some approaches may also attempt to predict future states based on historical data and the current state. Finally, in the "actuation" stage, the system may or may not perform certain actions based on the observed state. Some "open-loop" systems do not affect the system per se, that is their results are merely informative, without direct actuation. However, "closed-loop" systems actuate directly on the environment or the human, in order to influence the loop and achieve a given target.

From now on we will call this reference model the Internet of all (IoA), meaning that it includes not only (traditional) IoT but also humans as fundamental elements. In this way, we emphasize that this Internet is made by humans, for humans, and with humans.

IoA is built from spatially distributed devices that are considered by standard IoT, like laptops, mobile phones, computers, sensors, actuators, "classic" network elements (we mean all passive elements like routers, switches, access points, etc.), RFID tags, readers, cars, intelligent clothes, wearable devices, furniture, and home appliances. As can be inferred from the general model in Figure 3.1, IoA also includes robotics and its interaction with intelligent devices and sensors. However, on top of these man-made devices, we also consider human beings themselves as part of the system: their actions, drives, desires, and emotions.

A Practical Introduction to Human-in-the-Loop Cyber-Physical Systems, First Edition.
David Nunes, Jorge Sá Silva and Fernando Boavida.

Companion Website URL: www.wiley.com/go/nunesloop

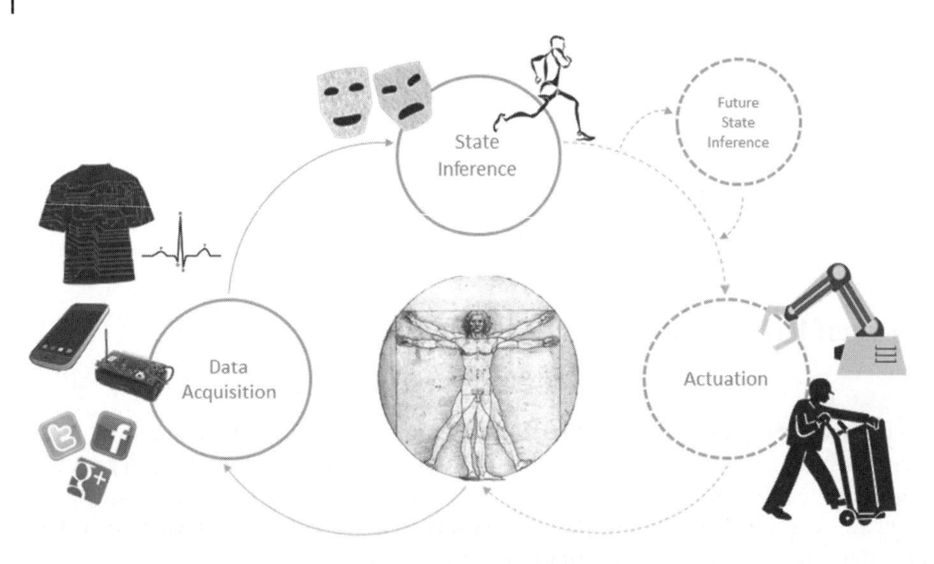

Figure 3.1 Basic processes of human-in-the-loop control.

How can we distinguish between different types of HiTLCPSs? In order to better comprehend the dimension of such an expansive field as HiTLCPSs, it is important to resort to taxonomies that allow us to better structure our ideas and concepts.

Some authors have previously proposed taxonomic distinctions and classifications for HiTL systems based on the type of exerted control. In an attempt to attain a greater understanding of the spectrum of HiTL applications, and of their underlying principles, requirements, and models, Stankovic *et al.* [45] began to establish a taxonomic foundation for HiTLCPSs applications. According to Stankovic *et al.* [45], it is possible to organize existing HiTL applications into three types: (1) applications where humans directly control the system, (2) applications where the system passively monitors humans and takes appropriate actions, and (3) a hybrid of (1) and (2). These three basic types are represented in Figure 3.2, and are described below.

Human Control: there are two main scenarios where humans directly control CPSs. In supervisory control scenarios, human operators oversee an otherwise mainly autonomous process. The operators are responsible for adjusting certain set points that may influence the system. This is the case of, for example, industrial scenarios where operators mainly set or adjust certain target parameter values that are then enforced by an autonomous robotic CPS. If humans have a more direct command over the process, we are in the presence of direct control scenarios. These are typical master slave scenarios where humans issue commands to the CPS, which then carries the necessary actions, and reports back the results. An example of such a system can be seen in [74], where a wheelchair-mounted robotic arm is controlled by a disabled person to retrieve objects.

Human Monitoring: applications that passively monitor humans, also known as people-centric sensing applications, use their monitoring data to take appropriate actions. In the scope of CPSs, these can be of two types: open-loop and closed-loop systems. Open-loop systems monitor information about humans regarding several aspects (e.g. sleep quality, physical activity, attention-level) and report these results.

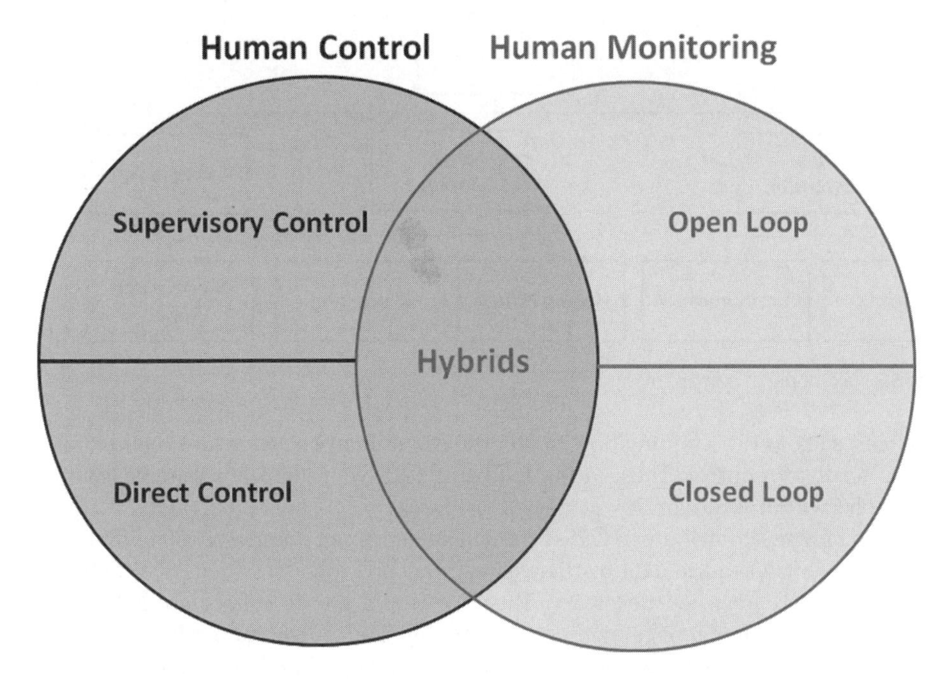

Figure 3.2 Taxonomy of human control.

One example is Look4MySounds, a remote monitoring platform for auscultation of cardiac sounds and automatic detection of pathologies [75]. The platform uses an integrated stethoscope with which auscultation sounds are recorded and processed for automatically detecting pathologies. The sound samples and obtained diagnosis are thereafter remotely sent to a clinician. Despite the human being in the loop, the system does not take any proactive actions and simply relays the results to a specialized medical practitioner. On the other hand, closed-loop systems use their sensory data and processing results in order to actively contribute to a specific goal. For example, a smartshirt may monitor a human's exercise levels at the gym, while a sensor placed on the wall monitors the room temperature. When the human is exercising, the HiTL control may signal the heating, ventilation, and cooling (HVAC) systems to reduce the room's temperature in order to make the exercise more pleasurable.

Hybrid Systems: hybrid systems take people-centric sensing information as feedback to their control-loops while also taking direct human inputs into consideration. Let us expand our smartshirt example to include a smartphone application that allows the user to keep track of his/her exercise and also set a desired room temperature. The hybrid system could take the user's desired temperature as input while using the activity information to fine-tune the absolute temperature value, or to control the rate of temperature change.

Stankovich *et. al.* [45] only classified the different types of HiTL applications according to how humans interact with them. In this book, we intend to go a little further and provide an alternative point of view of the HiTL process. We will expand this taxonomic exercise to also consider the possible roles of humans in these systems. We believe such a distinction is important, since it will allow us to better cope with some of the existing

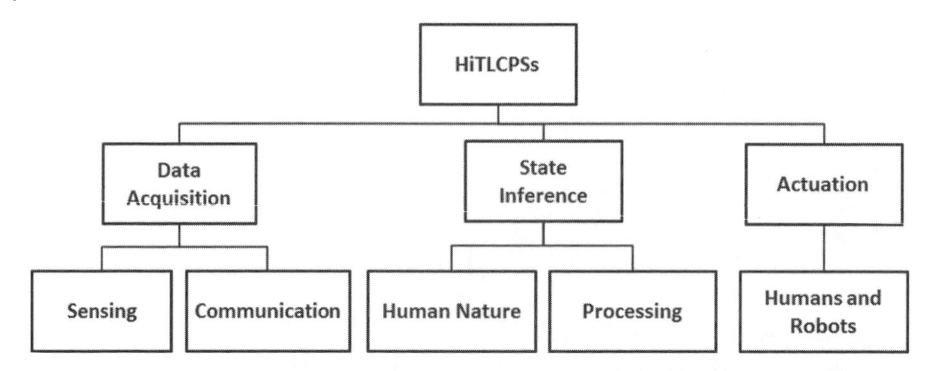

Figure 3.3 Taxonomy of human roles.

challenges, such as determining how to incorporate human behavior into the methodology of feedback control. Thus, in the following text, we establish a classification of human roles within future IoT.

How can humans contribute to CPSs? Human presence can manifest itself in different ways: humans may acquire data by themselves, may provide assessment of situations, and may also actuate when necessary. Thus, we would like to reflect not simply on where to place the model of human behavior within the control loop but also understand what roles a human may play within an HiTLCPS and how to best explore this resource; not as an external and unpredictable element but as an inherent part of the system. Figure 3.3 summarizes our understanding of the possible human roles within HiTLCPSs.

The next subsections detail the human role in each of the identified categories.

3.2 Data Acquisition

3.2.1 Humans as Sets of Sensors

There are several ways a human can act as a sensor, or even as a set of sensors. Whenever the human is capable of directly feeding the system with information, we can say that human is acting as a sensor. For example, each time a user indicates that he likes something on a social network, gives feedback for machine learning, or provides collaborative information for crowd sourcing (e.g. indicates that a road is blocked on a collaborative navigation app), that user is acting as a human sensor. Human-provided information has several advantages. It tends to be of a more abstract/complex type, and it may be easier and less expensive to obtain than the one provided by sensor machines, as most data is provided voluntarily and without the need for additional hardware. Taking our road example, it is difficult to reliably detect a blockage without a considerable amount of sensors (infrared proximity, cameras, etc.) and signal processing, which can end up being rather expensive even for very short sections of the road. However, a human with a smartphone can easily take pictures, comment, and report such blockages in a way that is useful for the rest of the HiTLCPS. On the other hand, this information is also more difficult to parse: rarely do people communicate in a machine-readable way. At the same time, this source of information is also more unreliable: unlike people, machines do not tend to lie or misinform (unless they are broken). Hence, most of the effort and cost of using human-derived information comes from its parsing and validation: depending on the use-case, it may be useful to use it or not.

Another way in which humans become an integral part of sensor networks is through the sensors that they carry. Wearable devices, such as smartphones, smartwatches, or intelligent clothing, can also become important elements in the future Internet. Some years from now, nano-technology might also become an important element in this regard; some researchers point out how this technology can bring intra-body elements into the IoA [76]. Nano-networks have been receiving a lot of attention from the scientific community and, in the near future, new studies and prototypes will begin to emerge that might result in very advanced applications in the biomedical area. HiTL concepts will certainly apply to these types of scenarios. The source of information is definitely human-influenced but, at the same time, machine-derived, as this mitigates many of the limitations associated with information derived exclusively from humans. Still, depending on the use-case, the usage of these sensors in a useful way may require a considerable amount of integration and processing.

The concept of humans as sensors extends beyond the use of direct human feedback and sensor devices. Social-networking, for example, also serves as a rich information source. In the not-too-distant future, this information could be combined from both sources, with sensor nodes placed in major shopping centers to, for instance, help and support the shopping of human beings. Smart glasses could overlay price-tags on the products of its user's interests (i.e. on things that they "liked").

We can talk about two types of sensing in these scenarios: direct sensing and indirect sensing. Direct sensing involves using sensors or human feedback that is directly related to the sensing target. On our shopping mall example, using GPS localization or a questionnaire asking the user which stores he visited are direct ways of determining his shopping habits. On the other hand, indirect sensing refers to a case where we infer desired information from other responses. For example, by using the shopping mall building's vibration sensors [77], or even by aggregating information from the shopping mall's information terminals, one can infer which floors or shops tend to attract most interest and customers.

While these sorts of applications are not unfeasible, concerns over the intrusiveness of such practices are more than justified. Thus, as we will see in later chapters, it is also the responsibility of HiTLCPSs to ensure the privacy of their users, enforcing strict control over the sharing of personal information.

3.2.2 Humans as Communication Nodes

As social creatures, humans are masters of communication. Many of the technologies that we see all around us were built specifically to improve this faculty: televisions, radios, telephones, the Internet; they all have the purpose of conveying messages. Thus, this ability of spreading information is not to be undervalued within HiTLCPS: messages, photographs, *tweets*, comments, and posts are all perfectly valid subjects of communication.

In fact, some researchers have pointed out that these data items may enable faster and more reliable communication than traditional "news" media outlets. Wang *et. al.* have explored these ideas in their 2015 book *"Social Sensing: Building Reliable Systems on*

Unreliable Data" [78]. According to the authors, human beings are "sensors engaging directly with the mobile Internet", emphasizing the key problem of extracting reliable information from data collected from largely unknown and possibly unreliable sources. The book explains how myriad societal applications can be derived from the massive amount of data collected and shared by average individuals. In other words, how can we know if this human communication is reliable? In the authors' opinion, the rate of human information generation has long outclassed humans' own cognitive ability for processing it. Thus, new algorithms are needed to preserve the quality of information as much as possible. Ascertaining the correctness of reported information is referred to as the "truth estimation problem", and it affects the ability of humans to act as both sensors and communication nodes.

Nevertheless, much like when acting as sensors, the ability of communicating information is greatly increased when humans and machines work together. Multi-hop is a very common technique used by tiny devices to save energy. Intermediate nodes can be used in a communication process between a sender node and a receiver node to reduce the required signal power. In this context, human devices such as smartphones and body-area sensors may also be used as intermediate nodes in the "hopping" process, taking advantage of human mobility and intelligence for distributing information more effectively in the network. This may be particularly useful in, for example, metropolitan-wide collaboration systems, where human presence and mobility may be crucial in re-passing non-critical information about the environment. Instead of using multi-hopping or long-distance communication between sensor nodes to, for example, monitor temperature, this information might be aggregated and stored by human-carried devices as people move around the city, opportunistically forwarding it when appropriate, thus, reducing the amount of energy required for communications.

Either way, be it through their own means or supported by machines, the human ability for transmitting information within HiTLCPS is undeniably important. Despite having discussed this communication ability under *Data Acquisition*, it remains important through all phases of HiTL control, particularly in *Actuation*, as discussed in Section 3.4.

3.3 State Inference

3.3.1 Human Nature

Human nature is a mysterious thing, and tremendously difficult for current machines to understand. Our best efforts at understanding what a person wants still reside in simply asking him/her directly. In fact, researchers are still trying to correlate smartphone sensing data features with human behavior, through sampling questions and surveys [11]. How can we hope to create machines that are capable of decoding such elusive but important aspects of existence that are so difficult to understand even for humans themselves?

The combination of sensors through body-area-networks may be able to alleviate this difficulty in gathering human information. This human body-area-network is composed of a variety of sensors (accelerometers, smartshirts, smartshoes, bracelets, watches, etc.) and is capable of measuring several different aspects of human activity, including vital signs (heart rate, ECG, EEG, movement, etc). More interestingly, we are continuously learning how to use this information for characterizing actions, and for detecting activities and even psychological states and emotions. Current research indicates promising leads to new powerful and complex machine learning solutions that are becoming increasingly more cognizant of "human nature" phenomena, making them an integral part of the control loop in IoA scenarios. For example, the attention level of a driver affects the cruise control mechanisms of an automobile, the user's exercise levels affect the air conditioning of his/her house, or a human's emotional state may affect the user interface of his/her smartphone application. Humans are no longer external entities that simply benefit from the system. Their presence, actions, and emotional states strongly affect how IoT things react. We will discuss some of these new research lines in greater detail in Sections 4.1.2 and 4.2.

3.3.2 Humans as Processing Nodes

As we have discussed in Section 3.2.1 (Humans as Sets of Sensors) and 3.2.2 (Humans as Communication Nodes), humans can also directly contribute to the processing of information. No machine has yet been capable of matching human capability for pattern recognition; thus, as previously discussed, human information should not be undervalued.

Still, there are other aspects of processing that are not directly related to human cognition and yet still greatly depend on human behavior. Since single individuals are now becoming equipped with a considerable number of mobile devices (smartphones, tablets, smart wearables, etc.), human behavior begins to have a significant impact on the availability of resources within an HiTLCPS. These resources contribute to the overall computation capability of the system: each of these individual devices represents an untapped computational resource that is available on site; by taking advantage of these devices, it is possible to reduce the need for distant service providers: direct communication with neighboring devices becomes key for handling local tasks and information. Thus, the traditional cloud is descending to the network edge and becoming diffused among the client devices in both mobile and wired networks. This concept has come to be known as the "fog of things" [79].

Although most devices carried by humans are very simple and have limited processing capabilities, the use of distributed algorithms can take advantage of the huge number of processing elements, and enable collaborative tasks that could not be fulfilled by any particular individual node. Smartphones can be major participants in this processing, but other, simpler, wearables and appliances can also become useful processing sub-nodes in the new IoA.

3.4 Actuation

3.4.1 Humans and Robots as Actuators

Humans already act as actuators and as a function of the medium. If a gas leak is detected in a factory, the responsible employee quickly goes to the control room and closes the respective valve. If in a hospital the blood pressure of a patient reaches a prohibited value, the nurse on duty, hearing the alarm signal, goes directly to the patient's room to administer a new drug. Unfortunately, current IoT systems are still mostly unprepared for handling human actuation as an inherent component of the system. In HiTLCPSs, human actions remain extremely important, since human conceptualization will continue to be unmatched by artificial intelligence (AI) for, most likely, many years to come. However, unlike most current CPSs, the IoA paradigm takes human action into consideration in the control-loop, in the sense that these systems are made for humans, with humans. Examples of this human role are industrial systems that may use WSNs and robots to monitor and detect problems, and then require specialized actuation of humans to fix the problem. On our social-networking shopping mall example, users may consider product suggestions from other clients with similar interests and psychological states, and collaboratively suggest products of their own interest.

In HiTLCPSs, human and machine actuation go hand-in-hand and can often complement each other. In this way, IoA systems are not "devoid of human soul" but make human actuation as an integral part of their functioning. As we will see in later chapters, particularly in Part III,the ability to work in combination with actuation machines, such as robots, will become increasingly important in the future.

3.5 In Summary...

In this chapter, we defined a general reference model, the Internet of all, where human actions become a fundamental part of the control loop of CPSs. We then organized the major ideas behind HiTLCPSs, starting with a previously proposed taxonomic classification, based on the type of exerted control.

In this classification, we saw that HiTL control can be classified into three types: (1) applications where humans directly control the system (*human control*), (2) applications where the system passively monitors humans and takes appropriate actions (*human monitoring*), and (3) hybrids of (1) and (2).

We then proposed an alternative taxonomic point of view of the HiTL process that highlights the roles a human may play within HiTLCPSs:

- **Data Acquisition**
  - *Humans as sets of sensors*: Humans can feed the system with information, either collaboratively or through the sensors they carry.
  - *Humans as communication nodes*: Humans are masterful communicators, able to quickly share information through social networks. Their body-area devices can also store and pass data as part of a "hopping" process.
- **State Inference**
  - *Human nature*: Understanding human nature is tremendously difficult, but a combination of body-area sensors and powerful machine learning solutions may alleviate the problem of recognizing human-centric states.
  - *Humans as processing nodes*: Machines have yet to match humans in their capability for pattern recognition. Additionally, human-carried devices (e.g. smartphones, smartwearables) have considerable amounts of processing power and may diminish the need for cloud-centric solutions in the near future.

- **Actuation**
  - *Humans and robots as actuators*: There is much to be said about human actuation in HiTLCPSs. Issues such as human motivation, robotic collaboration, and AI will have an important impact on future deployments. As we will see in later chapters, it is likely that human-machine collaboration will play an important role in future technologies.

Now that we have described HiTLCPSs from a theoretical perspective, we will focus on providing practical examples of existing deployments and technologies.

4

HITL Technologies and Applications

In an effort to gain an understanding of the existing types of solutions and methods, we will begin by analyzing the scope of HiTLCPSs from a practical perspective, first delving into the processes of data acquisition, then considering different solutions for inferring state, and finally, different techniques for actuation. In the second part of the chapter we will look at several experimental projects and research initiatives in this field.

4.1 Technologies for Supporting HiTLCPS

4.1.1 Data Acquisition

The acquisition of data through which a human's state may be inferred is a complex process, with a multitude of possible sources of information. HiTLCPSs have previously used physical properties, such as localization (e.g. GPS positioning), vital signals (heart rate, ECG, EEG, body temperature), movement (accelerometers), and sound (voice processing) among many other types of information that can be acquired directly from the physical reality. There are also many non-physical properties that may be derived, such as communication behaviors (e.g. phone calls, SMSs) or socialization habits (e.g. social networking data, friend lists). We will discuss some examples of the use of socialization habits and social networking in Sections 4.2.3 and 11.2. Here, we focus on physical properties, since they have greater technological requirements.

Most raw physical data comes from distributed sensor architectures, which are critically important for HiTLCPSs, since they allow the measurement of physiological changes, which may be processed to infer current environmental conditions and human activities, psychological states, and intent. In this regard, several types of architectures and technologies have been proposed.

One of the most important technologies for the process of data acquisition in HiTLCPSs is the WSN. These are networks of small, battery-powered devices with limited capabilities, wireless communication, and various sensors that have been applied in countless application scenarios worldwide. One highly debated challenge of WSNs applied to HiTLCPS is the integration of these tiny devices into the worldwide IoT. The ease of integration of these small devices is of particular importance for HiTLCPSs, since it would make these systems more distributed, open, interactive, discoverable, and heterogeneous, as envisioned in [33]. In WSNs, the use of the Internet protocol (IP) has

A Practical Introduction to Human-in-the-Loop Cyber-Physical Systems, First Edition.
David Nunes, Jorge Sá Silva and Fernando Boavida.

Companion Website URL: www.wiley.com/go/nunesloop

always raised some concerns, owing to the fact that it does not minimize memory usage or processing. Moreover, the use of the full transmission control protocol (TCP) and/or the IP stack is not possible because it requires more resources than the ones most of these devices can offer. However, the integration of IP has the advantage of offering a transparent communication between nodes while using a well-known protocol, providing interoperability and even Internet connectivity. While working towards employing IP in WSNs, the IETF created the 6LoWPAN group that has been working on standards for the transmission of IPv6 packets over low-capability devices in wireless personal areas, using IEEE 802.15.4 radios [80]. New drafts were also proposed for adapting 6LoWPAN to other technologies like *Bluetooth*. Unfortunately, 6LoWPAN cannot be applied to devices devoid of processing or memory capabilities, such as RFID tags. Gateway-based approaches are a possible solution to support IP functionality in these scenarios. The main advantage of these approaches is that terminal devices do not require any processing or communication capabilities. Moreover, they make the sensor and device networks transparent to external environments, and developers can use any protocols as long as they are suitable for their needs. However, one inherent problem of gateway-based approaches is that gateways are single points of failure. This problem is exacerbated in environments where devices present some type of mobility, i.e. moving from place to place while maintaining connectivity. In these mobile environments, all of the communication processes are more fragile, and failure of fixed gateways can compromise the integrity of the HiTLCPS. Another problem of gateway-based approaches is that sensor nodes are often required to format the data according to a specification defined by the provided drivers of the gateway, forcing the developer to create a software driver or analyzer for each specific data frame format, further reducing their interoperability for HiTLCPSs.

Some research work focused on some of the architectural issues of including WSNs into HiTLCPS. SenQ [9] is a data streaming service for WSNs designed to support user-driven applications through peer-to-peer (P2P) in-network queries between wearable interfaces and other resource-constrained devices. It introduced the concept of "virtual sensors", user-defined streams that could be dynamically discovered and shared. For example, in assisted-living scenarios for elderly people, a doctor could combine information from mobility speed, movement, and location (e.g. nearby stairs) to create a virtual sensor that alerted nearby healthcare staff of an elevated risk of falls. SenQ took into consideration several requirements that were not satisfied by existing query system designs at the time, such as heterogeneity of sensor devices, dynamics of data flow patterns, localized aggregation of sensor data, and in-network monitoring. The system supported hierarchical architectures but predominantly favored ad hoc decentralized ones, as the authors argued that ad hoc architectures with neighborhood devices and service discovery are better suited to supporting large-scale and open systems with many users and sensors [33]. Data and control logic was also kept close to the concerned devices, in order to save energy and preserve scalability by providing a stack with loosely coupled layers that were placed on devices according to their capabilities and by enabling in-network P2P query issue for streaming data. Figure 4.1 shows SenQ's query system stack and the topology of one of its prototype implementations.

Another type of communication paradigm that may benefit HiTCPSs is body-coupled communication (BCC) [45] for supporting low-energy usage, heterogeneity, and reduced interference. BCC leverages the human body as the communication channel,

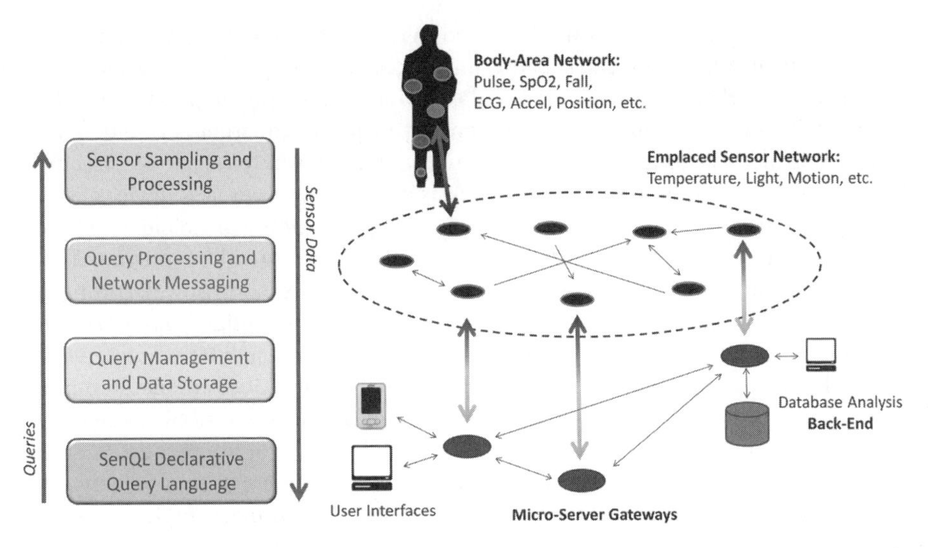

Figure 4.1 SenQ's query system stack shown side-by-side with the topology and components of AlarmNet, a prototypical implementation for assisted-living [9]. *Source:* Adapted from Wood 2008.

i.e. signals are transmitted between sensors as electrical impulses directly through human tissue to a point of data collection, a circuit-equivalent representation of the body channel in which different types of body tissue (skin, fat, muscles, and bone) are modeled with variable levels of impedance [81]. In particular, "Galvanic coupling" differentially applies the signal over two coupler electrodes, which is then received by two detector electrodes. The coupler establishes a modulated electrical field, which is sensed by the detector. Therefore, a signal transfer is established between the coupler and detector units by galvanically coupling signal currents into the human body [82]. There are several motivations for using this paradigm. First, the energy consumed in BCC is shown to be approximately three orders of magnitude less than the low-power classic RF-based network created through IEEE 802.15.4-based nodes. This technology is also bolstered by high bandwidth availability, of approximately 10 Mbps, which accommodates the needs of multiple sensor measurements. Finally, it offers a considerable mitigation of fading phenomena and overcomes typical interference problems of the industrial, scientific, and medical (ISM) which are usually affected by nearby devices (e.g. *Bluetooth*, WiFi, or microwave ovens) [45].

Conversely, we argue that much of the computational power and sensing capabilities for future HiTLCPSs will come from devices already existing in the environment. In particular, we believe that near-future HiTL systems will be heavily based on smartphone technology, owing to their rapidly expanding dissemination and powerful computation and sensing capabilities. A smartphone's sensors can be used by simple inference mechanisms to evaluate a human's psychological and physiologic states and integrate this information into HiTLCPSs. These sensors may include accelerometers, GPS positioning, microphones, or even the smartphone's camera.

In this line, some research attempted to combine smartphone data acquisition with social networking. The term "people-centric sensing" was used by the MetroSense project [83] to describe a vision where the majority of network traffic and applications are related to sensor and actuator data, applied to the general citizen. The MetroSense project envisioned collaborative data gathering of sensed data by individuals, facilitated by sensing systems that comprise cheap and easily accessible mobile phones and their interaction with software applications. The project proposed an architecture that supported heterogeneity by joining a variety of sensing platforms into a single architecture, while considering the communication limitations of low-power wireless devices [84].

MetroSense's architecture design followed a number of guiding principles: "network symbiosis" meant that traditional networking infrastructures would be integrated into the new sensing infrastructure in order to utilize the already existing power, communication, routing, reliability, and security resources; "asymmetric design" was a principle where differences in computational power and resources between different members of the network were exploited by pushing computational complexity and energy burden to devices with greater capacity; "localized interaction" implied that network elements should only interact with devices within a constrained "sphere of interaction", sacrificing network flexibility with the aim of increasing service implementation, simplicity, and communications performance. The architecture of one of its project implementations is shown in Figure 4.2.

Many of these guiding principles also apply to more general HiTLCPSs. We provide further insight into this matter in Chapter 10.

Several mechanisms have been proposed to support continuous sensing using smartphones. Jigsaw [85] is a continuous sensing engine that supports resilient accelerometer, microphone, and GPS data processing. It comprises a set of plug-and-play sensing pipelines that adapt their depth and sophistication to the quality of data as well as the mobility and behavioral patterns of the user, in order to drive down energy costs. This reusable and application agnostic sensing engine proposed solutions to problems that usually arise in mobile phone sensors, such as performing calibration of the accelerometer independently of body position, reducing computational costs of microphone processing, and reducing the GPS duty cycle by taking into account the activity of the user. Focusing more specifically on the microphone, as one of the most ubiquitous but least exploited of the smartphone's sensors, SoundSense [86] is a scalable sound sensing platform for people-centric sensing applications which classifies sound events. It is a

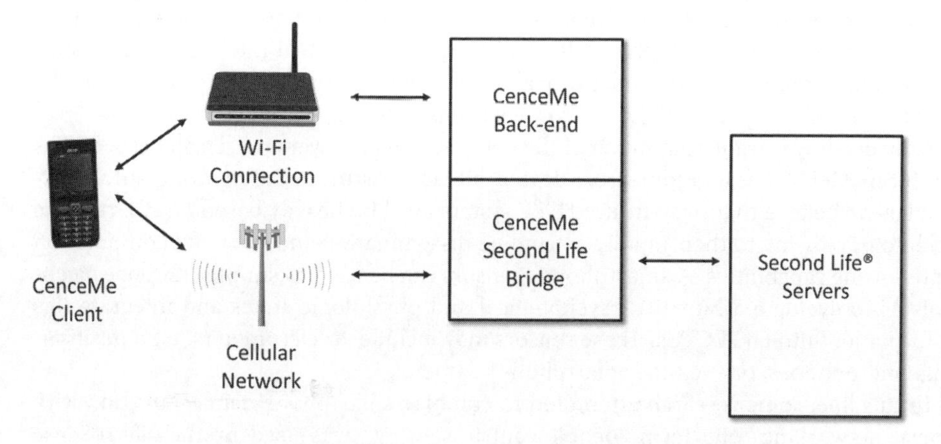

Figure 4.2 The architecture of CenceME [10], one of MetroSense's implementations.

general purpose sound sensing system for phones with limited resources that uses several supervised and unsupervised sensing techniques to classify general types of sound (music, voice) and discover novel sound events that are specific to individual users. These sorts of sensing architectures could be exploited to enable future continuous and ubiquitous smartphone sensing in HiTLCPSs.

4.1.2 State Inference

A recurring premise behind powerful HiTL systems is transparent interfaces that can infer human intent, physical and psychological states, emotions, and actions. While traditional interface schemes such as the mouse and keyboard have long been used to transmit human desires, they are not practical, involving series of key combinations or sequences of mouse clicks that are unintuitive and require practice and repetition in order to be learned and mastered. On the other hand, HiTLCPS applications are meant to react to natural human behavior and do not necessarily require direct human interaction. However, deriving advanced mathematical models or machine learning techniques that can reliably classify and possibly predict human behavior is a colossal challenge.

Many different methods for human activity classification can be found in the literature. One of the most successful and popular techniques is the use of Hidden Markov Models [87–89], but some approaches also use naive Bayes classifiers [90–93], Support Vector Machines [94], C4.5 [92, 94], and Fuzzy classification [95]. Research also uses different kinds of sensory data for activity detection: wearable sensor boards with many different types of sensors [87, 88], wearable accelerometers [89, 92, 94, 96] , gyroscopes [96], ECG [94], heart rate [92], smartphone accelerometer data and sound [97], and even RSSI signals [98]. The application of activity recognition is present in many areas, from sport solutions to social networking and health monitoring. Research in these topics is very active and presents very good results, in some cases achieving accuracy levels in the order of 90–95%.

The detection of a user's psychological state has been previously attempted. A communication framework for human–machine interaction that is sensitive to human affective states is presented in [99], through the detection and recognition of human affective states based on physiological signals. Since anxiety plays an important role in various human–machine interaction tasks and can be related to task performance, this framework was applied in [100] to specifically detect anxiety through the user's physiological signals. The presented anxiety-recognition methods can be potentially applied in advanced HiTLCPSs.

One example of such an application is the use of smartphones in experience sampling method (ESM) studies. ESM is a research methodology that requires periodical notes of the participant's experience. The notes can encompass the participant's feelings at a given moment and are best employed when the subjects do not know in advance when they will be asked to note their experience [101].

Different means can be used to signal a request for the participant's notes. Traditional methods include the use of preprogrammed stopwatches controlled by the researcher. However, through a prototype smartphone-based ESM system, named EmotionSense, it was possible to study the influence of different sampling strategies on the inferred conclusions about the participants' behavior [102]. This prototype system was based on an Android application that used both "physical sensors" (including accelerometer, microphone, proximity, GPS location, and the phone's screen status) as well as "software sensors" (capturing phone calls and messaging activity). These sensors were used to evaluate the context of users and to trigger survey questions about their feelings, namely how positive and negative they felt, their location, and their social setting. These short questionnaires gave insight into the participants' moods and behavior. The application could be remotely reconfigured to vary the questions, sampling parameters, and triggering mechanisms that notified users to answer a questionnaire. The results were used to empirically quantify the extent to which sensor-triggered ESM designs influence the breadth of behavioral data captured in this kind of studies.

EmotionSense was also used to enhance behavior change interventions (BCI) by using the devices capabilities to positively influence human behavior [11]. Traditional BCIs involve advice, support, and relevant information for the patient's daily activities, which are given during sessions by therapists and coaches. Smartphones with their powerful sensing and machine learning capabilities, ubiquity, and presence allow for behavior scientists to use directed, unobtrusive, and real-time BCIs to induce lifestyle changes that may help people coping with chronic diseases, smoking addiction, diets, or even depression. Information can be delivered and measured in the moments when the users need it the most. For example, people addicted to smoking usually suffer from detectable stress when feeling the need to smoke, creating an opportunity for the system to send a notification urging them not to do so. Thus, detecting the user's context and emotions allows for interventions to be delivered at the right time and place. The authors identified three key components of BCI using smartphones (shown in Figure 4.3). Interestingly, they closely match the basic processes of HiTL control (refer to Figure 3.1), where *Monitor* corresponds to *Data Acquisition*, *Learn* is a part of *State Inference* and *Deliver* is a form of *Actuation*.

The EmotionSense application uses the gaussian mixture model machine learning technique to detect ongoing conversations and their respective participants. An emotion inference component was also developed using a similar approach, training a background gaussian mixture model representative of all emotions through the Emotional Prosody Speech and Transcripts library [103]. This component allowed the application to infer five broad emotional states from the smartphone's microphone: anger, fear, happiness, neutrality, and sadness. The authors reported an accuracy of over 90% for speaker identification and over 70% for emotion recognition.

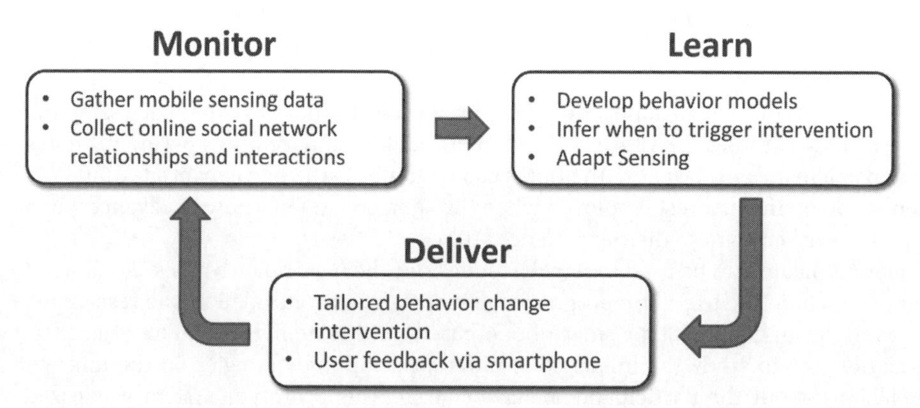

Figure 4.3 The three key components of BCI using smartphones [11]. *Source:* Adapted from Lathia et al. 2013.

Also with the objective of providing positive behavioral change, SociableSense [12] was a platform that monitored the user's social interactions and provided real-time feedback to improve their relations with their peers. In this work, the authors attempted to measure the "sociability" of users, which is an important factor in many behavioral disorders, ranging from autism to depression. The system then closed the loop by providing real-time feedback and alerts that aimed to make people more sociable. The sociability measurement was divided into two factors: collocation and interaction. Collocation was defined by the proximity between users, and it was inferred by a coarse-grained *Bluetooth*-based indoor localization mechanism. Interaction was derived from the speaking between users, and it was inferred via the microphone sensor and a speaker identification classifier, in a fashion similar to EmotionSense. Active socialization was promoted through a gaming system which classified the most sociable persons as "mayors" of the social groups. Results showed that such feedback mechanisms influenced users and increased their sociability. SociableSense's architecture is shown in Figure 4.4.

Several ongoing challenges for mobile sensing were also identified in [11], including:

- Energy constraints associated with continuous sensing, which require intelligent mechanisms that dynamically adapt sampling rates depending on the user's context.
- Data processing and inference mechanisms that can accurately extract information on human behavior from raw sensor data, and the importance of balancing the distribution of this computation among smartphone sensors and cloud-based back-ends.
- Generality of classification mechanisms that need to make uniform inferences regarding widely different populations of users.
- Privacy concerns about the acquisition of sensitive data (locations, activities) and the recording of data without people's informed consent (e.g. inadvertently capturing the voice of an external person through the microphone of a smartphone user).

The detection of emotion is not restricted to voice pattern recognition, however. Other interesting ways of inferring the user's psychological state have been proposed. The touch interface and movement of a smartphone were used in [104] as a way

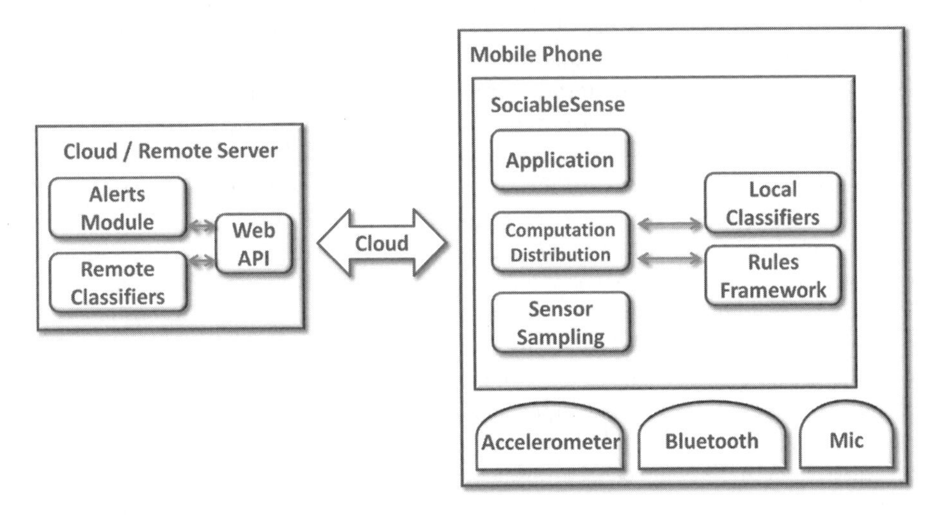

Figure 4.4 SociableSense architecture [12]. *Source:* Adapted from Rachuri 2011.

of inferring emotional states. The proposed framework consisted of an emotion recognition process and an emotion preference learning algorithm which were used to recommend smartphone applications, media contents, and mobile services that fit the user's current emotional state. The system collected data from three sensors: the touch interface, accelerometer, and gyroscope, classifying it into types (e.g. touch actions could be divided into tapping, dragging, flicking, and multi-touching). The processed data was used to quantify higher level emotions, such as "neutral", "disgust", "happiness", or "sadness", through decision tree classification methods. By analyzing communication history and application usage patterns, MoodScope [105] also inferred the mood of the user based on how the smartphone was used. The system passively ran in the background, monitoring application usage, phone calls, email messages, SMSs, web browsing histories, and location changes as user behavior features. With daily mood averages as labels and usage records as a feature table, the authors applied a regression algorithm to discern a mood inference model.

More recently, image-based processing of facial expressions was used to detect irritation in drivers [44] and to improve driving safety. The developed system was non-intrusive and ran in real-time. Through an NIR camera mounted inside the dashboard, a near frontal view of the driver's face was captured. A face tracker was applied to track a set of facial landmarks used to classify the facial expressions. Experimental results demonstrated that the system had a detection rate of 90.5% for in-door tests and 85% for the in-car tests.

4.1.3 Actuation

Actuation has a very broad definition in the field of HiTLCPS. For example, applications that passively monitor a human being's sleep environment to give information about potential causes of sleep disruption [106] or that record a human's cardiac sounds to detect pathologies [75] do not directly influence the associated environment, nor do they attempt to achieve a certain goal and yet they still "actuate" by providing information.

More direct actuation with the physical world can be achieved through specialized devices, such as robots [16]. Historically, robots were designed and programmed for relatively static and structured environments. Once programmed, it was usually expected for the robot's environment and interactions to remain within a very constrained range of actions. Anything unaccounted for in the robot's configuration is essentially invisible and only minimal feedback is traditionally available, such as joint position measurements. These primitive sensory capabilities require robots to operate in isolated "work cells", free from people and other interferences. Thus, current robots, including mobile ones, far from being integrated into HiTLCPSs, continue, on the whole, to use collision sensors that halt operation whenever something unaccounted for happens or whenever somebody enters their workspace, to prevent accidents. This continues to enforce the need for having areas for workers and areas for robots, which are mutually exclusive and preclude any type of human–robot cooperation typically found in HiTLCPSs [107]. Apart from safety reasons, there is also the lack of trust of workers in robots. People prefer to work alongside teleoperated robots than with autonomous ones [108]. The reason for this mistrust is that people cannot predict the robot's intentions or behavior, owing to the lack of body language signs, common in humans. A second reason for mistrusting robots is that people do not know if the robot "sees them" (lack of presence awareness).

Without HiTL behavior modeling, robots in many automated factories remain isolated in both physical and sensorial senses [109].

Yet, this is about to change. While robots were initially used in repetitive tasks where all human commands were given a priori, the next generations of advanced robots is envisioned to be mobile and operating in unstructured or uncertain contexts.

Achieving "human-awareness" requires robots to have sensing capabilities greater than mere joint position measurements. Interestingly, in recent years there has been a combination of two important technologies—robots and WSNs—that can complement each other in this respect. WSNs can assist in the process of discovering the environment where robots actuate; the detailed level of information provided by sensors may be essential for the tasks to be undertaken by a robot. On the other hand, robots can be used as mules that collect and forward information from several sensor nodes spread in the environment. Thus, the energy needed for long-distance or multi-hop transmissions can be reduced. Robots can also perform the calibration of sensors and support their recharging process when energy levels are low. Robots and wireless sensor technologies can be exploited to support remote monitoring in dangerous environments under maintenance, using a set of sensors to measure, for example, gas levels. They can also be applied in the monitoring of environmental parameters, such as in wastewater treatment facilities or for measuring air pollutants, allowing a proactive implementation of a social responsibility culture. Using wireless technology and robotic mobile inspection for the monitoring and surveillance of wide areas, where diagnosis and intrusion detection are critical, is also more reliable and cost-efficient than traditional methods.

While WSNs offer the sensorial capabilities necessary for robots to perform the desired tasks, humans provide the necessary management of their operation. Thus, robots are capable of performing missions in hazardous environments in cooperation with humans, taking into consideration the psychological state of humans, while using data from WSNs to scan both humans and the environment. In fact, the human–WSN–robot combination has huge potential in the perspective of actuation in HiTLCPSs, since advanced industrial automation can strongly benefit from distributed sensing capabilities. Robots, humans, and WSNs can be deployed to support personnel safety, by complementing human work in hazardous contexts, with wireless sensor networks collecting and processing information. Mobile workers and robots can be equipped with multiple sensory systems that send information to a control center, accessible and monitored by safety and health-control personnel. This allows workers to safely and remotely control operations, and to take faster decisions. The combination of these technologies allows us to envision highly advanced HiTLCPSs applied to many different scenarios. As an example, flying inspection robots could be used to navigate interactively and inspect power plant structures (including various components within and around boilers, environmental filters, or cooling towers), as well as oil and gas industrial structures (inside and outside large-scale chimneys, inside and outside flare systems, inside button part of refining columns, as well as pipelines and pipe webs). On the other hand, workers in the field may collaborate with these robots in their inspection tasks, in the management of the whole operation, and in the deployment and collection of sensor networks. HiTL controls allow for this collaboration to be safe for humans, since their presence, actions, and intentions are made known to individual robots as well as to the entire system.

There are several projects that specifically study and evaluate the integration of WSNs with robots. For example, the Robotic UBIquitous COgnitive Network (FP7-ICT-269914) [110] was a project that aimed to create autonomous and auto-configured systems by combining WSNs, multi-agents, and mobile robots. The proposed mechanisms reduced the complexity and the time needed in deployment and reconfiguration tasks. However, the main objective of this project was to remove, as much as possible, the human from the configuration and maintenance processes. According to the authors, this meant that the quality of service that was offered by robot WSNs was significantly improved, without the need for extensive human involvement. Considering that these technologies coexist with human beings, why were humans excluded from control loop decisions? Why not take advantage of the human potential to create immersive HiTLCPSs?

Other research initiatives focus more on this human–robot cooperation. For instance, the NIFTi European FP7 project (natural human–robot cooperation in dynamic environments, FP7-ICT-247870) [111], which ran from January 2010 until December 2013, proposed new models for cooperation between robots and humans when they work towards a shared goal, cooperatively performing a series of tasks. However, this project required a lot of direct instructions from human to robots, and WSNs were not used to dynamically contribute and adapt to these systems. PHRIENDS: Physical Human–Robot Interaction: Dependability and Safety (FP6-IST-045359) [112] was a project that aimed to propose the co-existence of robots and humans. One of its main objectives was to find the strictest safety standards for this coexistence. Later, this project resulted in a new PF7 project, SAPHARI [113], which maintained its main objective but now used soft robotics, combining cognitive reaction and safe physical human–robot interaction. In contrast with its precursor project, SAPHARI intended to provide reliable, efficient, and easy-to-use functionality. There are also other projects on the topic of safety in interaction between humans and robots. CHRIS (Cooperative Human Robot Interaction Systems; FP7-ICT-215805) [114] evaluated a mapping mechanism between robots and humans. This project also aimed at studying the safety of cooperative tasks between humans and robots. However, once again, these environments did not assume the existence and participation of WSNs. On the other hand, humans were just seen as end users and they were not integrated into the system. SWARMANOID: Towards Humanoid Robotic Systems (FP6-IST-022888) [115] and SYMBRION (Symbiotic Robotic Organisms; FP7-ICT-2007.8.2) [116] were two similar projects that aimed to find strategies to achieve collaborative work between robots. SWARMANOID proposed joint mechanisms both by air and land to achieve search tasks. The latter project intended to optimize energy by sharing policies. Robot-Era [117] was a project that started in 2012 and finished in 2015. It intended to implement and integrate advanced robotic systems and intelligent environments in real scenarios for the aging population. Some of these intelligent environments were based on WSNs, and their role was to support the quality of life and independent living for elderly people.

In Table 4.1, we summarize the main technologies/solutions that are discussed in this section.

Despite all of these efforts, much work still has to be done, in particular for robotic actuation that considers the human state. Thus, the role of robotics in future HiTLCPSs cannot be yet fully understood. In addition to the unmet technical challenges, there are

Table 4.1 Summary of some of the technologies/solutions that support HiTLCPS.

Technology	Description	Examples of applied research
Data Acquisition		
Wireless Sensor Networks	Networks of small, battery-powered devices	6LoWPan [80] and SenQ [9]
Body-Coupled Communication	Sensors that leverage the human body as their communication channel	[81] and [82]
Smartphones	Devices with powerful computation and sensing capabilities that accompany users throughout their days	MetroSense [84], Jigsaw [85], and SoundSense [86]
State Inference		
Physical Activity Classification	Detection of human activities (e.g. walking, brushing teeth,)	Hidden Markov Models [87–89], naive Bayes classifiers [90–93], Support Vector Machines [94], C4.5 [92, 94], and Fuzzy classification [95]
Psychological State Classification	Detection of human affective states (e.g. happiness, anger)	Frameworks such as [99], EmotionSense [11], SociableSense [12], smartphones' touch interface [104], MoodScope [105], and facial recognition [44]
Actuation		
Human–WSNs–Robots	WSNs offer the sensorial capabilities necessary for robots to perform the desired tasks, while humans provide the necessary management of their operation	RUBICON [110], NIFTi [111], PHRIENDS [112], SAPHARI [113], CHRIS [114], SWARMANOID [115], SYMBRION [116], and Robot-Era [117]
Notifications and Suggestions	Smartphone-based systems often show suggestive notifications	EmotionSense [11], SociableSense [12, 118], and HappyHour [119]

also questions of an ethical nature that will also need to be considered. We will identify some these matters in Section 11.2.

4.2 Experimental Projects

As previously mentioned, the area of HiTLCPSs is vast. Not only can they be applied to many different areas, the variety of their possible configurations and technologies makes it very difficult to establish well-defined borders for classification. Nevertheless, in this section we present several examples of some of the works in the area of HiTLCPSs that apply the various technologies discussed in Chapter 2. By no means do we intend to provide an extensive review on the state of the art; the purpose of this section is to give the reader a better idea of the applicability of HiTLCPS concepts.

4.2.1 HiTL in Industry and at Home

To contribute towards a better understanding of the spectrum of HiTL applications, the work in [13] provided its own implementation of an HiTL system that attempted to reduce the energy waste in computer workstations by modeling human behavior and detecting distractions. While current practices for reducing energy consumption are usually based on fixed timers that initiate sleep mode after several minutes of inactivity, the proposed system uses adaptive timeout intervals, multi-level sensing, and background processing to detect distractions (e.g. phone calls, restroom breaks) with 97.28% accuracy and cutting energy waste by 80.19% [13]. The control architecture of the system is shown in Figure 4.5. The proposed "distraction model" comprised two main sources of information: user activity and system activity. At user-activity level, the authors used a "gaze tracker", which evaluated the user's gazing at the computer's screen through a webcam. At system-activity level, the system evaluated keyboard and mouse events, CPU usage, and network activities to infer the machine's level of use. The control loop combined both types of information to determine the distraction status of the user, with some self-correcting measures. For example, if the user resumed the system shortly after it was put to sleep, the control loop took this as a negative feedback event, and subsequently adjusted the timeout interval.

At the same time, there is an increasing concern of corporations with the well-being and happiness of employees, since a happy employee is a more productive one. An article published in the December 2012's edition of *IEEE Spectrum* [120] discusses how the same technology advances in computers and telecommunications that have brought about tremendous gains in productivity may also be applied to increase happiness, instead of stress. The work of engineers and psychologists over the last few decades has

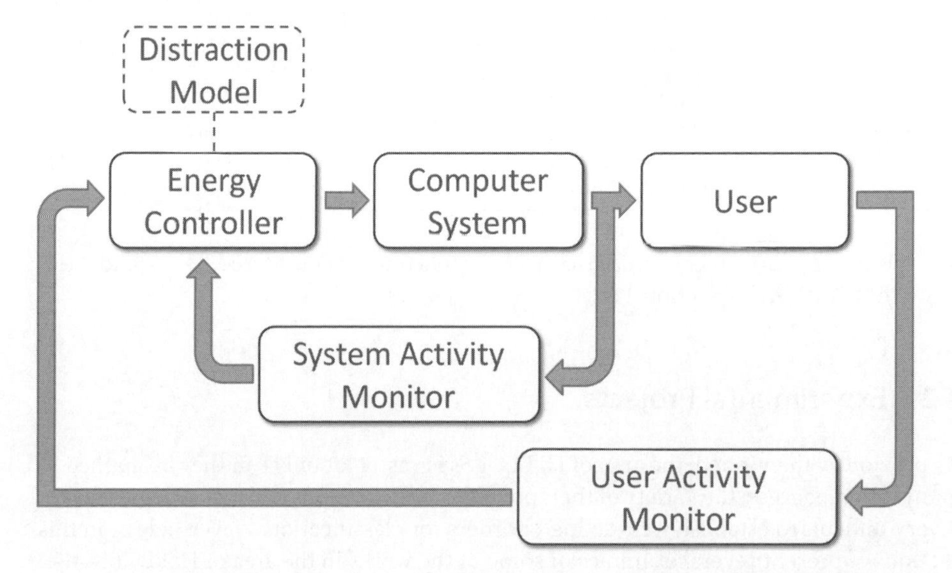

Figure 4.5 Control architecture for energy saving with HiTL [13]. *Source:* Adapted from Liang 2013.

allowed us to infer a person's level of happiness by monitoring and analyzing a person's sleep patterns, exercise and dietary habits, as well as vital statistics like body temperature, blood pressure, and heart rate. This technology might be used to improve the overall environment of the workplace, resulting in better communication, teamwork, and job satisfaction. The Hitachi Business Microscope is a small wearable device containing six infrared transceivers, an accelerometer, a flash memory chip, a microphone, a wireless transceiver, and a rechargeable lithium-ion battery that allows the badge to operate for up to two days at a time. It measures the wearer's body movements, voice level, and location, as well as the ambient air temperature and illumination. When these transceivers detect another badge within two meters, the two badges exchange IDs and each badge then records the time, duration, and location of the interaction. This allows the collection of data on the type of social exchanges that take place in the workplace. The Hitachi Business Microscope is being used by hundreds of organizations (banks, design firms, research institutes, hospitals, etc.) to collect behavioral data. This data is then used in conjunction with studies from the field of positive psychology, which focuses on desired mental states (including happiness), to improve people's personal and professional lives. Happy people tend to be more creative, more productive, and feel more fulfilled by their work. Happy people also tend to more easily achieve a state of full engagement and concentration. Interestingly, the research presented in [120] suggests that it is possible to infer when a person has reached this state by analyzing the consistency of their movements: for some people, that consistent movement is slow, while for others it is fast. The time of day during which people tend to experience the "flow" is also highly variable and individual, some people favor mornings while others favor afternoons or evenings. Regardless of when participants experience "flow", their motions become more regular as they immerse themselves in the activity at hand. One advantage of measuring the worker's activity is that once people become aware of their daily patterns they can better schedule their work to take advantage of times when they are most likely to be in this focused mental state. Documenting social interactions can also help to identify the areas in an office which tend to host the most frequent and active discussions, helping in the restructuring of office layouts to foster more fruitful collaboration.

The area of human-computer interaction has long studied the concept of HiTL. Humans prefer to attend to their surrounding environment and engage in dialog and interaction with other humans rather than to control the operations of machines that serve them. Thus, in [121] it is suggested that we must put computers in the human interaction loop (CHIL), rather than the other way around. In this line, a consortium of 15 laboratories in nine countries has teamed up to explore what is needed to build usable CHIL services. The consortium developed an infrastructure used in several prototype services, including a proactive phone/communication device; the Memory Jog system for supportive information and reminders in meetings, collaborative supportive workspaces, and meeting monitoring; and a simultaneous speech translator for the lecture domain. These projects led to several advances in the areas of audiovisual perceptual technologies, including speech recognition and natural language; person tracking and identification; identification of interaction cues, such as gestures, body and head poses, and attention; as well as human activity classification.

HVAC systems have also been endowed with HiTL control. For example, in [122] the authors implemented a system that used cheap and simple wireless motion sensors and door sensors to automatically infer when occupants were away, active, or sleeping. The

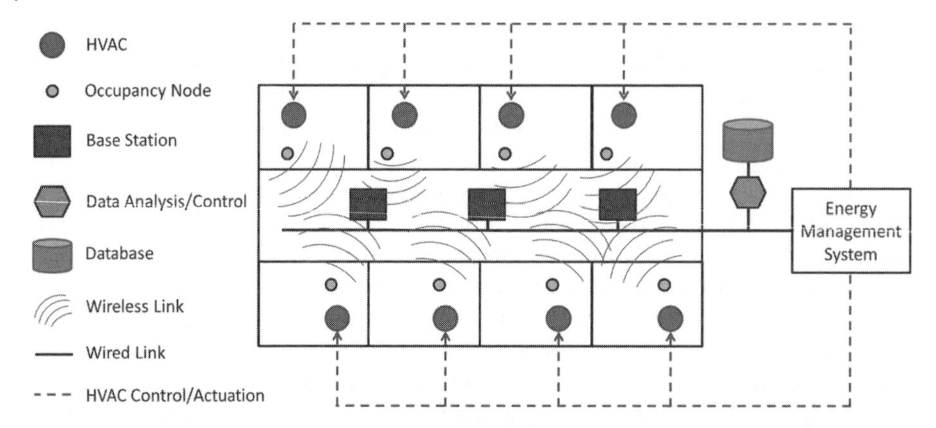

Figure 4.6 Architecture of an HiTL HVAC system [14]. *Source:* Adapted from Agarwal 2011.

system used these patterns to save energy by automatically turning off the home's HVAC system as much as possible without sacrificing occupant comfort, effectively creating a HiTLCPS. Another example can be found in [14], where an occupancy sensor network was deployed across an entire floor of a university building together with a control architecture (see Figure 4.6) that guided the operation of the building's HVAC system, turning it on or off to save energy, while meeting building performance requirements.

4.2.2 HiTL in Healthcare

CAALYX [15] was a research work that intended to develop a wearable light device directed towards the monitoring of elderly people that could measure specific vital signs in order to detect falls and to automatically communicate in real time with assistance services in case of emergency (Figure 4.7). A number of wireless sensors that detected several vital signs (blood pressure, heart rate, temperature, respiratory rate, etc.) were used. These wireless sensors communicated within a body-area-network and with a mobile phone with GPRS access. The mobile phone was able to analyze the data and detect emergency situations, during which it would contact an emergency service, regardless of the elderly person's location. The CAALYX project also developed an initial simulation of its workings in the Second Life® virtual world, as a means of disseminating and showcasing the project's concepts to wider audiences [123]. There are two interesting aspects of this work: its use of mobile phones as gateways for ubiquitous communication and its early attempts at integrating health monitoring with virtual environments.

Schirner *et al.* [16] stress the existing multidisciplinary challenges associated with the acquisition of human states in HiTLCPS. For example, embedded systems are key components used in these systems and, as such, they propose a holistic methodology for system automation in which designers develop their algorithms in high-level languages and fit them into an electronic system level tool suite which acts as a system compiler, producing code for both the CPU and the field-programmable gate array. This automation allows researchers to more easily test their algorithms in real-life scenarios and to focus on the important task of algorithm and model development. Schirner *et al.* also used an EEG-based brain–computer interface for context-aware sensing of a human's status, which influenced the control of an electrical wheelchair. To improve the intent inference accuracy, the authors suggest that inference algorithms should adapt to the

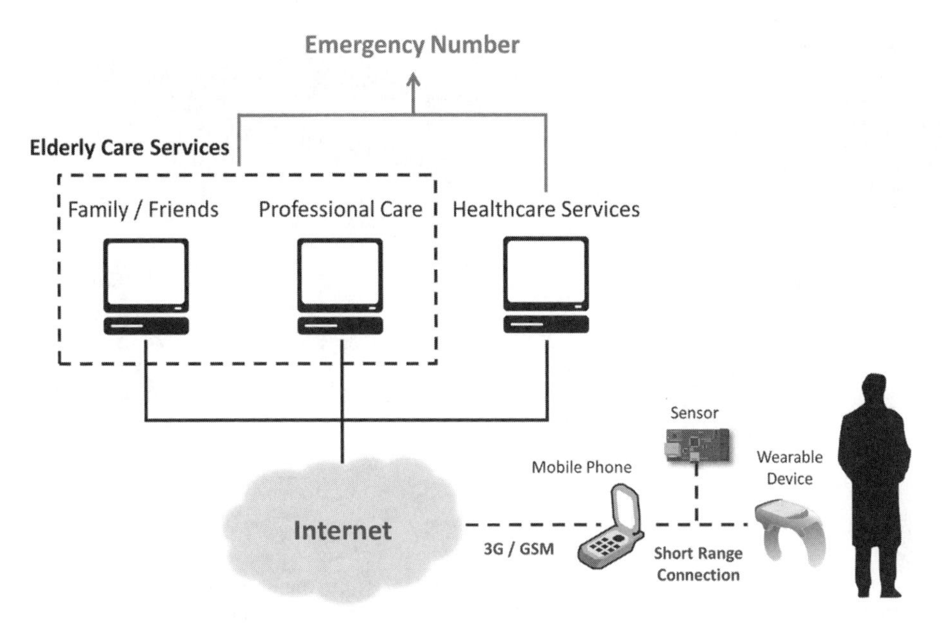

Figure 4.7 Diagram showing the main components of CAALYX's roaming monitoring system [15]. *Source:* Adapted from Boulos *et al.* 2007.

current application as well as to the user's preferences and historical behavior, that is they should use application-specific history and contextual information. The field of robotics is also addressed, as robots are the primary means for actuation and interaction with the physical world in CPSs. Their semi-autonomous wheelchair interpreted brain signals that translated high-level tasks such as "navigate to kitchen", and then executed path planning and obstacle avoidance [16]. However, important research questions are still unresolved in this area, namely the problem of dividing control between human and machine, as well as the modularity and configurability of such systems. Distributed sensor architectures are also very important for HiTL since they allow the measurement of physiological changes, which may be processed to infer current human activities, psychological states, and intent. In this regard, BCC was presented as a means of supporting low-energy usage, high bandwidth, heterogeneity, and reduced interference.

At the Worcester Polytechnic Institute [124], a HiTLCPS prototype platform and open design framework for a semi-autonomous wheelchair was developed. The authors considered disabled individuals, namely those suffering from "locked-in syndrome", a condition in which an individual is fully aware and awake but all voluntary muscles of the body are paralyzed. To improve the life condition of these individuals, they created a HiLCPS wheelchair system which used infrared and ultrasonic sensors to navigate through indoor environments, enabling the user to share control with the wheelchair in an HiTL fashion. This allowed handicapped individuals to live more independently and have mobility. The resulting prototype used modular components to provide the wheelchair with a degree of semi-autonomy that would assist users of powered wheelchairs to navigate through the environment. This work was extended in [16], where the user could interface with the wheelchair through a brain–computer interface based on steady-state visual-evoked potentials induced by flickering light patterns in

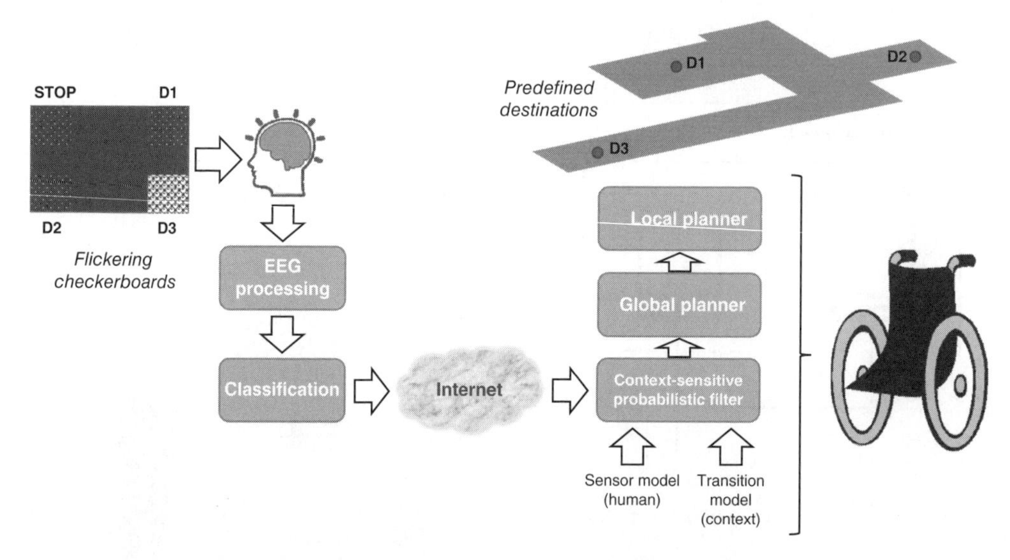

Figure 4.8 A semi-autonomous wheelchair receives brain signals from the user and executes the associated tasks of path planning, obstacle avoidance, and localization [16]. *Source:* Adapted from Schirner 2013.

the operator's visual field (see Figure 4.8). A monitor showed flickering checkerboards with different frequencies. Each checkerboard and frequency corresponded to one of four desired locations. When the operator focused on a desired checkerboard on the monitor, his/her visual cortex predominantly synchronized with the checkerboard's flickering harmonic frequencies. These frequencies were detected through an electrode on the scalp near the occipital lobe, where the visual cortex is located.

On the other hand, there are other projects that focus on the development of intelligent wheelchairs with HiTL to assist disabled people. The work "I Want That" [74] proposes a system that controls a commercially available wheelchair-mounted robotic arm. Since people with cognitive impairments may not be able to navigate the manufacturer-provided menu-based interface, the authors improved it with a vision-based system which allows users to directly control the robotic arm to autonomously retrieve a desired object from a shelf. To do so, they use a touchscreen which displays a shoulder camera view, an approximation of the viewpoint of the user in the wheelchair. An object selection module streams the live image feed from the camera and computes the position of the objects. The user can indicate "I want that" by pointing to an object on the screen. Afterwards, a visual tracking module recognizes the object from a template database while the robotic arm grabs the object and gives it to the user.

A vision-based robot-assisted device to facilitate daily living activities of spinal cord injured users with motor disabilities is also proposed in [125], through an HiTL control of a robotic arm. The research objective was to reduce time for task completion and the cognitive burden for users interacting with unstructured environments via a wheelchair-mounted robotic arm. Initially, the user needed to indicate the approximate location of a desired object in the camera's field of view using one of a number of diverse user interfaces, including a touchscreen, a trackball, a jelly switch, and a microphone.

Afterwards, the user could order the robotic arm to center the object of interest in the visual field of the camera, and then grab the desired object.

A model-driven design and validation of closed-loop medical device systems is presented in [126]. The safety of a closed-loop control system of interconnected medical devices and mechanisms was studied in a clinical scenario, with the objective of reducing the possibility of human error and improving patient safety. A patient-controlled analgesia pump delivered a drug to the patient at a programmed rate while a pulse oximeter received physiological signals and processed them to produce heart rate and peripheral capillary oxygen saturation outputs. A supervisor component got these outputs and used a patient's model to calculate the level of drug in the patient's body. This, in turn, influenced the physiological output signals through a drug absorption function. Based on this information, the supervisor decided whether to send a stop signal to the pump. The main contribution of the project was the methodology for the analysis of safety properties of closed-loop medical device systems.

4.2.3 HiTL in Smartphones and Social Networking

Exploiting the line of people-centric sensing, the authors of [10] propose the use of sensors embedded in commercial mobile phones to extrapolate the user's real-world activities that in turn can be reproduced in virtual settings. The authors' goal was to go further than simply representing locations or objects; they intended to provide virtual representations of humans, their surroundings, and their social interactions. The proposed system prototype implementation was named CenceMe, and allowed members of social networks, namely Second Life®, to process the information sensed by their mobile phones and use this information to extrapolate the user's surroundings and actions (refer to Figure 4.2). The authors proposed to use mobile phone sensors, such as microphones or accelerometers, to infer the user's current activity. In fact, today's mobile phones are powerful enough to run activity recognition algorithms, and the results can be sent to virtual worlds thanks to the phone's mobile Internet capabilities. Activity recognition algorithms extracted patterns from the obtained data, such as the current user status in terms of activity (e.g., sitting, walking, standing, dancing, or meeting friends), and logical location (e.g., at the gym, coffee shop, work, or other). This information was reflected in the virtual world, where the user's friends could see what activities he/she was performing at a given moment and his/her current geographical position. Real-world activities could be mapped to different activities in the virtual world; for example, the user could choose to have real-world running represented by flying on the virtual world. The avatar's clothes and accessories could also be changed according to a user's location; the avatar's shirt could display a logo of the user's current location (cafe, home, school, etc.). The authors also used external sensors to complement the ones provided by the mobile phone, namely they suggested the use of galvanic skin response sensors to infer emotional states and stress levels. The activity recognition algorithms were performed on the mobile device in order to reduce communication costs and computational burden on the server. However, in cases where the algorithms were too intensive for handheld devices, the computation was partially or even completely performed on the server side. The use of mobile phones as means of sensing and communicating with virtual realities is an important aspect of this work, as relying on common and easily accessible technologies fosters the adoption of these new systems by more users.

A few more recent mobile applications attempted to combine context awareness with user social connections. Highlight [127] is a social application that allows users to learn more about the people around them by displaying profiles of nearby users. The application presents several data items, including names, photos, mutual friends, and other information users have chosen to share, as well as a tiny map that shows their recent location (in a fashion similar to the mockup shown in Figure 4.9). The closer a person is to the user (the more interests, friends, or history they have in common), the more likely the user will be notified of their presence. According to Paul Davison, Highlight's chief executive officer, the application "started with the idea that if you can just take two people and connect them, you can make the world a better place" [128]. Thus, Highlight hopes to increase synchronicity and reduce the friction in meeting new people, allowing users to know a few things about each other in advance.

The SceneTap [129] application is an even more flexible and complex example of detection of people for social networking purposes. The application uses anonymous facial detection software to approximate the age and gender of people entering a nightclub environment. By counting the number of people entering and leaving a venue, the application can estimate and report crowd size, gender ratios, and the average age of people in a given location. This information is shared among users, allowing them to better plan their nights out and decide which nightlife establishments are a better fit for their desires.

HiTL concepts have also been applied to smartphone data usage. In fact, HiTL has previously been proposed as a solution for addressing the increasing demand for wireless data access [118]. Since the wireless spectrum is limited and shared, and transmission rates can hardly be improved solely with physical layer innovations, a "user-in-the-loop" mechanism was proposed that promoted spatial control, in which the user is encouraged to move to a less congested location, and temporal control, in which incentives, such as dynamic pricing, ensure that the user reduces or postpones his/her current data demand in case the network is congested. This closed loop

Figure 4.9 A mockup of a map interface similar to the Highlight application.

controlled user activity itself through suggestions and incentives, influenced by the current location's signal-to-interference-plus-noise ratio and traffic situation. The authors propose that users receive control information in the form of a graphical user interface, showing a map and directions towards a better location and a better time to start the user traffic session (e.g. outside busy hours).

Another area that has been closely linked with smartphones and HiTLCPS is the area of recommendation systems. On this information era, the ability to quickly and accurately understand consumers' desires allows companies to timely control supply and demand, and cope with quick changes in consuming trends. There has been a considerable amount of research in the area of data mining to derive intelligence from large amounts of transaction records, so that individual consumer marketing strategies can be developed [130]. Since consumers are also influenced by relevant information provided by retailers, context-aware recommendation systems can have a considerable effect on consumerism dynamics. In particular, smartphones are amazing candidates for sensing and understanding consumer context (*Data Acquisition* and *Inference*) and powerful dissemination vectors for recommendations (*Actuation*). In a sense, these systems can be considered open-loop HiTLCPSs as they usually do not directly affect the environment or the consumer, but merely suggest products and services. A very large body of research work has studied how smartphones can be used in this context. However, we will only present a few examples of such research work, for the sake of brevity.

In [130], context-aware recommendation systems for smartphones were divided into two modules to provide product recommendations. First, a simple RSSI indoor localization module located the user's position and determined his/her context information. RFID readers would be equipped in shopping centers with a consumer location mechanism. Consumers, would place their RFID tags close to the readers and let them recognize their identity and location (based on which reader was accessed). Second, a recommendation module provided directed product information to users, through association rules mining. The system performed recommendation calculations pertaining to merchandise in the region of the user, and passed on this information through the smartphone.

Another example can be found in [17], where the authors managed to create a recommendation system to suggest smartphone applications. The motivation behind this work is the huge number of available apps; Google's own Android *Play Store* currently has over 1,600,000 applications [131]. Bayesian networks processed data from several of the smartphone's sensors, including accelerometer, light, GPS, time of day, and date, to perform context inference (see Figure 4.10). From this information, the recommendation system was able to associate the user's context with application categories of interest, based on the author's domain knowledge. The categories included communications, health and fitness, medical, media and video, news and magazines, weather, business, social, games, or traffic information.

In a similar vein, the authors in [132] proposed AppJoy, a system that also made personalized application recommendations; however, instead of using the smartphone's sensors to understand context, the system actually analyzed how the user interacted with his/her installed applications. AppJoy measured usage scores for each app, which were then used by a collaborative filter algorithm to make personalized recommendations. What the user did directly affected his/her application profiling. AppJoy followed

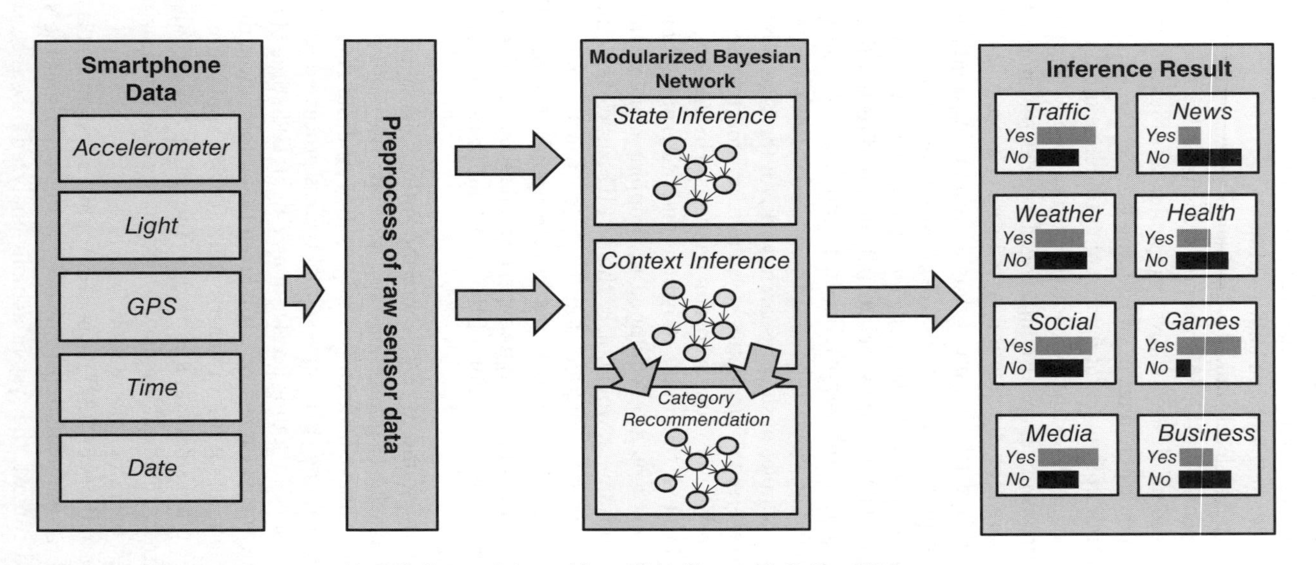

Figure 4.10 Overview of the system proposed in [17]. *Source:* Adapted from W.-H. Rho and S.-B. Cho 2014.

Table 4.2 Summary of experimental HiTLCPS projects.

Project	Main objectives and features
Industry and home	
Reducing Energy Waste for Computers by Human-in-the-loop Control [13]	Attempted to reduce the energy waste in computer workstations by detecting distractions
Can Technology Make You Happy? [120]	Used the Hitachi Business Microscope to acquire workspace behavioral data. This data was then used in conjunction with positive psychology to improve people's mood, creativity, and productivity
Handbook of Ambient Intelligence and Smart Environments [121]	Focused on human–computer interaction within smart environments and described several prototype services (proactive phone/communication, information reminders, collaborative supportive workspaces, speech translators, etc.)
The Smart Thermostat [122]	Smart HVAC systems that used HiTL control to improve performance and save energy
Duty-cycling buildings aggressively [14]	
Healthcare	
CAALYX [15]	Used wearable devices to measure specific vital signs and detect falls of elderly people. Mobile phones with GPRS access were also used to analyze data and automatically communicate with assistance services
Modular Designs for Semi-autonomous Wheelchair Navigation [124]	Presented a semi-autonomous wheelchair that navigated through indoor environments and was controlled through a brain–computer interface based on flickering light patterns
The Future of Human-in-the-loop Cyber-physical Systems [16]	
I Want That [74]	HiTL control of wheelchair-mounted robotic arms through touchscreens, visual tracking of objects, trackballs, jelly switches, and microphones
HiTL control of an assistive robotic arm in unstructured environments for spinal cord injured users [125]	
Toward patient safety in closed-loop medical device systems [126]	Design and validation of a closed-loop medical device system that consisted of an HiTL-controlled analgesia pump

(Continued)

Table 4.2 (Continued)

Project	Main objectives and features
Smartphones and social networking	
CenceMe [10]	Second Life® users could use their smartphone's sensors to automatically detect and share their location and actions in the virtual world
Highlight [127]	Mobile social application that allows users to learn more about the people around them by displaying profiles of nearby users
SceneTap [129]	Application that used anonymous facial detection software to approximate the age and gender of people entering a nightclub
User-in-the-loop: spatial and temporal demand shaping for sustainable wireless networks [118]	HiTL control of the wireless spectrum through suggestions and incentives that encouraged less congested locations and the reduction of data demand
Recommendation-aware Smartphone Sensing System [130]	HiTL recommendation systems for smartphones that used location, sensors, and interaction to understand user context
Context-aware smartphone application category recommender system with modularized Bayesian networks [17]	
AppJoy [132]	

a ubiquitous usability approach, being completely automatic, without requiring manual input, and adapted to changes of the user's application taste.

Let us summarize the projects presented in this section by looking at Table 4.2.

4.3 In Summary...

The objective of this chapter was two-fold. In Section 4.1, we presented some of the technologies (summarized in Table 4.1) that can support current and future HiTLCPSs in terms of *Data Acquisition*, *State Inference*, and *Actuation*. This section was meant to give the reader an overview of the tools currently available to him.

On the other hand, Section 4.2 presented several projects and scientific prototypes (summarized in Table 4.2) to give the reader a better idea of how the previously presented HiTL concepts and technologies can be applied in real-world scenarios.

As such, we hope that, by now, the reader has enough awareness of the theoretical concept behind HiTLCPSs to begin a practical exercise. In fact, in the next part of this book we will attempt to strengthen this theoretical understanding through a hands-on approach.

Part II

Human-in-the-Loop: Hands-On

In this part of the book, we will perform a step-by-step tutorial on how to create a simple, collaborative HiTL Android application, named HappyWalk. Our application will be a BCI system that will roughly estimate the user's current mood to improve their physical and mental well-being. This sample application requires some knowledge of Android programming and the Java programming language, as well as some notions about databases and RESTful web services.

Our main goal with this part of the book is to guide the reader through the creation of a simple HiTLCPS. It is not our intention to provide in-depth knowledge about Java or Android programming or the necessary machine learning algorithms to create complex HiTL systems. Instead, we aim at giving the reader some hands-on experience that might be helpful in consolidating some of the theories ideas presented in the previous chapters.

This part of the book is composed of Chapters 5 through 9. In Chapter 5 we will describe the objectives and concepts of our sample app, its base architecture, and techniques to be used. Chapter 6 explains how to install the necessary software, libraries, dependencies, and development environments. Chapter 7 focuses on the *Data Acquisition* process of the application and describes how to acquire and pre-process data from the smartphone's sensors. Chapter 8 is dedicated to the process of *State Inference* and explains how to implement a machine learning technique and acquire user feedback. Finally, chapter 9 discusses *Actuation* and explains how to handle the results of human-centric data and provide feedback to the user.

A Practical Introduction to Human-in-the-Loop Cyber-Physical Systems, First Edition.
David Nunes, Jorge Sá Silva and Fernando Boavida.

Companion Website URL: www.wiley.com/go/nunesloop

5

A Sample App

This chapter presents general architectural aspects of the sample HappyWalk Android HiTLCPS application and, in doing so, some of its underlying concepts and technologies, from a practical perspective. We encourage avid readers to further explore the presented technologies and solutions by complementing this presentation with material from books focusing on each of the addressed topics.

HappyWalk's base architecture will be presented, comprising an Android client application and a server-side application. Also, the technologies that will be used for the application development are briefly identified. Subsequently, the main classes that constitute this sample app will be listed and succinctly explained, both for the client side and the server side. Finally, the architectural options concerning emotion awareness will be presented and justified, and some initial implementation aspects will also be discussed.

5.1 A Sample Behavior Change Intervention App

As previously mentioned in Section 4.1.2, BCIs are therapeutic systems that focus on providing advice, support and relevant information to patients. Traditionally associated with presential therapeutic consultations, BCIs have recently begun to be delivered through the Internet and smartphones [11]. Using smartphone sensors to monitor humans with BCIs not only provides more effective feedback to help users in adapting or controlling some aspects of their behavior but also helps behavioral scientists' research. The journey towards our first HiTL system begins with a simple Android application, named HappyWalk, which we will modify throughout the book to introduce several HiTL capabilities, turning it into a full BCI system. HappyWalk will be an HiTL BCI application that attempts to positively influence its user's mood through moderate physical exercise.

In fact, recent research work has found evidence that moderate walking exercise and a change of environments can contribute to the improvement of mental health, providing several cognitive benefits such as improved memory, attention, and mood [133, 134]. Other studies suggest that contact with natural environments not only makes people feel better but also makes them behave better, thus presenting both personal health benefits and broader social benefits [135].

A Practical Introduction to Human-in-the-Loop Cyber-Physical Systems, First Edition.
David Nunes, Jorge Sá Silva and Fernando Boavida.

Companion Website URL: www.wiley.com/go/nunesloop

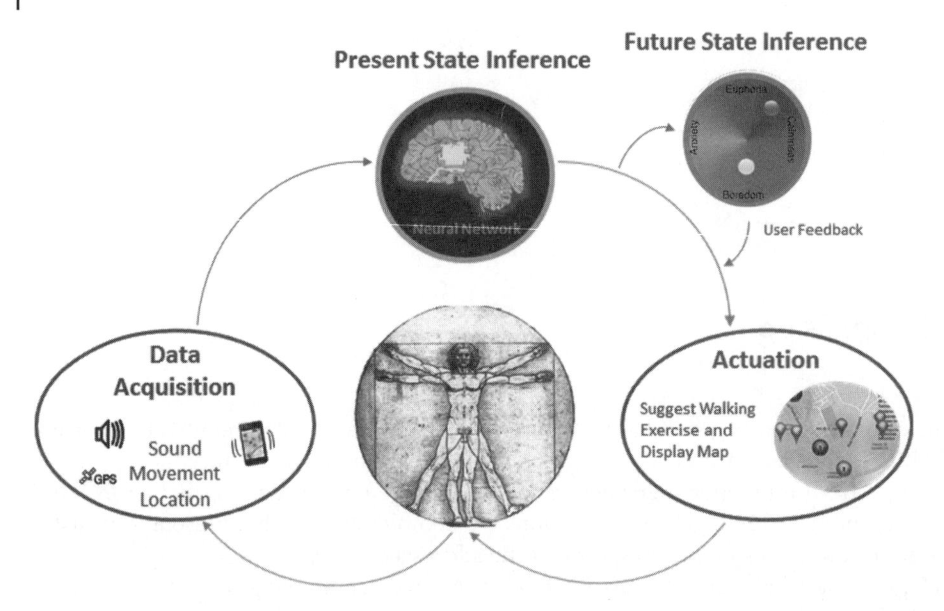

Figure 5.1 HappyWalk HiTL control.

Thus, in this part of the book we will develop and extend an Android app that promotes walking with the aim of improving the user's mood. Our objective is to use the smartphone's sensors and a machine learning algorithm to trigger positive feedback notifications that suggest walking exercise whenever the data from the smartphone's sensors indicates a negative state of mind. Collaborative data gathering is also employed to show heatmaps representing the near real-time context of nearby points of interest (POIs) that might be of interest to visit. Thus, HappyWalk will employ a full closed-loop HiTL control, as shown in Figure 5.1.

5.2 The Sample App's Base Architecture

Since the objective of this book is not to teach Android programming, we will not be developing HappyWalk from scratch. Instead, we will be enhancing an existing base app, capable of showing POIs on a map, with HiTL control. A high-level architecture of HappyWalk can be seen in Figure 5.2, which shows that the basis of the system is composed of an Android client application and a server-side application.

5.2.1 The Android App

The focus of this chapter is on HappyWalk's Android App, responsible for the interaction with the end user. It displays a map where the relevant POIs are shown, as well as menus that show information about them.

Android is an ecosystem supported by the Open Handset Alliance made up of devices and, primarily, an open-source operating system (OS) designed for smartphones, tablets, and other embedded devices. While Android applications are written in the **Java programming language**,[1] they are not run within a traditional Java virtual

1 See *The Java® Language Specification*; https://docs.oracle.com/javase/specs/jls/se8/jls8.pdf.

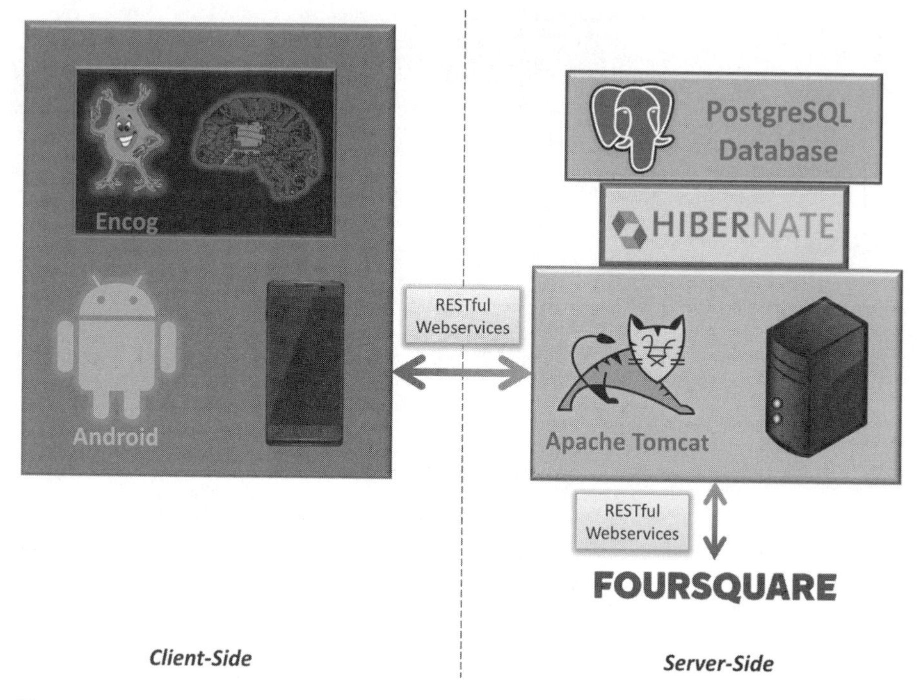

Figure 5.2 HappyWalk's architecture.

machine. Android has its own runtime (Android Runtime and, in older devices, its predecessor Dalvik) and performs its own management of the application's life cycle. The end user is not concerned about which apps are running. Android is optimized for low-power, low-memory devices, and is capable of closing and opening processes as the device's capabilities dictate. This directly translates into a programming in which the developer has to always be aware of the possibility of the app suddenly being shut down because of processing or memory constraints.

For the didactic purposes of our book, we would like to ensure that our application is compatible with as many versions of Android as possible. Thus, as our minimum Android SDK we chose API 10, corresponding to Android 2.3.3 (Gingerbread). This ensures compatibility with 99.5% of Android handsets, according to Google's statistics.[2] This tutorial uses Android API level 21 as the compile SDK; since Android is an ever-evolving development environment, the reader might be tempted to use a more recent compilation API. However, to ensure that the tutorial can be smoothly followed, we ask the reader to refrain from doing so, since it would certainly imply the adaptation of various parts of the code.

Additionally, during this part of the book we **strongly** recommend using a real device to develop the application, as virtual devices cannot be used to debug the usage of sensors, such as microphone and accelerometer.

Most Android apps are constructed by linking several building blocks known as *Activities*. In their essence, an activity can be described as a "thing" that the user can do.[3] This implies that, for the most part, an activity has a way of interacting with the

2 https://developer.android.com/about/dashboards/index.html?utm_source=ausdroid.net
3 http://developer.android.com/reference/android/app/Activity.html

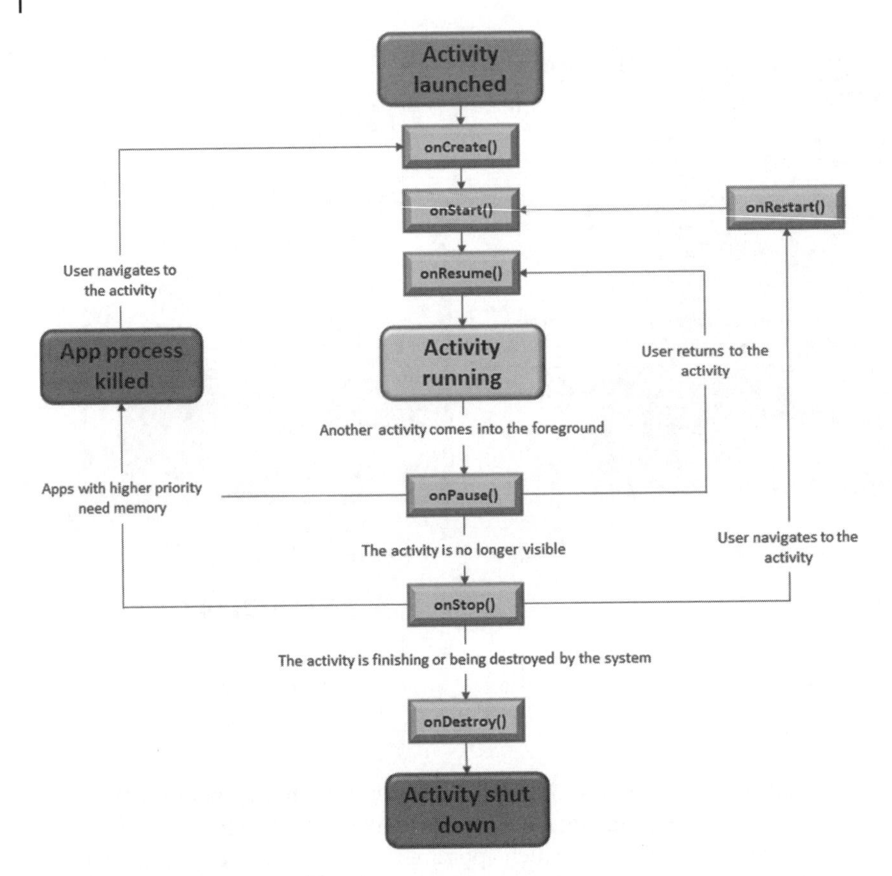

Figure 5.3 Android's *activity lifecycle*.

user (typically, a graphical user interface). The managing of an Android application is inherently tied to the lifecycle of its activities. During their lifecycle, Android activities can be in one of several states, as shown in Figure 5.3, which displays Android's *Activity Lifecycle*. Activity states are managed by the OS itself; the developer has the responsibility of handling the transitions between each state. This is done through special method calls, which should be overridden whenever necessary.

The entire lifetime of an activity occurs between the first call to *onCreate()* and a single final call to *onDestroy()*:

- ***onCreate()*** is called when the activity begins. It is typically used to perform initialization tasks and the global setup of the activity.
- ***onRestart()*** runs when a previously stopped activity restarts and is always followed by *onStart()*.
- ***onStart()*** marks the beginning of the *visible lifetime* of an activity. It is called to indicate that an activity is about to be displayed and can be used to maintain the necessary visual resources.
- ***onResume()*** is called when an activity is about to become interactable. It is important to note that this method may be called multiple times, since an activity may frequently

waver between the resumed and paused states. Thus, this is a good place to update visual elements or start animations and music.

- ***onPause()*** is a very important method and is called whenever an activity is about to go into the background. The importance of this activity pertains to the fact that this is the last *non-killable* method in older Android versions (pre-3.0); that is, after this method returns, the process hosting the activity may be killed by Android *at any time*, following memory or processing constraints. This is important to us, since we will be targeting Android 2.3.3 (Gingerbread). Therefore, *onPause()* should be used to write any persistent data to storage. Additionally, this method is also typically used to stop animations, music, etc.
- ***onStop()*** is called when the activity is no longer visible to the user. Starting with Android 3.0: Honeycomb, an application is not in the killable state until this method has returned. The method can be followed by either *onRestart()* (activity goes back to user interaction) or *onDestroy()*.
- ***onDestroy()*** marks the end of the activity's lifecycle, called right before it is destroyed. This is triggered by specific methods (e.g. *finish()*) or because the system is destroying the activity to save space.

An overview of HappyWalk's class structure is shown in Figure 5.4, and the relationship between its main classes is detailed in Figure 5.5.

HappyWalk's primary class is named *MapsActivity*, which encompasses the handling of the application's main activity: the one containing the map. HappyWalk uses the

Figure 5.4 HappyWalk's Android class structure.

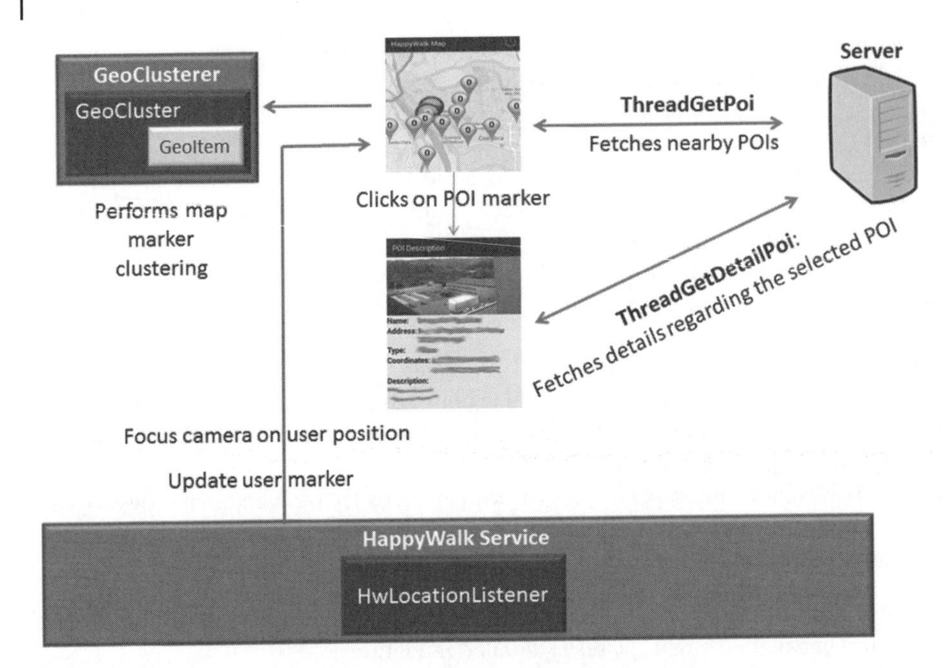

Figure 5.5 An overview of HappyWalk Android app's main classes.

Google Maps Android API v2 to show its map and POIs.[4] The *MapsActivity* class is responsible for managing the Google Maps object and the user position marker. The clustering of individual marker POIs on larger icons that occurs when the user zooms out is performed by the *GeoClusterer* class, in conjunction with the *GeoCluster* and *GeoItem* helper classes.

Whenever the user taps on a POI marker, the *POIDescription* activity is called. This activity is responsible for displaying information regarding the selected POI, such as a name, address, description, coordinates, and an illustrative image.

Another essential component of HappyWalk is its background service, providing various functionalities to other classes. It runs on the background in a separate thread, showing a notification while it is running, and also provides a handler object that can be used to perform background tasks. The background service supports the *HwLocationListener* class, which is responsible for acquiring and managing the user's location.

All of these functions are supported by helper *Thread* classes, which perform various tasks in parallel with the main application (e.g. on their on thread). Namely, *ThreadGetPoi* and *ThreadGetDetailPoi* handle the fetching of POI information.

Finally, the *utilities* package contains several useful classes that are used throughout the application. *CalcDistance* is used to calculate the distance between two points from their latitude and longitude coordinates. The *CommunicationClass* contains several methods that facilitate communicating with the server. Lastly, the *GlobalVariables* class is used to store values that are used by most other classes. This class allows it to easily tune various aspects of the application (server URL, when to update POIs, etc.).

4 https://developers.google.com/maps/documentation/android-api/

5.2.2 The Server

HappyWalk's server-side application is responsible for the provision, management, and fetching of POI information. It is a Java EE web application implemented through the Java Servlet API[5] that runs on the Apache Tomcat™ 7 open-source web server.[6] It communicates in the form of Representational state transfer (RESTful) web services, the communication style of the the World Wide Web. This form of communication typically occurs over HTTP and uses the usual HTTP verbs (*GET, POST, PUT, DELETE,* etc.). Services are identified through a URI, and data is encapsulated within an Internet media type, which in our case is JSON. These RESTful web services and the JSON encapsulation are supported by the Jersey Java library.[7]

HappyWalk's POI information is retrieved from the well-known Foursquare® database.[8] Foursquare® is a location-based mobile social network that takes into consideration the position of its users to provide suggestions of places to visit. It allows users to discover places that fit their interests based on the advice of other users they trust. Foursquare®'s POI database is very complete and available free of charge through a web API with a limit of 5000 requests per hour, which is more than enough for our educational purposes. To facilitate the integration with Java, our server uses the *foursquare-api-java* library.[9]

The server also communicates with a PostgreSQL database,[10] where the records of emotions and POI locations are kept, through Hibernate[11]. PostgreSQL is an open-source object-relational database system that has earned a reputation for reliability and correctness. On the other hand, Hibernate is an Object/Relational Mapping framework concerned with data persistence in relational databases. Hibernate allows us to cleanly map Java objects with PostgreSQL tables, greatly simplifying HappyWalk's database management.

Figure 5.6 presents an overview of the server's packages and classes. Each package holds a specific purpose:

- The *Model* package contains classes that are representative of the JSON communication messages, providing encapsulation through the Jersey library. As convention, each class has either a "Request" or a "Response" prefix.
- The *Web* package contains classes that implement the RESTful web services' interfaces. Using Jersey, two POST services were implemented: *GetDetailPoi*, which returns detailed information regarding a certain POI, and *GetListPoi*, which returns a list of POIs around a certain location.
- The *Com* package contains the actual intelligence of the server. The classes within these packages tend to communicate with other entities, such as the Database and Foursquare®. The *GetDetailPoi* and *GetListPoi* web service requests are processed by the classes *ComGetDetailPoi* and *ComGetListPoi*, respectively. The classes

5 http://docs.oracle.com/javaee/6/api/javax/servlet/Servlet.html
6 "Apache", "Apache Tomcat", and "Tomcat" are trademarks of the Apache Software Foundation - https://tomcat.apache.org
7 https://jersey.java.net/
8 https://developer.foursquare.com/
9 https://github.com/clinejj/foursquare-api-java
10 http://www.postgresql.org/
11 http://hibernate.org/

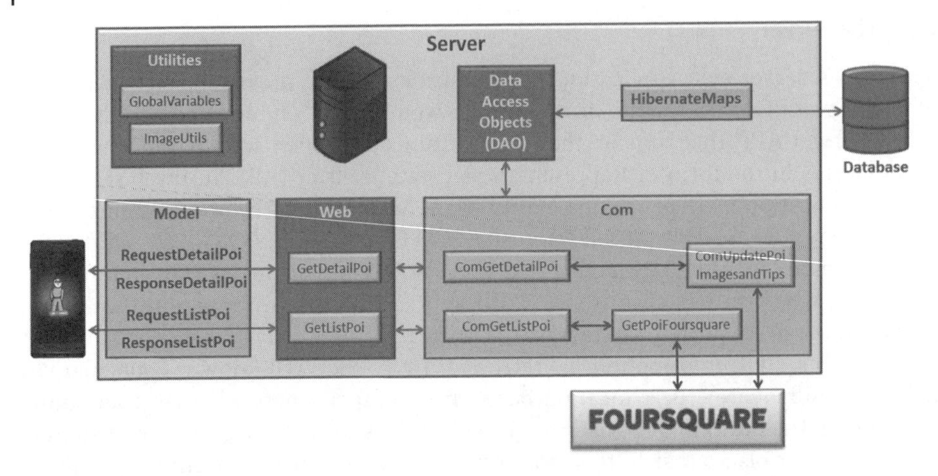

Figure 5.6 An overview of HappyWalkServer's main classes.

ComUpdatePoiImagesandTips and *ComGetPoiFoursquare* retrieve the necessary information from Foursquare®.

- Within the *DAO* package rest several data access objects that facilitate communication with the database. These are, in turn, supported by the *HibernateMaps*, which provide an interface with the database's tables.
- Finally, the *Utilities* package contains an *ImageUtils* class that processes images and a *GlobalVariables* class, similar to the one used in the Android app, for storing important variables that are used throughout the server.

Since the focus of this book is on HiTL control, which rests mostly on the Android app, as far as the server is concerned we will focus mostly on the intelligence associated with handling emotional information. Database structure and communication is already implemented and ready for handling HiTL information. The tutorial sections will simply use the *Dao* and *HibernateMaps* helper classes to save and update emotions. We will explain how to use these classes in Section 9.1.4.

We will also detail how to fetch and deploy the server using the Eclipse Mars IDE for Java EE Developers.[12] Despite the fact that the scope of this book does not include many of the server's inner workings, its source code is openly available for the inquisitive reader who might desire to tinker with it.

5.3 Enhancing the Sample App with HiTL Emotion-awareness

From this base map application, we will now take steps to introduce HiTL control. Since we will be dealing with possibly sensitive data, such as location and mood, one of the fundamental requirements of the design of our HiTL application will be to respect the privacy of users. We will consider this requirement through data anonymization, by generating a pseudo-random identifier (we will discuss this in greater detail in Section 8.3). While the app will be responsible for acquiring and processing GPS

12 https://eclipse.org/downloads/

positions, as well as accelerometer and microphone data, the resulting emotion will be **periodically sent in an anonymous way to the server**. This allows HappyWalk to display a near real-time average of the mood at each POI, through heatmaps with different colors. This information allows users to pick areas which are either livelier (euphoric mood) or calmer (relaxed mood). All of this real-time information may provide the necessary motivation for walking and visiting places that the user feels are better suited to his/her current mood.

5.3.1 Choosing a Machine Learning Technique

In HappyWalk, the core of our emotion-awareness rests on our ability to associate sensory input with certain emotions. It would be arguably possible to simply periodically ask the user how they are feeling. However, such an approach would go against the principles of “calm” computing and non-intrusiveness inherent to HiTLCPS design. Thus, HappyWalk will employ a state inference mechanism based on machine learning which will attempt to automate the detection of mood. Nevertheless the detection of emotions is an extremely complex issue and, as such, we will still require the use of supervised learning systems which rely on direct user feedback. Even so, we will attempt to reduce the amount of required feedback whenever the state inference component is performing well enough.

As mentioned in the introduction to this part of the book, our objective with HappyWalk is not to propose/develop robust methods for mood detection, but instead to present a practical “proof of concept” that can show the reader how HiTL concepts can be applied to create a simple, smart HiTLCPS. Thus, in order to determine a good machine learning technique for our application, let us consider previous comparisons between the different possibilities. Previous research work [18], the results of which are shown in Table 5.1, has scored different classification algorithms in terms of correct classification rate and in terms of CPU time needed for the classification. The latter is of particular importance for smartphone HiTLCPSs, since these are limited in terms of available processing power and energy. Based on this study, we will be using an artificial neural network (ANN) as our mood inference tool, since it offers a reasonable classification rate while being one of the least time-consuming techniques. Other good alternatives could be C4.5 decision trees and Support vector machines, but ANNs make it significantly easier to update our inference model with new data provided by the user.

Table 5.1 Machine learning approaches for sensing context in smartphones [18]. *Source:* Adapted from Guinness 2013.

Machine learning techniques	Correct classification rate	CPU time requirements ranking
Random Forest	96.5%	5th
Support vector machines	80.2%	1st
Naive Bayes classifiers	81.5%	6th
Bayesian networks	90.9%	4th
Logistic regression	83.4%	3rd
Artificial neural networks	87.2%	2nd

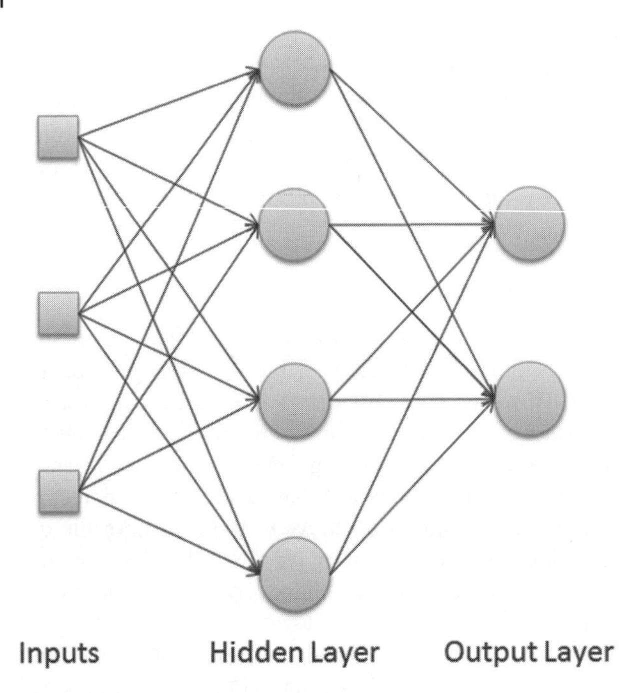

Figure 5.7 A typical artificial neural network architecture.

ANNs are machine learning techniques based on biological neurons, the cells that are the most representative of the thinking function of animal brains. As put by Haykin in his book *Neural Networks: A Comprehensive Foundation* [136], the brain is a highly complex, nonlinear and parallel computer that processes information in an entirely different way from the conventional digital computer. Much like biological brains, ANNs are systems of interconnected "neurons" which exchange messages with each other. They possess *plasticity*, that is the ability to adapt the "strength" of the connections between neurons, through numeric weights that can be tuned based on experience. This endows ANNs with the ability to learn from training data. In Figure 5.7 we can see a typical ANN architecture. The neurons of an ANN are usually grouped in layers; the first layer receives the input, while the last layer transmits the final output. In between these layers are the *hidden* layers, which allow the ANN to extract higher-order information from the data, by providing additional transformations and processing. ANNs are usually considered *black-box* systems, in the sense that their functioning is opaque: studying an ANN's inner structure does not provide any logical insight into the function being approximated. All the processing and memory of the ANN rests within the weights of the connections between its neurons, which, by themselves, do not mean much to the ANN designer.

5.3.2 Implementing Emotion-awareness

An important matter to decide on how to implement our emotion-aware ANN is which sources of input should be used to teach it. In order to avoid obligating the reader to use additional hardware and perform particularly complicated integration tasks, we want to limit our choice of sensors to those already provided by the smartphone device.

Current scientific knowledge does not yet have an exhaustive picture of all the factors influencing a person's emotions [137]. Nevertheless, in the case of our sample application, we intend to consider at least three general sources of data: contextual information, environmental clues, and body movement information.

In terms of context, most smartphones are equipped with GPS, allowing us to know where the user is located. Regarding environmental clues and body movement, the accelerometer and microphone have been previously identified as effective sensors for identifying human context [97]. Thus, our application will acquire raw data from these sensors, and will process it through a simple classifier.

Our classifiers will use some concepts of "signal processing". In particular our accelerometer processor will use a "Fourier Transformation": a powerful signal processing technique. In fact, signals can be understood from two different perspectives. The time perspective is the way we instinctively perceive our reality: things happen and vary as time passes. However, every signal in nature can also be described as a frequency spectrum and is determined by its inherent *frequencies* [138]. Thus, we can analyze a signal either in the time (or spatial) domain or in the frequency domain.

A Fourier transformation is a mathematical process which converts a finite signal, acquired at a certain frequency for a certain duration, into a series of *coefficients*, which represent a finite combination of complex sinusoid functions [139]. This combination of sinusoid functions represents that same signal on its *frequency domain*, as shown in Figure 5.8[13]. Our accelerometer classifier will use this type of analysis by performing a fast Fourier transformation (FFT) and summing the resulting fourier coefficients. This sum gives the neural network an idea on the amount of movement detected by the smartphone.

The user's emotion will be inferred once or twice an hour. The time between two sensory acquisitions will be **randomly determined within these constraints**, in order to avoid user habituation. In our example implementation, we will consider four

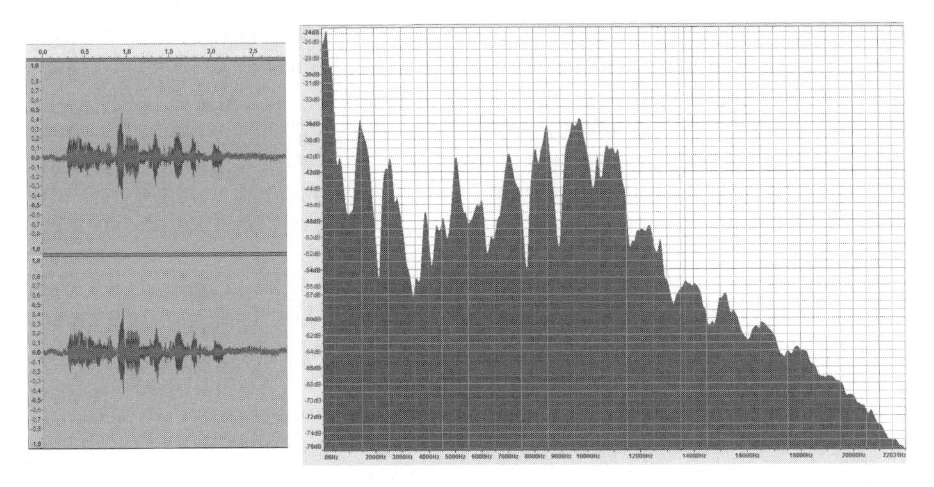

Figure 5.8 Sound signal in the time domain (left side) analyzed through a Fourier transformation to show its frequency domain (right side).

13 Obtained from the Audacity software, www.audacityteam.org

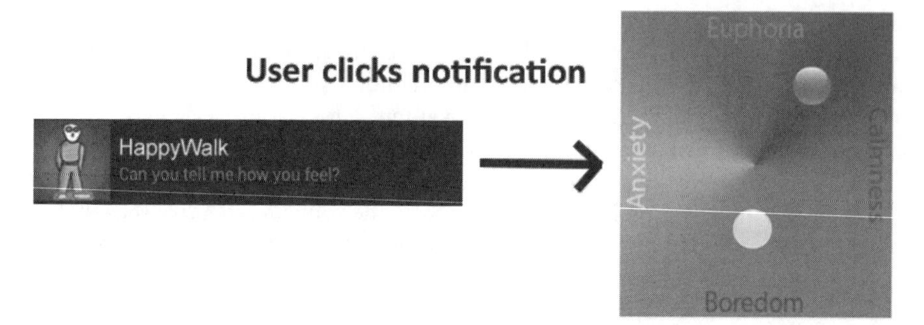

Figure 5.9 HappyWalk's Emotional Feedback.

distinct moods: euphoria, calmness, boredom, and anxiety. Boredom and a Anxiety are considered negative emotions, whereas euphoria and calmness are considered their positive counterparts. Users receive a notification when an emotion is detected, and, by selecting it, the application will open and display a feedback screen. The output representing the inferred emotion will be shown as a yellow circle in a two-dimensional space containing the four emotions (Figure 5.9). The user can provide corrective feedback by dragging the yellow circle to a new position, now shown in green. This feedback initiates an ANN re-training process, which will reflect the correction in future inference tasks. After some training, and when the neural network begins to become accurate, the feedback notifications will be progressively replaced with notifications suggesting walking exercise, whenever negative emotions are detected.

The ANN will be implemented using the Encog[14] [140] machine learning framework. Designing an ANN architecture is a challenging task. How should we select a number of neurons that provides minimal error and highest accuracy?

Previous research has shown that excessive hidden neurons will cause over fitting; that is, the neural networks overestimate the complexity of the target problem. A rule of thumb is selecting a size between the number of input neurons and number of output neurons [141]. Thus, it was decided to test two possibilities: using a hidden layer with four neurons or using two hidden layers, three neurons in the first and two neurons in the second. Both of these configurations fit the rule of thumb without becoming overly complex and taxing smartphone hardware. In the decision process, two major requirements were considered: the amount of effort required for training the network (which is important in terms of processing power and battery drain) and the accuracy of the network.

In order to test the training effort, simulated emotions were generated. For each type of emotion, a probability value for different ranges of its input components (movement, background noise, etc.) was empirically defined. Through this method, 150 simulated emotions were generated; while not valid for testing accuracy, these are sufficient for testing training performance. Thus, the number of epochs necessary to successfully train the network for each configuration was counted. The results are shown in Table 5.2. These show that using two hidden layers increases the training effort significantly. Therefore, it was also necessary to test if using more layers would bring any benefits in terms of accuracy. A test subject used HappyWalk for a week, during

14 Encog Framework: http://www.heatonresearch.com/encogin

Table 5.2 Testing training performance (150 emotions).

Configuration	Number of epochs
One hidden layer	100
Two hidden layers	3000

Table 5.3 Testing neural network accuracy (41 emotions).

Configuration	Sensitivity	Specificity
One hidden layer	0.679	0.766
Two hidden layers	0.720	0.830

which his sensory data and emotional feedback were recorded for a total of 41 records. Using this data, both neural network configurations were tested to evaluate their sensitivity and specificity. Considering that negative emotions are the events of interest, performance was evaluated through two statistical measures known as "sensitivity" and "specificity". In our case, sensitivity is the proportion of negative emotions that were correctly identified as such; in other words, it measures when our system was capable of detecting that it was necessary to actuate. Specificity, on the other hand, measures the proportion of correctly identified positive emotions. A perfect emotional predictor would present the maximum value of 1 for each of these metrics.

The results shown in Table 5.3 suggest that using a two-layer configuration leads to considerably better results. After pondering over the results, it was decided that, despite being more demanding, a two-hidden-layer configuration, the first containing three nodes and the second two nodes, constitutes a good compromise in terms of training time and accuracy. Thus, this is the configuration used for our sample application. The proposed neural network's architecture is presented in Figure 5.10.

5.4 In Summary...

In this chapter we have seen a high-level overview of the HappyWalk app, which we will use in this part of the book for illustrating the development of HiTLCPS applications. This is a kind of BCI application that provides feedback to the users and may influence their behavior, through a *Data Acquisition, State Inference,* and *Actuation* cycle.

HappyWalk is a BCI system that attempts to positively influence its user's mood throughout moderate physical exercise. The app's base architecture consists of a client application that requires at least Android Gingerbread. It already has classes for controlling POI clustering, the map interface, detecting location, and showing POI information. Many of its tasks are run under a background service to avoid encumbering the user interface. As for the server-side, it runs on Apache Tomcat™ 7 and uses the Jersey Java library to handle RESTful web service communication.

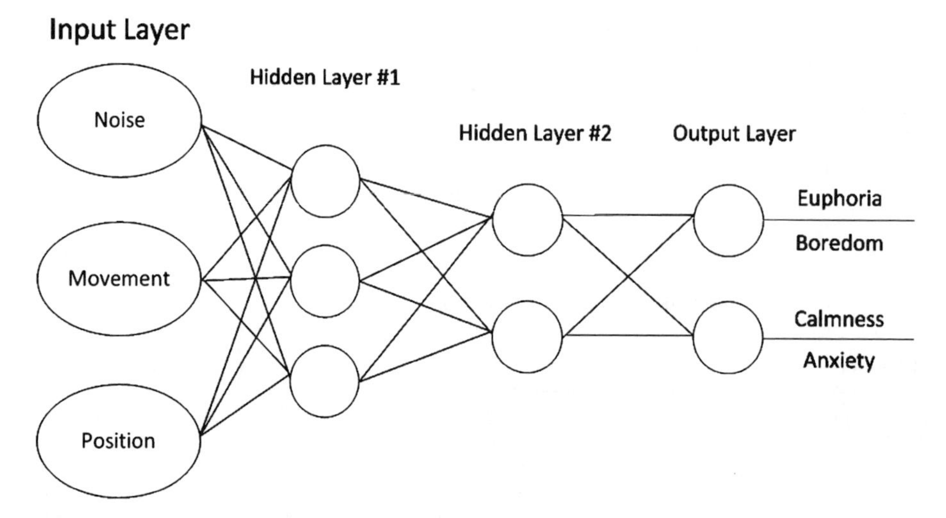

Figure 5.10 HappyWalk's neural network design.

The server also communicates with Foursquare® to retrieve POI information, and a PostgreSQL database to permanently store data.

The architectural options concerning emotion awareness were also presented and justified. We will be using the Encog library to implement a simple ANN with two hidden-layers as our mood inference tool. This network will be fed with various sources of data provided by the smartphone, including location, movement, and noise. We will pre-process this data before feeding it to the network through simple signal processing techniques, including frequency analysis. The user's emotion will be inferred once or twice an hour, within four possible moods: euphoria, calmness, boredom, and anxiety.

6

Setting up the Development Environment

Now that we have settled on the major ideas behind our HiTL application, it is time to begin the actual implementation. To do so, we first have to set up the proper development environment. Please note that this tutorial was devised from within a *Windows 7* OS and, as such, most screenshot images refer to this OS. Nevertheless, the tools used for these tutorials should also support most *Linux* distributions, *Windows 7, 8, and 10,* as well as *MacOSX*.

In the current chapter we will go through the various phases needed for setting up the development environment. These comprise installing the Android software development kit, cloning the HappyWalk Android project, deploying the server, and testing the basic sample app. The following sections describe each of these phases in detail.

6.1 Installing Android Studio

Android applications are developed through Android Studio (AS). This tutorial was written using version 2.1.3 of AS and, as such, we strongly recommend downloading and using this version since newer versions may introduce discrepancies and incompatibilities. In this section we will perform the necessary tasks to properly install this IDE, including installing the Java SE Development kit, AS, and Android SDK.

At the time of writing, AS 2.1.3 requires the Java Development Kit (JDK) 7; in particular, we used *Java SE Development Kit 7u79,* which can be downloaded from *http://www.oracle.com/technetwork/java/javase/downloads/jdk7-downloads-1880260.html*. From the JDK, we will need the *Development Tools* and the *Public JRE,* as shown in Figure 6.1.

After installing the JDK, the installation of AS 2.1.3 requires the reader to visit the page *http://tools.android.com/download/studio/builds/2-1-3*, and download the *Windows bundle with SDK* installer package. Alternatively, it is also possible to download a zip package appropriate for other operating systems; however, in doing so, the reader becomes responsible for manually setting up the necessary environmental variables.

After downloading and running the bundle's executable, AS should begin its installation. As shown in Figure 6.2, the reader needs to install the Android SDK (by ticking the appropriate checkboxes, if needed). The Android virtual device, while useful for testing many types of applications, is not sufficient for this tutorial. Since we require the accelerometer, location and microphone sensors, the use of a real device is strongly encouraged.

A Practical Introduction to Human-in-the-Loop Cyber-Physical Systems, First Edition.
David Nunes, Jorge Sá Silva and Fernando Boavida.

Companion Website URL: www.wiley.com/go/nunesloop

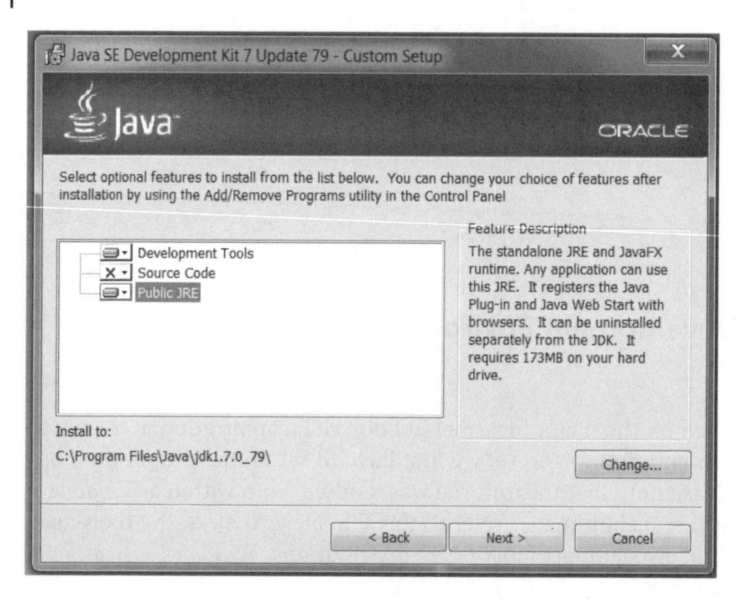

Figure 6.1 Installing Java SE Development Kit 7u79.

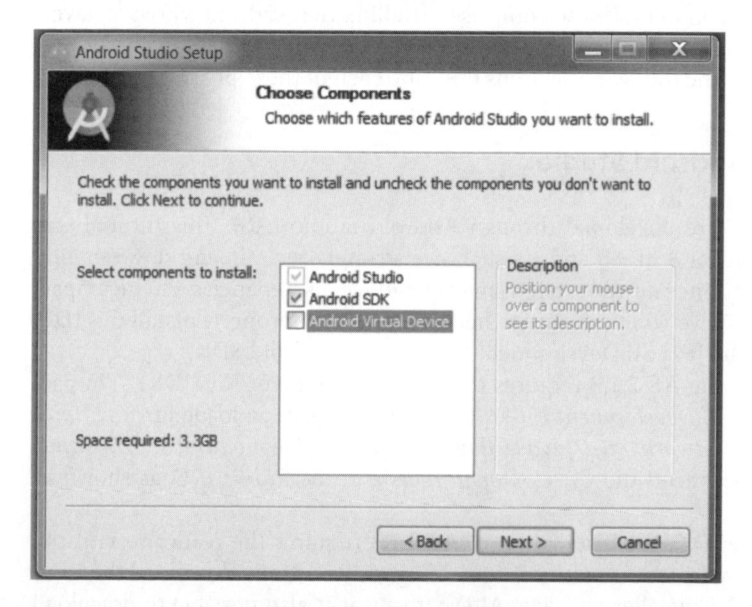

Figure 6.2 Installing Android Studio and Android SDK.

After its installation is complete, AS may greet the reader with a *Missing SDK* window. As shown in Figure 6.3, we can safely cancel this setup wizard since we will be installing the necessary SDK by ourselves. Notice the update notification in the upper-right corner of the window; we advise the reader *not* to update, as newer versions introduce incompatibilities.

When you finally see AS's welcome screen, open the Android SDK manager by clicking on *Configure* and then *SDK manager*, as shown in Figure 6.4.

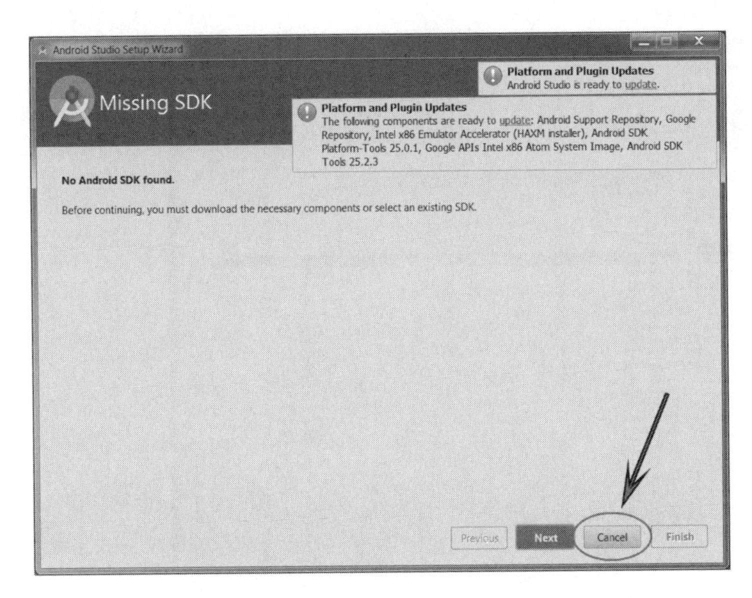

Figure 6.3 Canceling the setup wizard.

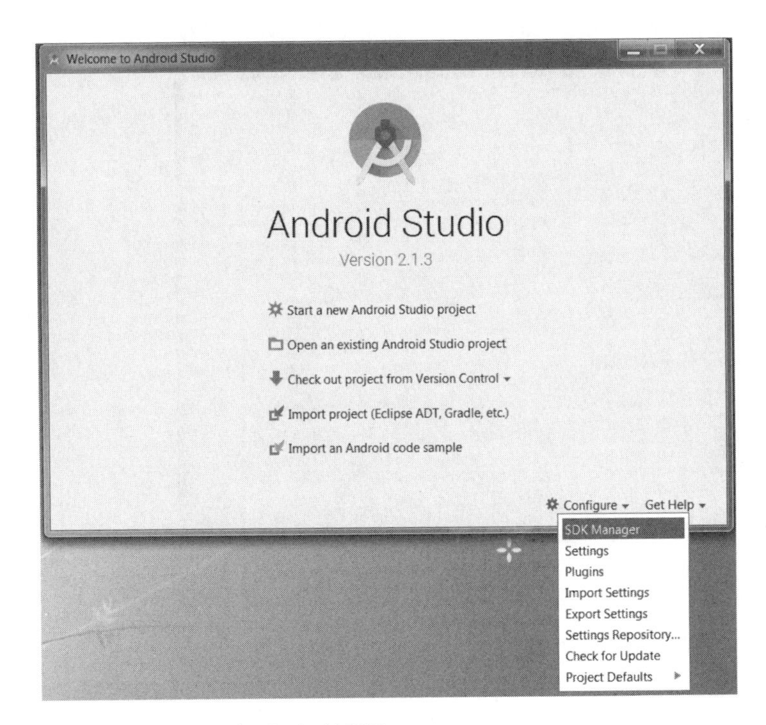

Figure 6.4 Opening the Android SDK manager.

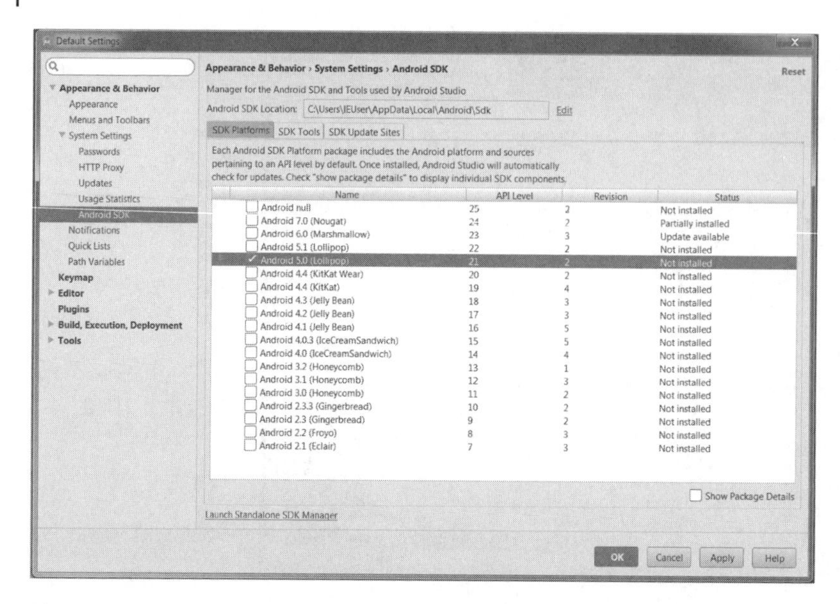

Figure 6.5 Installing Android API 21.

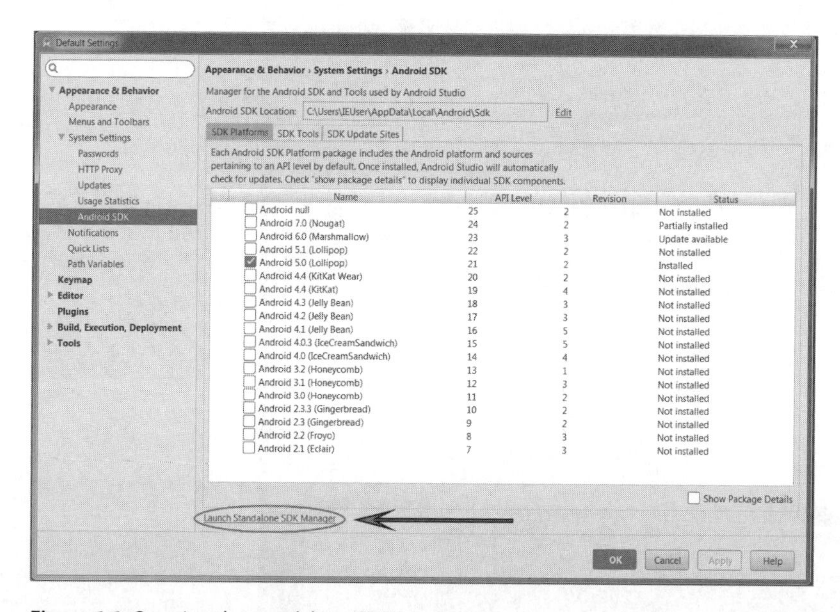

Figure 6.6 Opening the standalone SDK manager.

You should see a window similar to the one shown in Figure 6.5. Since in this tutorial we will be using Android API level 21, tick the corresponding checkbox as shown in Figure 6.5, and click the *OK* button to let AS install the associated components.

Afterwards, open the same window and click on *Launch Standalone SDK manager*, as highlighted in Figure 6.6.

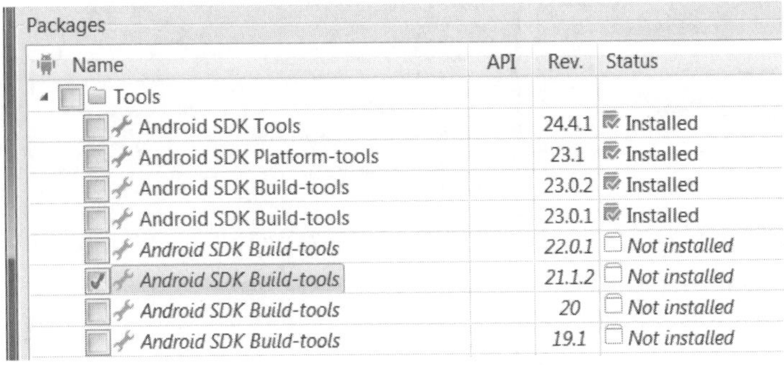

Packages

Name	API	Rev.	Status
Tools			
Android SDK Tools		24.4.1	Installed
Android SDK Platform-tools		23.1	Installed
Android SDK Build-tools		23.0.2	Installed
Android SDK Build-tools		23.0.1	Installed
Android SDK Build-tools		*22.0.1*	*Not installed*
Android SDK Build-tools		*21.1.2*	*Not installed*
Android SDK Build-tools		*20*	*Not installed*
Android SDK Build-tools		*19.1*	*Not installed*

Figure 6.7 Installing *Android SDK Build-tools 21.1.2.*

Table 6.1 Summary of the steps necessary to install AS 2.1.3.

Step	Summarized Objective
Install JDK	Installed JDK 7u79 from http://www.oracle.com/technetwork/java/javase/downloads/jdk7-downloads-1880260.html
Install AS 2.1.3	Downloaded and extracted the AS package from http://tools.android.com/download/studio/builds/2-1-3
Install Android SDK	Used the SDK Manager to install Android API 21 and Build-Tools 21.1.2

On the Standalone SDK manager, check the box corresponding to *Android SDK Build-tools 21.1.2*, just as Figure 6.7 shows. We will need these Build-tools to compile HappyWalk properly.

Table 6.1 summarizes the steps we have taken so far.

6.2 Cloning the Android Project

In this section we will fetch and set up HappyWalk's base client application code. To do so, we will install Git, check out the HappyWalk client project, run the application for the first time, discover our Android debug key, and obtain a Google Maps Android API key.

The base HappyWalk app is available through Git[1]. As such, you will need to install an appropriate Git distribution for your OS. You can download the appropriate Git release for your OS from http://git-scm.com/downloads. This tutorial was developed using Git version 2.7.1.2 for Windows, although it should also work with newer versions. We will now demonstrate how to properly install Git on *Windows 7*. Make sure to adapt these steps to your own OS.

After running the installation package, choose a folder to install Git on. Make sure to **add Git to your PATH system variable** so that AS can find and use it! On the Git Windows installation, this can be done easily by choosing the option *Use Git from the Windows Command Prompt* (see Figure 6.8a). We also recommend choosing the options

1 https://git-scm.com/

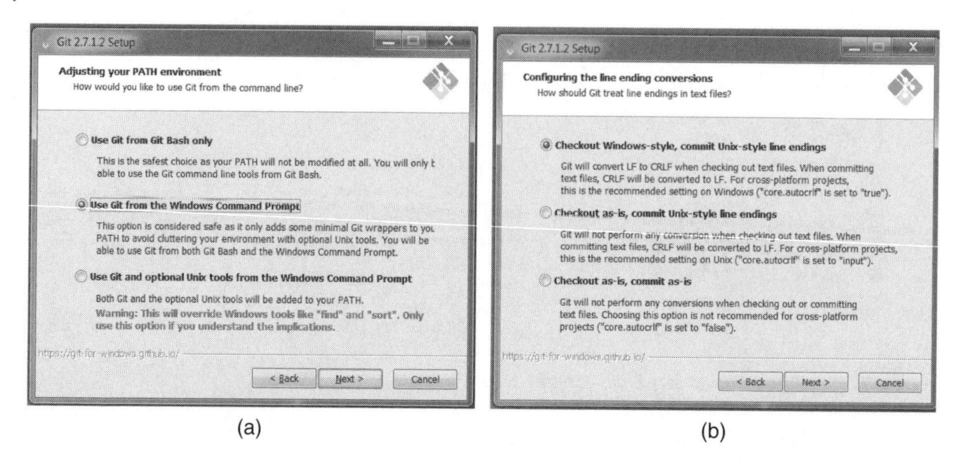

Figure 6.8 Installing Git #1. (a) Adding Git to the PATH, on Windows (b) Choose *Checkout Windows-style.*

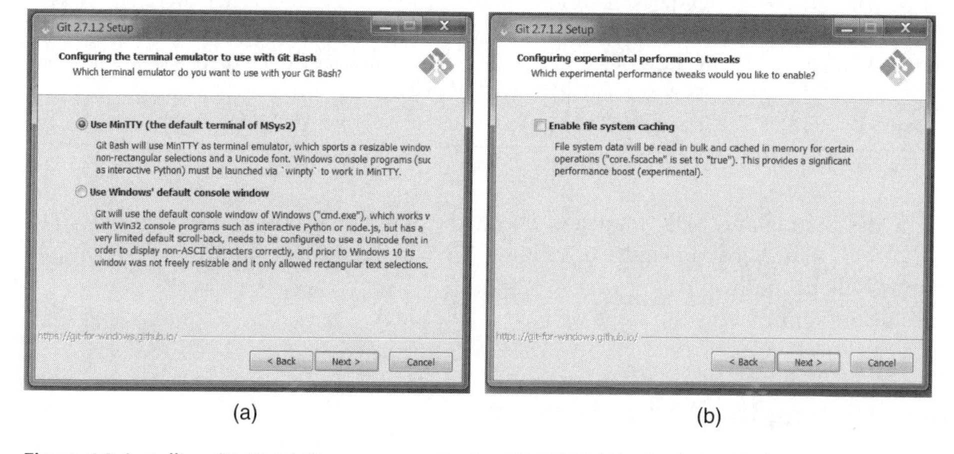

Figure 6.9 Installing Git #2. (a) We recommend using MinTTY (b) Uncheck *Enable file system caching.*

Checkout Windows-style, commit Unix-style line endings (see Figure 6.8b), *Use MinTTY* (see Figure 6.9a), and uncheck *Enable file system caching* (see Figure 6.9b).

Now that Git is installed, let us import the HappyWalk project in AS. From the welcome screen, click on *Check out project from Version Control* → *Git* as indicated in Figure 6.10 and fill the correct repository URL, as shown in Figure 6.11. Pick a *Parent Directory* of your choosing, but please take note of its location.

After AS clones the project, it should ask you if you want to open the checked-out Studio project file. As shown in Figure 6.12, respond "No" and, instead, select "Open an existing Android Studio Project"). Select HappyWalk's directory from the previously chosen *Parent Directory*, as shown in Figure 6.13. As AS loads the project it is likely that a "Gradle Update" and a "Gradle Plugin Update" window to appear, in which case you should answer "Don't remind me again for this project" to both (see Figure 6.14).

After AS finishes the project load process we can begin our deployment. The project's structure follows the general architecture presented in Section 5.2. As previously

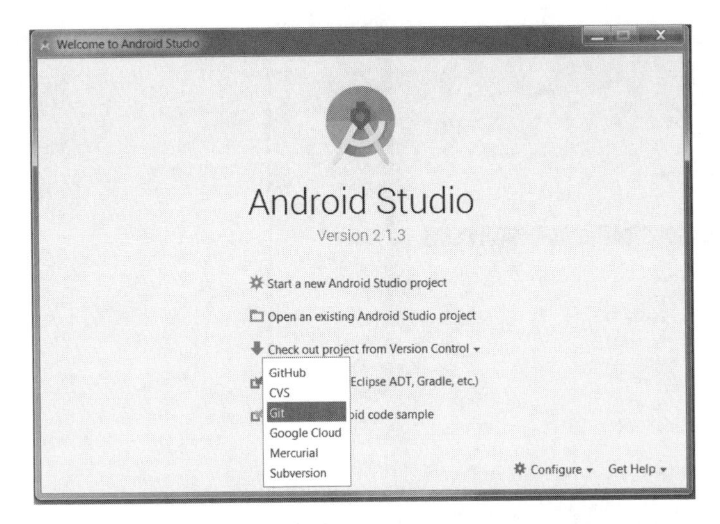

Figure 6.10 Importing HappyWalk from Git.

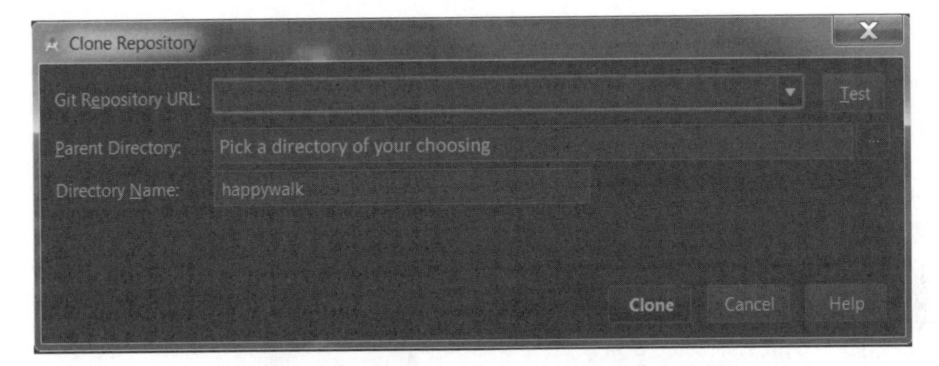

Figure 6.11 Cloning the HappyWalk project.

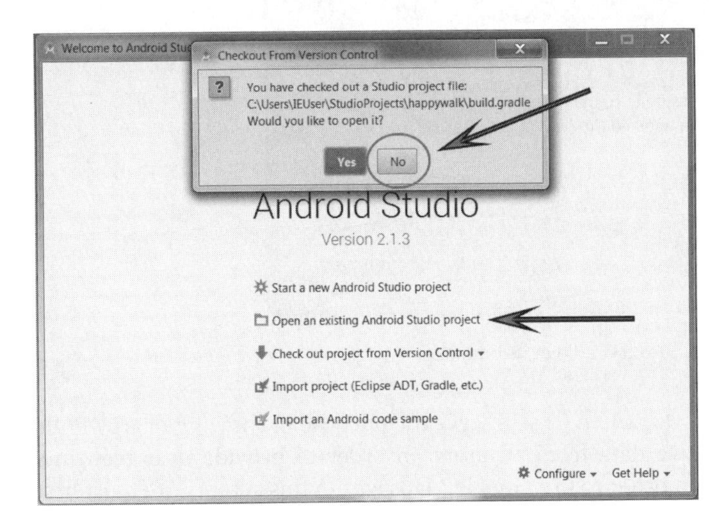

Figure 6.12 Opening the HappyWalk project.

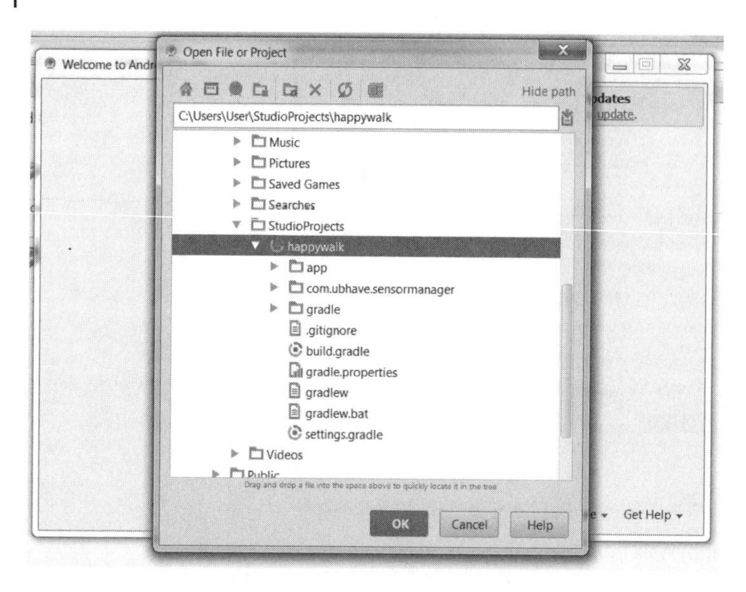

Figure 6.13 Choosing HappyWalk's project folder.

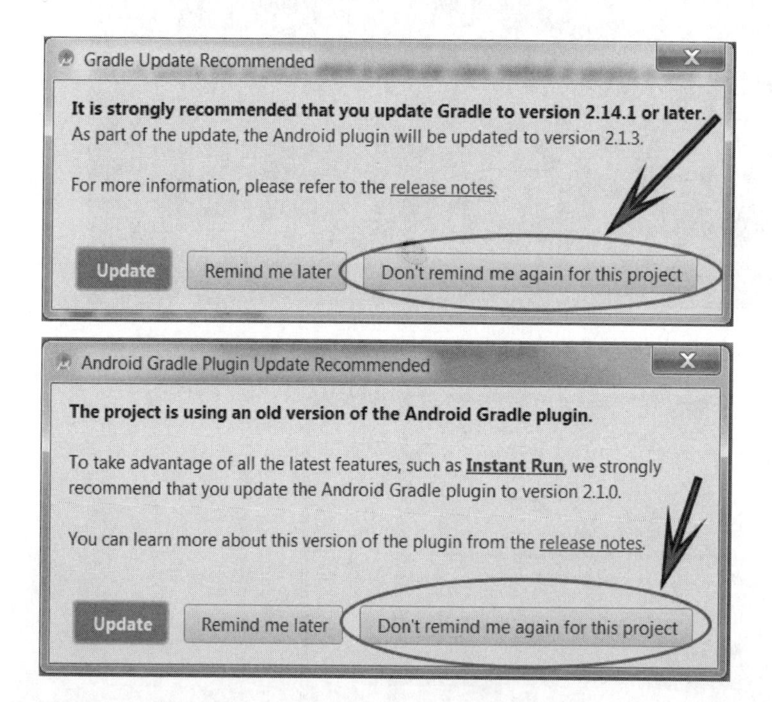

Figure 6.14 Do not upgrade Android Gradle or its plugin.

mentioned, HappyWalk's *MapsActivity* will serve the purpose of showing nearby POIs and displaying collaborative data from all users, in order to provide near real-time mood context. However, in order to use Google Maps within this activity, we must first perform some additional steps.

First of all, we first need to run our application at least once. This will allow AS to generate the necessary debug certificates. We can launch our Android application by making sure that *app* is selected from the dropdown list near the small Android logo and by clicking on *Run app* (the green "play" button), as shown in Figure 6.15.

Make sure your device appears on the *Select Deployment Target* window. In the Windows operating systems, this may require installing the appropriate Android Debug Bridge drivers for your device. For more information on this, check the Android Developer's website.[2] If AS was able to properly install the application on your device, you should see a screen similar to what is shown in Figure 6.16.

Notice that Figure 6.16 shows an empty map screen. To use Google Maps, the reader needs to acquire his/her own *Google Maps Android API* key. To do so, a

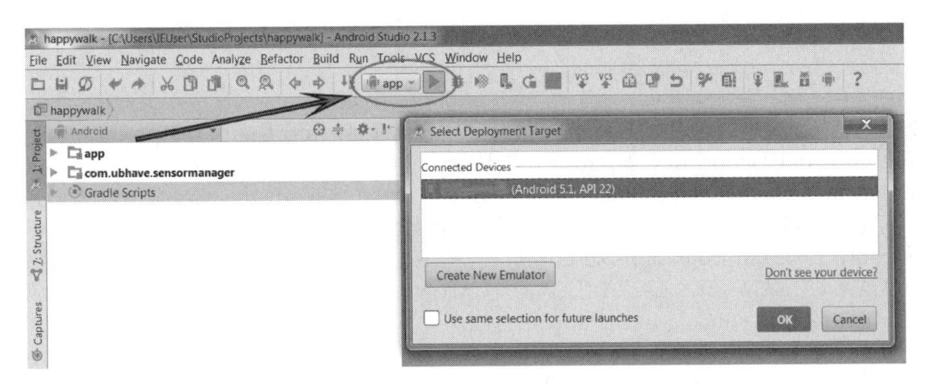

Figure 6.15 Running HappyWalk.

Figure 6.16 HappyWalk's first launch.

2 https://developer.android.com/studio/run/oem-usb.html

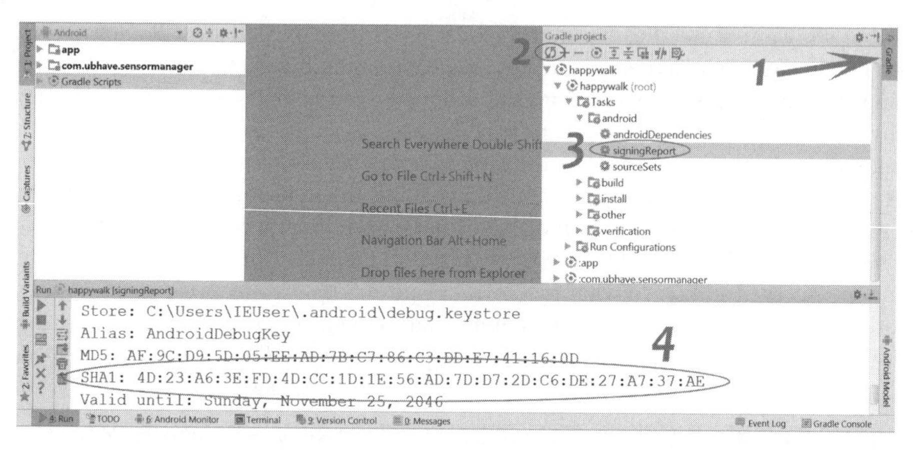

Figure 6.17 Obtaining the Android debug key.

Google Account is required and it is also necessary to find the development computer's Android Debug Key. Figure 6.17 shows how this may be accomplished from within AS. With the HappyWalk project open, first open up the *Gradle* right-side panel (1). Click on its refresh button (2) and double-click on "happywalk/happywalk(root)/Tasks/android/signingReport" (3). The required SHA1 key takes the form *XX:XX:XX:XX:XX:XX:XX:XX:XX:XX:XX:XX:XX:XX:XX:XX:XX:XX:XX:XX* and shall appear on the *Run* console, at the bottom (4).

We cannot guarantee accurate instructions for the next few steps, since they may have changed by the time the reader is following this sentence. At the time of writing, the reader could follow the link below, replacing the *XX:XX:XX:XX:XX:XX:XX:XX:XX:XX:XX:XX:XX:XX:XX:XX:XX:XX:XX:XX* on the URL with the SHA1 acquired above:

https://console.developers.google.com/flows/enableapi?apiid=maps_android_backend&keyType=CLIENT_SIDE_ANDROID&r=XX:XX:XX:XX:XX:XX:XX:XX:XX:XX:XX:XX:XX:XX:XX:XX:XX:XX:XX:XX%3Bhitlexamples.happywalk

Here, the reader should be able to log in to his/her own Google Account and follow the provided directions to acquire a *Google Maps Android API* key. This involves creating a project (Figure 6.18), creating a new Android API key (Figure 6.19), and obtaining the mentioned key (Figure 6.20).

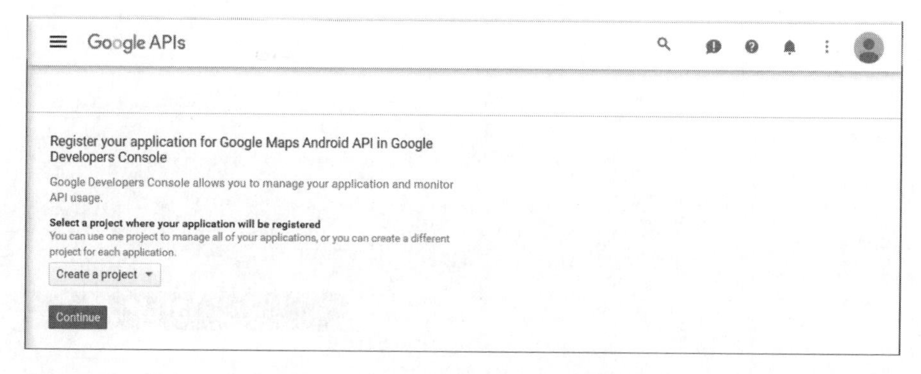

Figure 6.18 Creating a project to obtain a Google Maps Android API key.

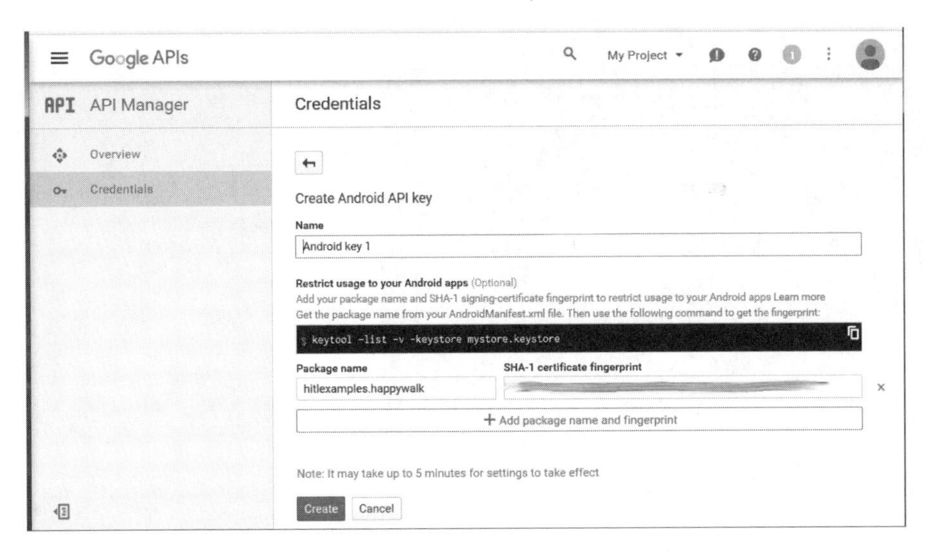

Figure 6.19 Creating the Google Maps Android API key.

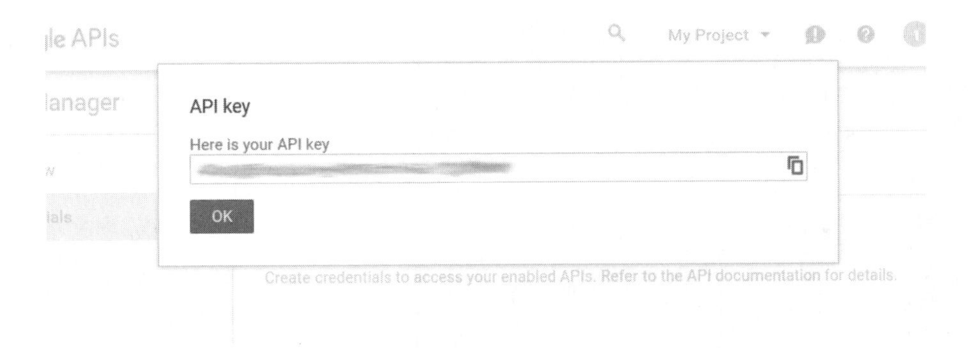

Figure 6.20 Obtaining the Google Maps Android API key.

In case the above instructions do not work, we advise the reader to search for the *Google Console Developers page*.[3] Either way, the reader should be able to log in to his/her own Google Account and follow the provided directions to acquire a *Google Maps Android API* key. At the end, Google should provide a development key free of charge (it starts with "AIza").

Let us now use the key by looking into the contents of the HappyWalk's file *app/debug/res/values/google_maps_api.xml*. To do so, it is easier to view the entire project using the project tab's *Project* view. By default, it should be set in *Android* view; you can change the project tab's view by clicking on the buttons shown in Figure 6.21. In this perspective, navigate to and double-click the *app/debug/res/values/google_maps_api.xml* file, as shown in Figure 6.22. As soon as you double-click *google_maps_api.xml*, you should see the following contents:

3 https://console.developers.google.com

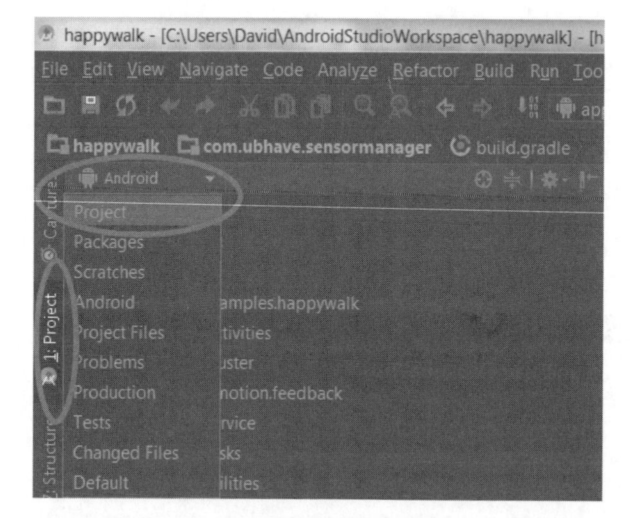

Figure 6.21 Changing into the project's view.

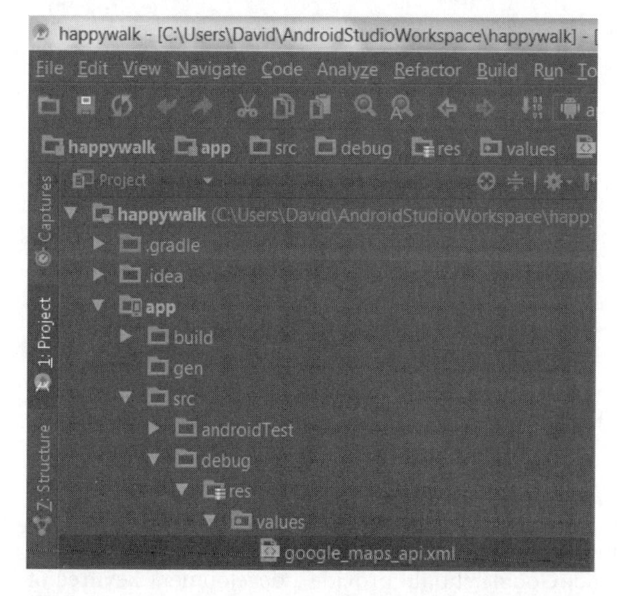

Figure 6.22 Opening *app/debug/res/values/google_maps_api.xml*.

```
<resources>
    <!--
    TODO: Before you run your application, you need a Google Maps API key.
    -->
    <string name="google_maps_key" translatable="false"
     templateMergeStrategy="preserve">
        YOUR_KEY_HERE
    </string>
</resources>
```

Table 6.2 Summary of the steps necessary to set up HappyWalk's Android project.

Step	Summarized Objective
Install Git	Installed Git for Windows from *http://git-scm.com/downloads*
Clone HappyWalk's Android project	Imported and opened HappyWalk's client project from *https://git.dei.uc.pt/dsnunes/happywalk.git* using AS
Launch HappyWalk	Used AS to run HappyWalk's Android application on a real device
Discover the debug key	Used AS's *signingReport* task to discover the computer's SHA1 Android debug Key
Acquire a *Google Maps Android API* key	Logged into the *Google Console Developers* page and used the debug key to acquire a *Google Maps Android API* key
Place the *Google Maps Android API* key	Navigated to the *google_maps_api.xml* file and replaced the *Google Maps Android API* key

In place of *YOUR_KEY_HERE*, your google_maps_api.xml should contain your own development key. Once the proper Google Maps API key is in place, we can fully utilize the Maps interface on our application.

Table 6.2 summarizes the steps we have taken in this section.

6.3 Deploying the Server

As previously discussed in Section 5.2.2, to compile and use our server, we will be using the following technologies:

- PostgreSQL 9.3, for hosting our database
- Eclipse Mars IDE for Java EE Developers
- Apache Tomcat 7.

As we also mentioned in Section 5.2.2, the server needs to communicate with Foursquare® to acquire POI information. Therefore, the machine hosting it *needs to have an active Internet connection.* This is essential for the proper functioning of the HappyWalk system.

In the next section we will perform the necessary tasks to properly install Eclipse Mars and PostgreSQL, import HappyWalk's server project, obtain a Foursquare® client ID and secret, set up the database, and deploy the server on Tomcat 7.

6.3.1 Installing the Software and Cloning the Server's Project

Installing the Eclipse Mars is trivial; visit *https://eclipse.org/mars/*, download the *Eclipse IDE for Java EE Developers* package, and extract it to a location of your choice.

As for PostgreSQL, browse to *http://www.postgresql.org/download/* and download the appropriate 9.3 installer for your OS. Several installers are available in *http://www.enterprisedb.com/products-services-training/pgdownload.* During installation, make sure to take note of the superuser password that is requested in the screen shown in Figure 6.23, we will need it later on. Uncheck *Launch Stack Builder at exit,* as it is not necessary (Figure 6.24).

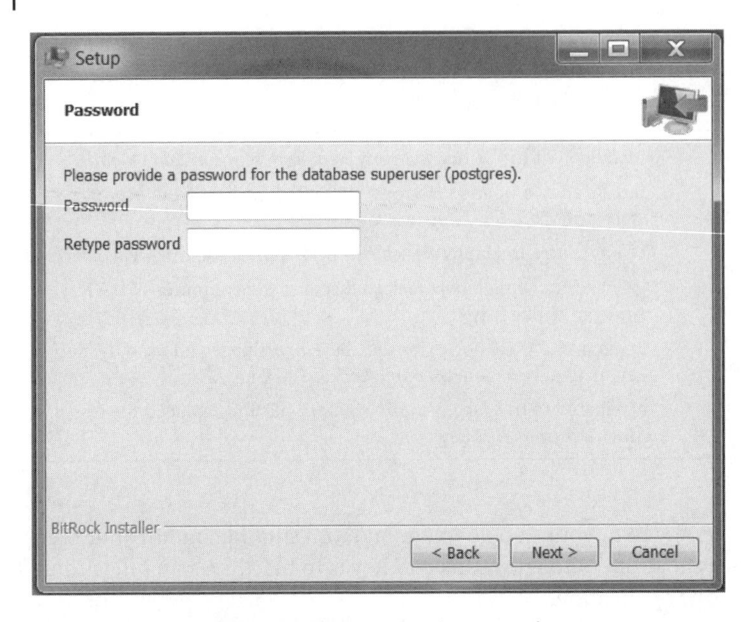

Figure 6.23 Choosing PostgreSQL superuser's password.

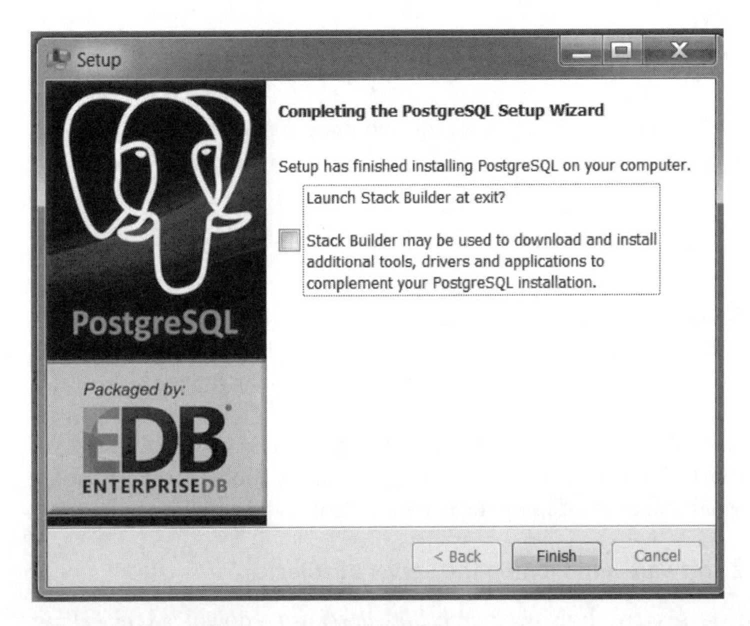

Figure 6.24 No need to launch Stack Builder.

After installing both of these tools, open the Eclipse Mars IDE and select a path of your choice for the workspace. Then, import a new Git project by clicking on *File → Import → Git → Projects from Git*. You should see a window similar to the one in Figure 6.25.

Select *Clone URI* and then fill the *URI* field, as shown in Figure 6.26. The *Host* and *Repository path* fields will then be automatically filled.

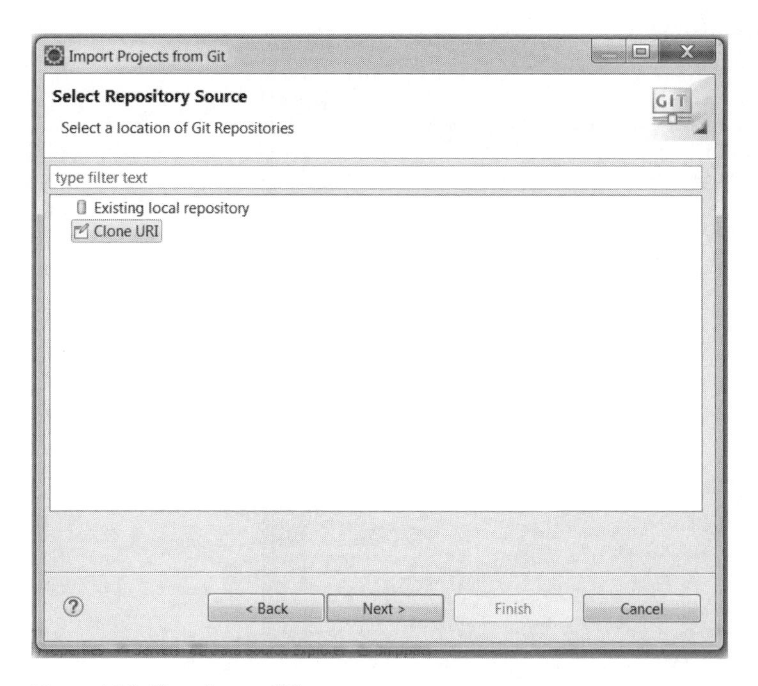

Figure 6.25 Clone from a URI.

Figure 6.26 Introduce the URI corresponding to HappyWalk's server.

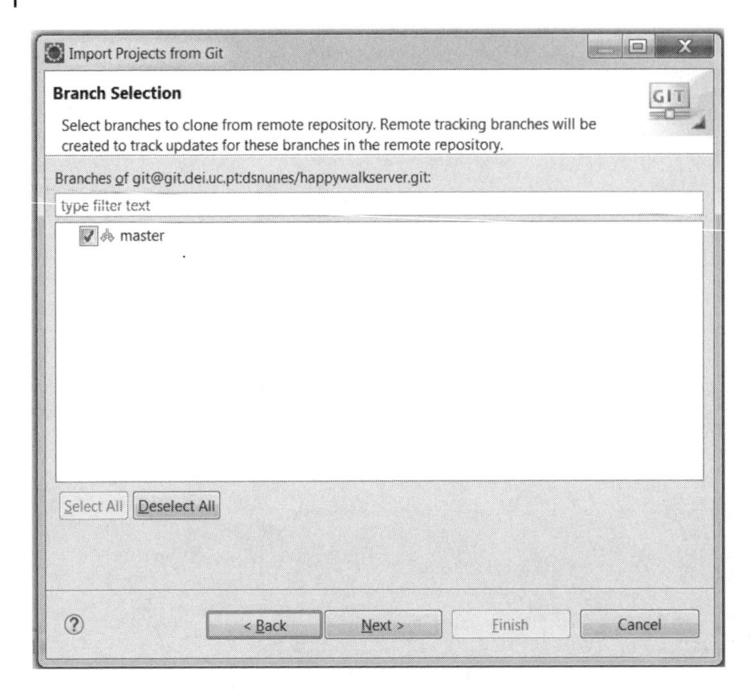

Figure 6.27 Select the *master* branch.

Afterwards, just select the *master* branch (Figure 6.27) and select *master* as the initial branch, *origin* as the remote name, and choose a directory to save the project (Figure 6.28).

Eclipse will begin to receive the objects associated with our server. We now need to tell it to import the Eclipse project (Figures 6.29 and 6.30).

6.3.2 Obtaining a Foursquare®'s Client ID and Client Secret

Now that we have cloned the HappyWalk Server project, we will need to make a few modifications to the server's code. First, we will need to create a Foursquare® Client ID and a Client Secret.

Browse to the address *https://developer.foursquare.com/*. The steps herein described are valid for the version of the Foursquare® website available in the beginning of 2016. We cannot account for future changes in website design but, hopefully, the functionality will remain the same for the foreseeable future.

First, click on *My Apps* and create an account (if you do not already possess one). After that, click on the button that says *Create a new app*. Fill the *Your app name* and *Download / welcome page url* fields as you desire. You can leave the other fields empty, as they are not relevant for our purposes. Afterwards, click on *Save changes*, shown in Figure 6.31. Now, when you navigate into the *MyApps* menu, Foursquare® will provide you with a Client ID and a Client Secret (Figure 6.32), which we will need in our server.

To do so, use Eclipse's Project Explorer to browse and double-click the *src/utilities/-GlobalVariables.java* class, as shown in Figure 6.33.

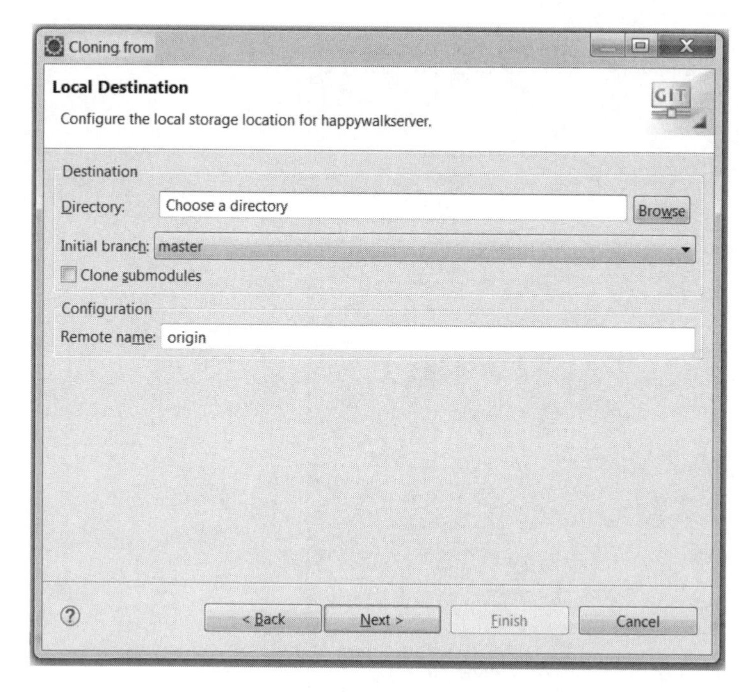

Figure 6.28 Selecting the local storage directory.

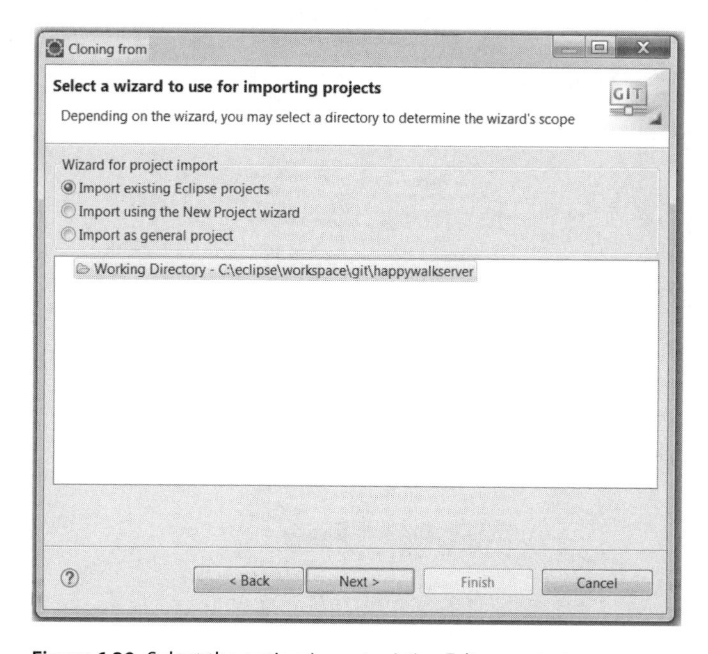

Figure 6.29 Select the option *Import existing Eclipse projects*.

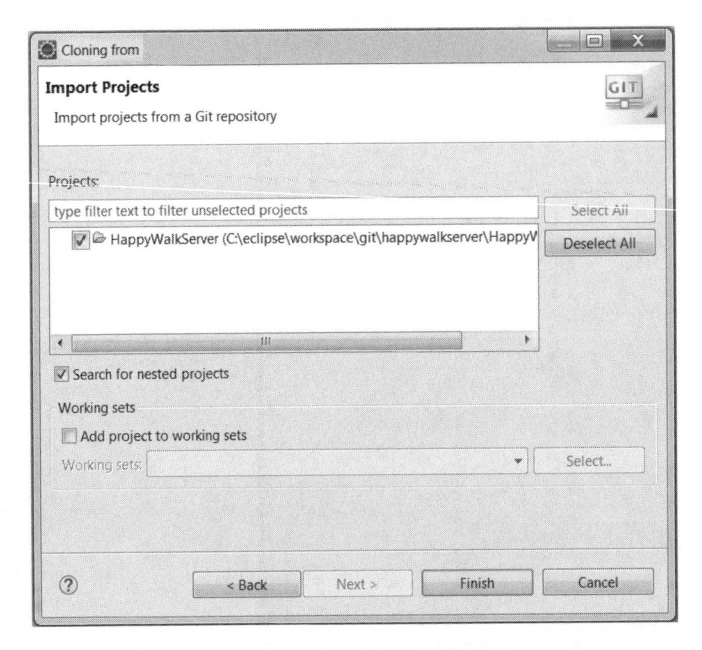

Figure 6.30 Tick the checkbox of the *HappyWalkServer* project.

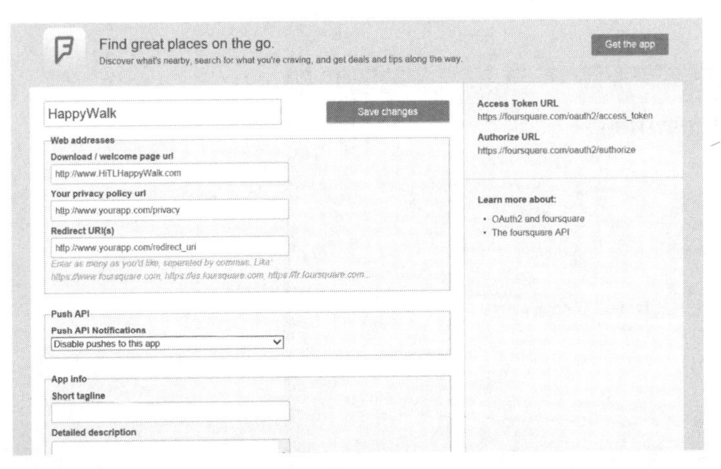

Figure 6.31 Creating a Foursquare® app.

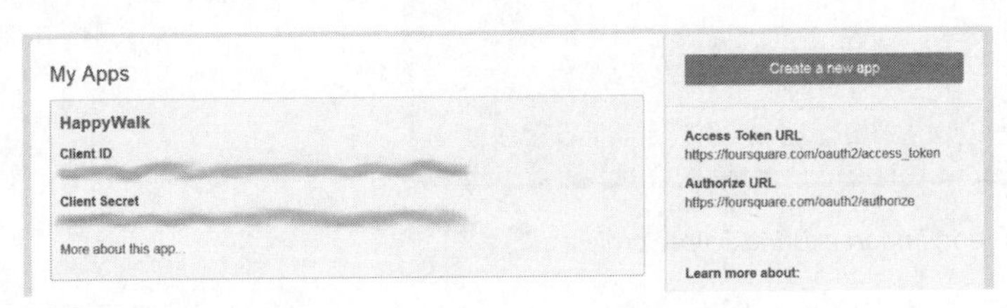

Figure 6.32 Foursquare®'s Client ID and Client Secret.

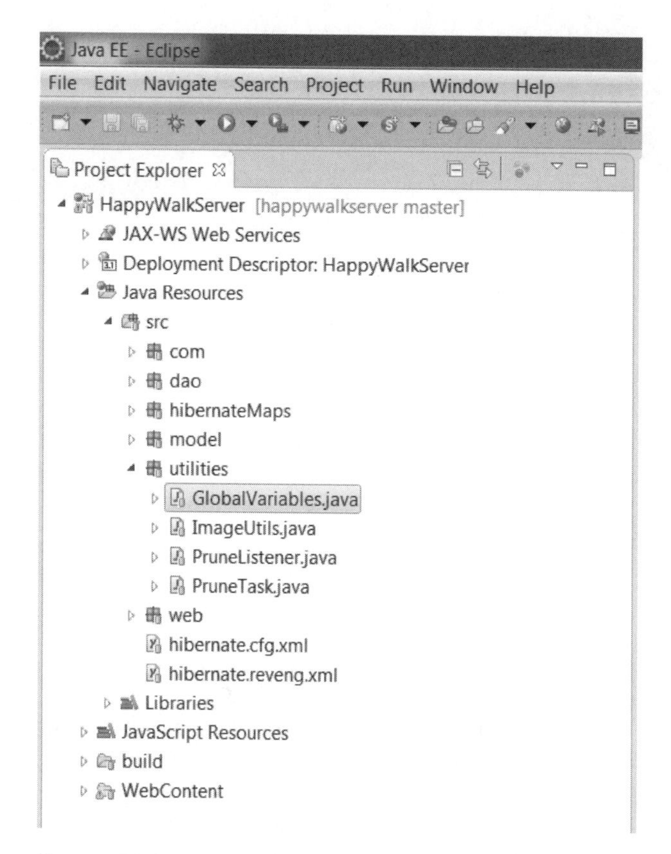

Figure 6.33 Navigating into the server's GlobalVariables.

```
public class GlobalVariables {
2
  //Foursquare keys
4 public static final String FOURSQUARE_CLIENT_ID = "YOUR_CLIENT_ID_HERE";
  public static final String FOURSQUARE_CLIENT_SECRET = "
    YOUR_CLIENT_SECRET_HERE";
6 public static final String FOURSQUARE_REDIRECT_URL = "https://api.
    foursquare.com/v2/";
```

Here, it is necessary to replace the *FOURSQUARE_CLIENT_ID* and *FOURSQUARE_CLIENT_SECRET* strings with the values provided by Foursquare®, from the page shown in Figure 6.32.

6.3.3 Setting up the Database

Now, let us set up our PostgreSQL database. To do so, we will use the pgAdmin III tool which usually accompanies a PostgreSQL installation. If, for some reason, you are unable to use pgAdmin, it is also possible to create our database using the command line as we will show later on.

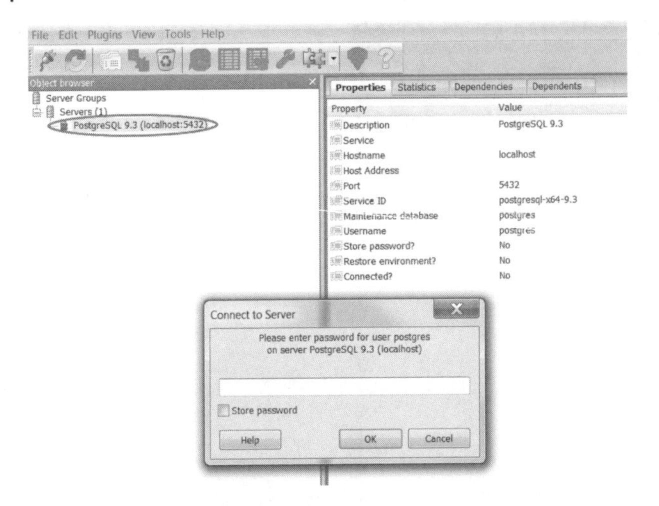

Figure 6.34 Log in to the PostgreSQL 9.3 server.

Using pgAdmin III, we first login by double-clicking on the *PostgreSQL 9.3* server, highlighted by the circle in Figure 6.34. When the login menu pops up, use the superuser credentials obtained during the installation step shown in Figure 6.23.

Afterwards, right-click on *Databases* and select *New Database*, as shown in Figure 6.35. Name the database *"happywalk"* and set its owner as your superuser (by default, it should be *postgres*; see Figure 6.36).

After creating the database, we need to populate it. To do so, we will need to provide the database creation SQL file contained within the server's git repository. Since we have already fetched the project files using Eclipse, the root of our server's project files rests on the directory selected in the step depicted by Figure 6.28. To begin, click on the SQL button while having the *happywalk* database selected, as shown in Figure 6.37.

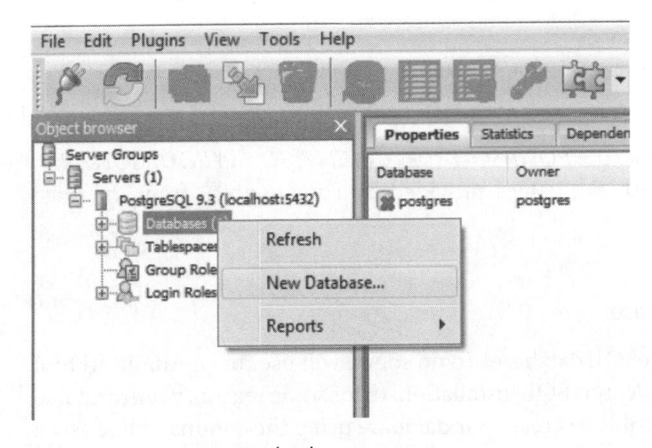

Figure 6.35 Create a new database.

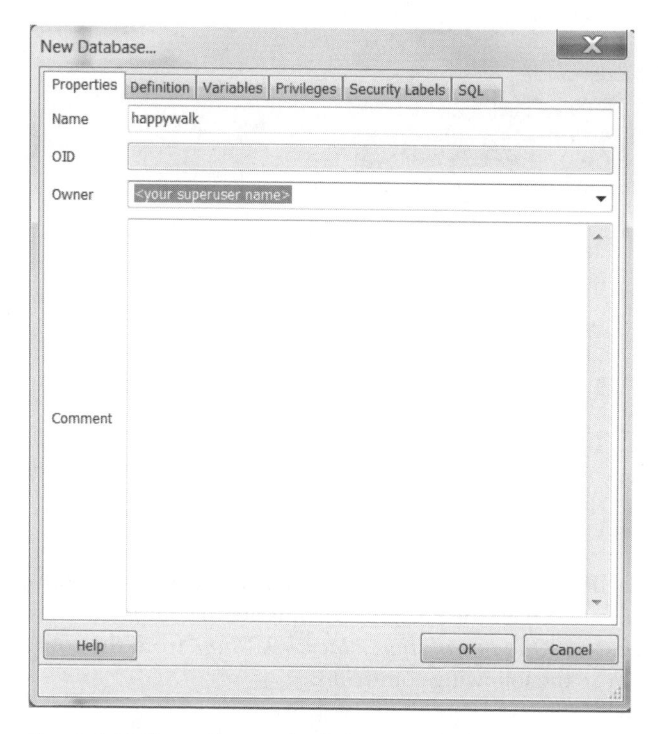

Figure 6.36 Name the new database as *happywalk*.

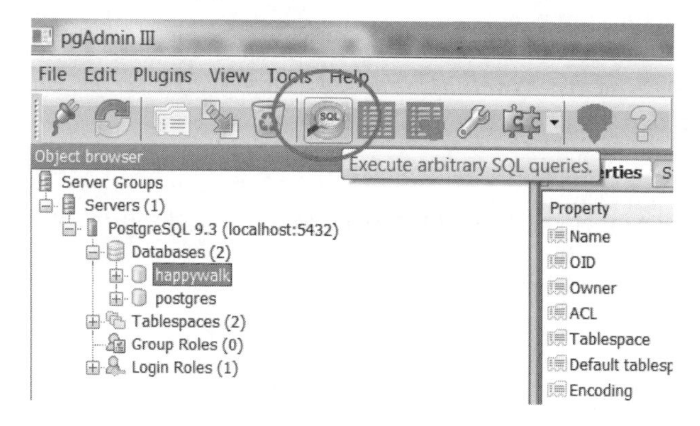

Figure 6.37 Select the correct SQL script.

On the *Query* window that appears, go to *File* → *Open* and select the file located in[4] *<root of HappyWalk's server Git>/Database/HappyWalkDB.sql*. Afterwards, click on the *Execute Query* button (shown in Figure 6.38). If everything went well, the server should output something akin to *Query returned successfully with no result in xx ms.* With this, our database is fully operational.

---

4 From here on, most of the text between < and >is merely indicative and should be replaced with the reader's own paths/filenames, etc.

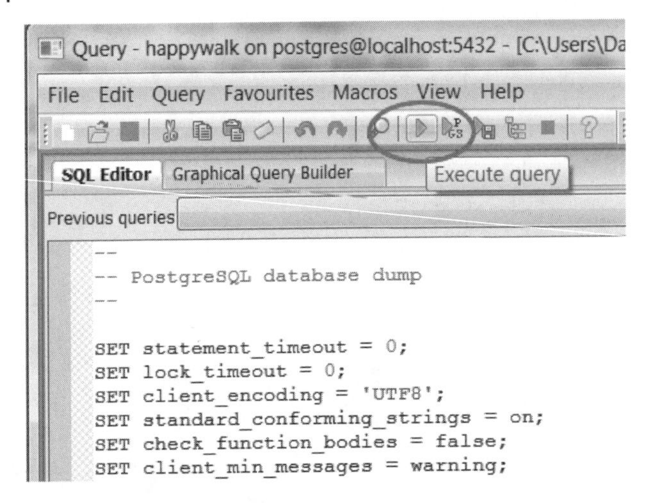

Figure 6.38 Populating the database.

As an alternative to pgAdmin III, you can also create the database through the command line. To do so, you need to either have the PostgreSQL binaries in your *PATH* system variable or navigate to *<PostgreSQL's installation folder>/9.3/bin/* from the command prompt. Afterwards, just use the following command:

createdb -U <your superuser> -O <your superuser> happywalk

to create the database, and then the command:

psql -U <your superuser> -d happywalk -f <root of HappyWalk's server Git >/Database/HappyWalkDB.sql

to populate it.

Now that the database has been created, let us return to our server. In Eclipse, browse to the *src/hibernate.cfg.xml* file:

```
<session-factory>
    <property name="hibernate.connection.driver_class">org.postgresql.
Driver</property>
    <property name="hibernate.connection.password">postgres</property>
    <property name="hibernate.connection.url">jdbc:postgresql://
localhost:5432/happywalk</property>
    <property name="hibernate.connection.username">postgres</property>
```

Here, you will need to replace the *hibernate.connection.password* and *hibernate.connection.username* properties with your PostgreSQL superuser credentials (chosen in the step shown in Figure 6.23). If necessary, also adjust the *hibernate.connection.url* property to match your server's URL and PostgreSQL port. If you have been following this tutorial from a single computer without changing default installation values, you should be connecting to a local PostgreSQL server and, thus, the default URL and port should be *localhost:5432*. This information should match the one shown in pgAdminIII, next to the *PostgreSQL 9.3* server (see Figure 6.34).

6.3.4 Deploying the Server on Tomcat 7

Finally, we need to deploy our server on *Tomcat 7*. To do so, let us define a new server on Eclipse through the menu *File* → *New* → *Other* → *Server* (Figure 6.39), where we choose a new *Tomcat v7.0 Server* (Figure 6.40).

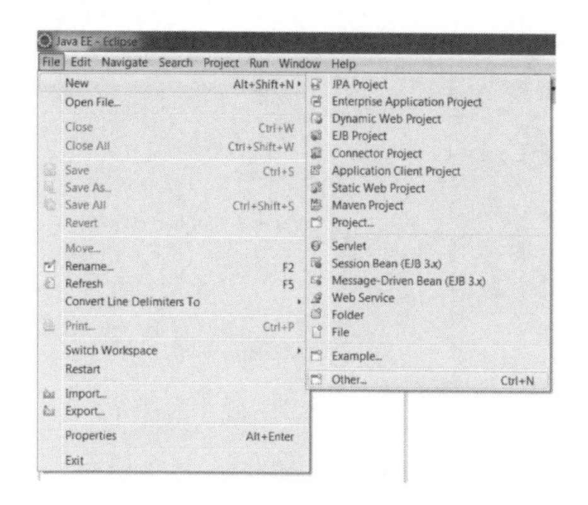

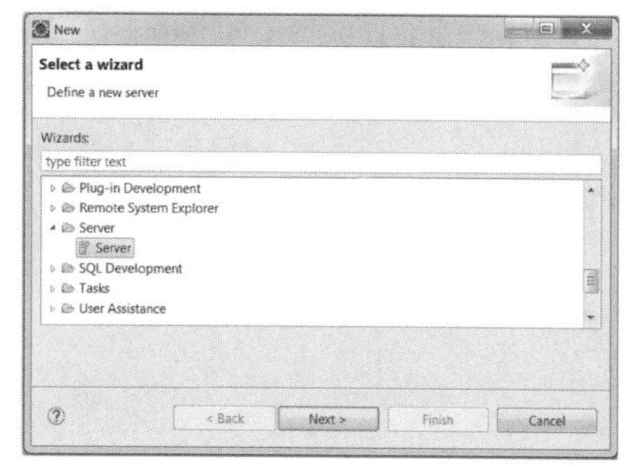

Figure 6.39 Create a new server.

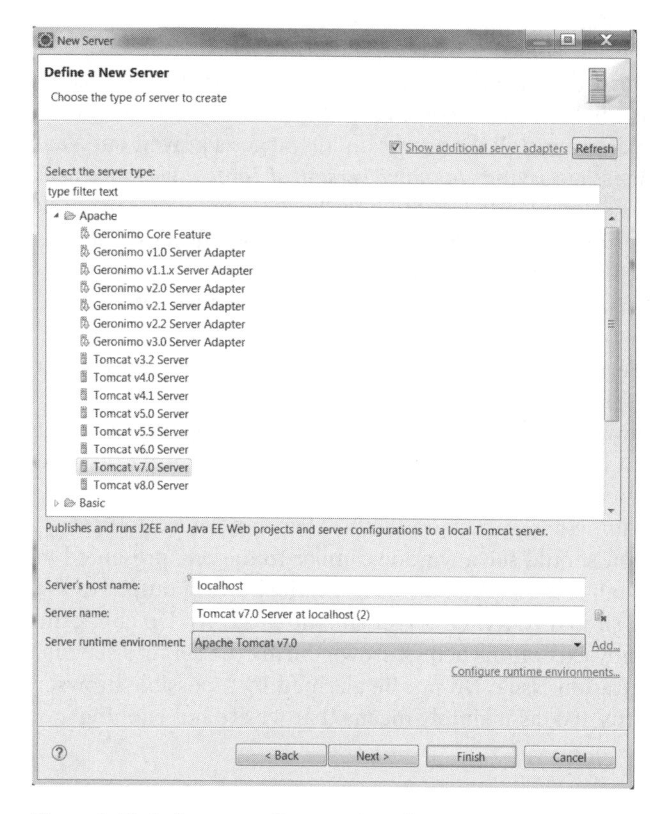

Figure 6.40 Define a new Tomcat 7 installation.

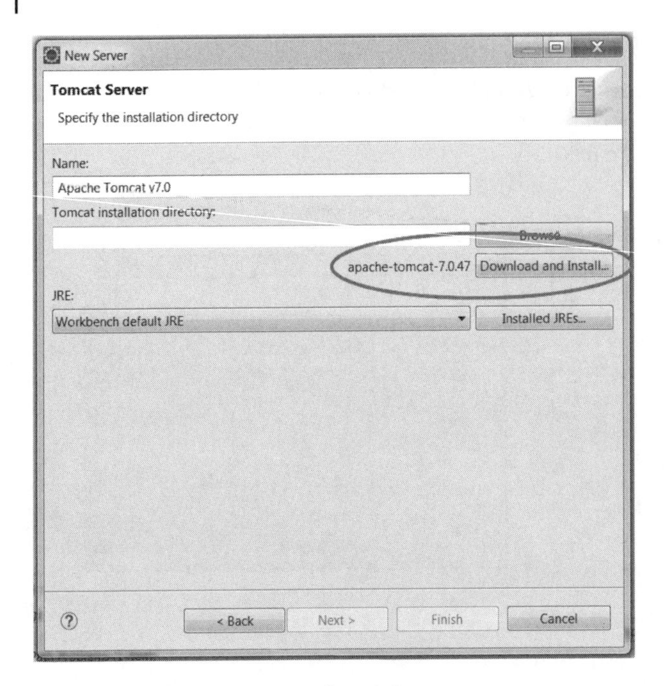

Figure 6.41 Installing Tomcat 7 from Eclipse.

Eclipse allows you to download and install the latest version of Apache Tomcat 7 from within the IDE. To do so, click on the *Download and Install...* button, as shown in Figure 6.41, and choose a folder to install the server on. If you get a grayed out *Next* button together with an error message saying *Unknown version of Tomcat was specified*, try to wait for a little while. Eclipse downloads Tomcat in the background (you can check the download progress in the bottom-right corner of Eclipse's main window) and it does not recognize the server until it has been fully downloaded.

Alternatively, you can also manually download *Tomcat 7*[5] and indicate its path to Eclipse.

Finally, add our HappyWalkServer project to the newly created server by selecting it and pressing *Add >* (figure 6.42) and then *Finish*.

Happywalk's server should now be ready to be run. To test the server, try to run it by pressing the *Run as...* button on Eclipse, choosing *Run on server* (Figure 6.43) and selecting the newly created Tomcat 7 server (Figure 6.44).

At the bottom of Eclipse's main window, click on the tab that says *Console*. If everything is correctly configured, you should see a window similar to the one presented in Figure 6.45, where the *Console* tab is highlighted by a circle. The bottom output of the console should read something similar to *INFO: Server startup in xx ms*. Do check the rest of the console's output for any exceptions being thrown during the server's startup, which would indicate a configuration issue. Do not be alarmed by a possible browser window showing an HTTP status 404, as it simply means that we are not running any web interface within our server.

5 http://tomcat.apache.org/download-70.cgi

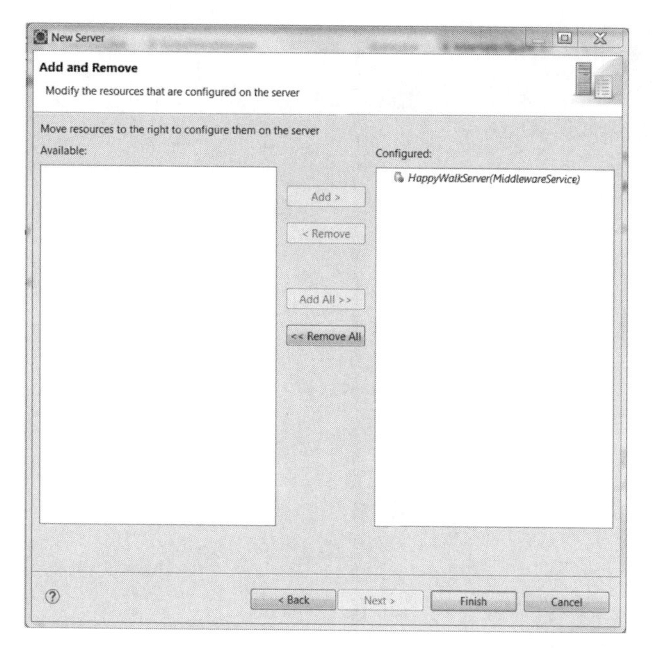

Figure 6.42 Adding HappyWalk to Tomcat 7.

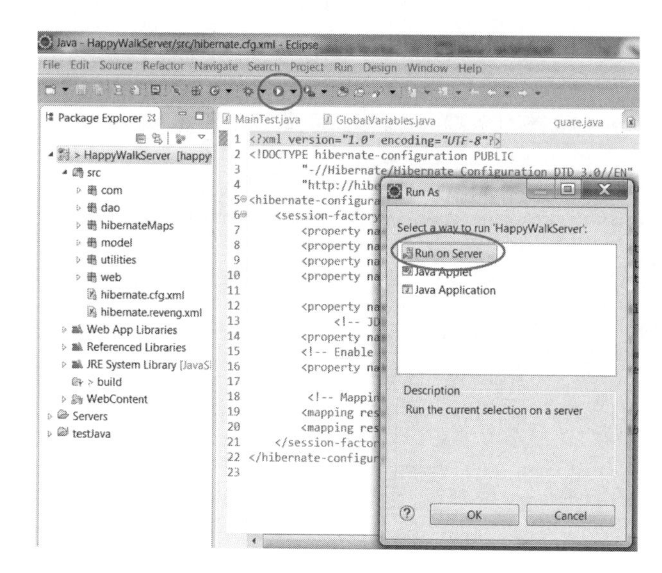

Figure 6.43 Running the HappyWalk server.

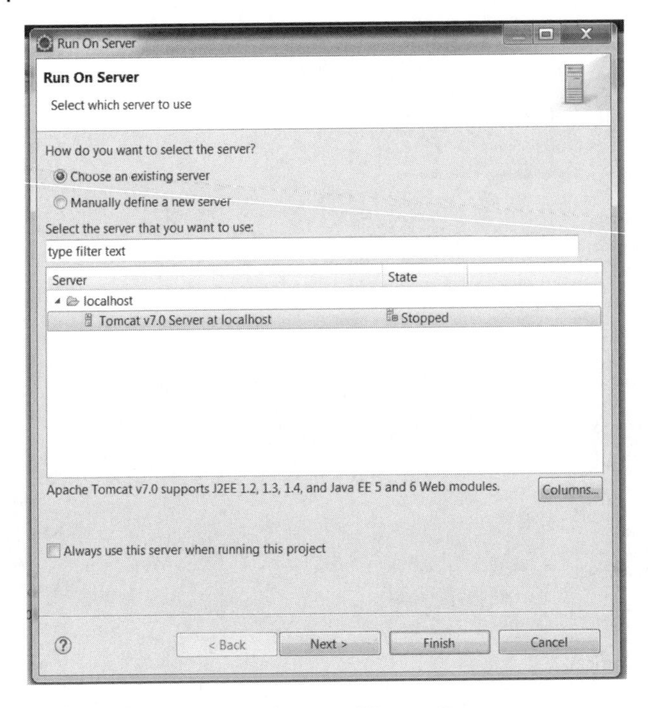

Figure 6.44 Select the newly created Tomcat 7.

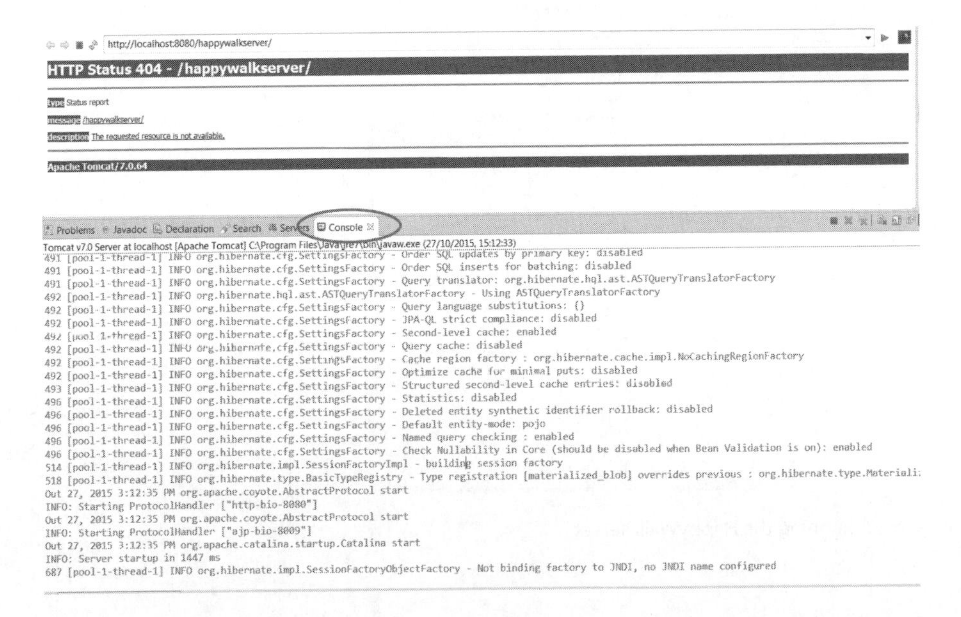

Figure 6.45 The HappyWalk server is up and running.

Table 6.3 Summary of the steps necessary to deploy HappyWalk's server.

Step	Summarized Objective
Install Eclipse Mars	Downloaded and extracted the *Eclipse IDE for Java EE Developers* package from *https://eclipse.org/mars/*
Install PostgreSQL 9.3	Downloaded a PostgreSQL 9.3 installer by visiting *http://www.postgresql.org/download/*
Import HappyWalk's server	Used Eclipse to import HappyWalk's server project from *https://git.dei.uc.pt/dsnunes/happywalk.git*
Obtain a Foursquare® client ID and secret	Visited *https://developer.foursquare.com/*, created an account and a new app to obtain a Client ID and a Client Secret, and replaced the appropriate variables within the server's *GlobalVariables.java* class
Set up the database	Used pgAdmin III to create a new *"happywalk"* database, populated this database using the project's *HappyWalkDB.sql* query, and updated the server's *hibernate.cfg.xml* file
Deploy the server on Tomcat 7	Used Eclipse to install Tomcat 7, added the HappyWalkServer project to it, and tested the server

Table 6.3 summarizes the steps we have taken in this section.

6.4 Testing the Sample App

To finalize our preparatory work, we will now attempt to launch HappyWalk's base system. This involves allowing communication between the server and the Android app. As such, we will need to discover the server's IP address and set that information in the client application.

The exchange of information between the server and the Android client requires a communication channel between the machine hosting the server and your Android phone. This may be achieved within a local network (e.g. both devices connected to the same wireless network) or by having both devices connected to the Internet and knowing the server's public IP. Either way, as mentioned in Section 6.3, **the server needs an active Internet connection**, to communicate with Foursquare®.

In Windows 7 (and most other Windows operating systems), knowing your computer's IP address can be easily achieved through the command line. Click on the *Start* menu, on the search box type *cmd*, and open *cmd.exe*. On the command prompt type *ipconfig* and press the *ENTER* key. As shown in Figure 6.46, the *command* will display several IP addresses; in the example, this computer was connected to a router giving access to a WiFi network to which our Android phone was connected. Thus, in this case, we were interested in the IPv4 address, which represents the address of the computer within the local area network.

The *Linux* and *MacOSX* operating systems have their own equivalent command for displaying IP information; typically, one can execute the *ifconfig* command in the terminal.

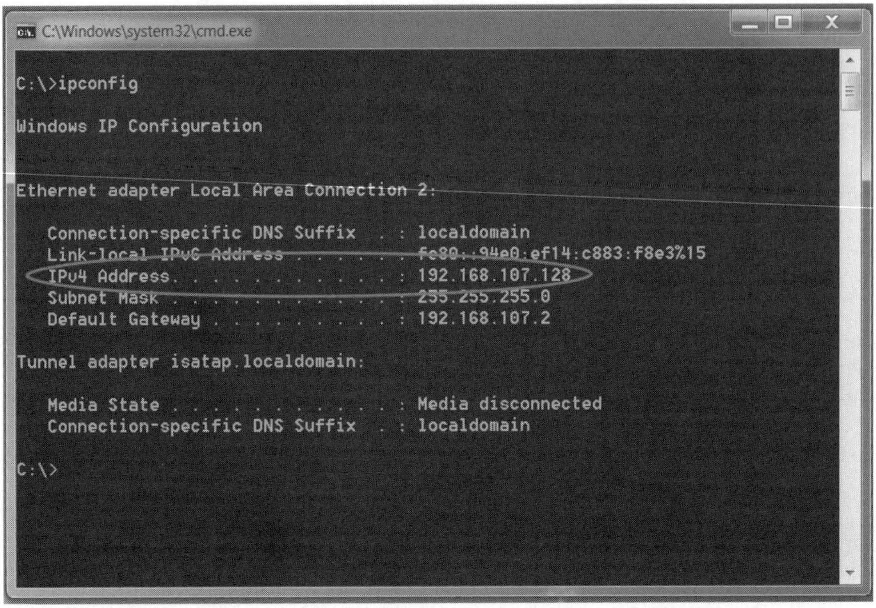

```
C:\Windows\system32\cmd.exe

C:\>ipconfig

Windows IP Configuration


Ethernet adapter Local Area Connection 2:

   Connection-specific DNS Suffix  . : localdomain
   Link-local IPv6 Address . . . . . : fe80::94e0:ef14:c883:f8e3%15
   IPv4 Address. . . . . . . . . . . : 192.168.107.128
   Subnet Mask . . . . . . . . . . . : 255.255.255.0
   Default Gateway . . . . . . . . . : 192.168.107.2

Tunnel adapter isatap.localdomain:

   Media State . . . . . . . . . . . : Media disconnected
   Connection-specific DNS Suffix  . : localdomain

C:\>
```

Figure 6.46 The *ipconfig* command.

Figuring out the correct IP address to provide to HappyWalk is a process that is highly dependent on your network configuration. If your server and your Android phone can communicate through the Internet, you can use websites such as *www.myipaddress.com* to discover your server's external IP.

After taking note of the server's IP, we will need to set this information in our Android app. In AS, change back to the *Android* view, just like we did a while back to change to *Project* view in Figure 6.21, and open the file *hitlexamples.happywalk/utilities/GlobalVariables.java*:

```
package hitlexamples.happywalk.utilities;
2
public class GlobalVariables {
4     //Server Location
      final static public String URL= "http://<IP_of_Server>:8080/
      happywalkserver/rest/";
```

Change the field <IP_of_Server> within the *URL* variable to match the location of your server. Now, we should attempt to run HappyWalk once more by clicking on the *Run app* button (refer to Figure 6.15 back on page ??).

If both the server and the app have been properly configured and assuming that your Android device has an **active Internet connection** and **location services enabled**, you should see an image similar to the one presented in Figure 6.47.

During development and testing, remember that your device needs to have an *active connection to the server and location services enabled*. If it doesn't, the app will not work properly.

Table 6.4 summarizes the steps we have taken in this section.

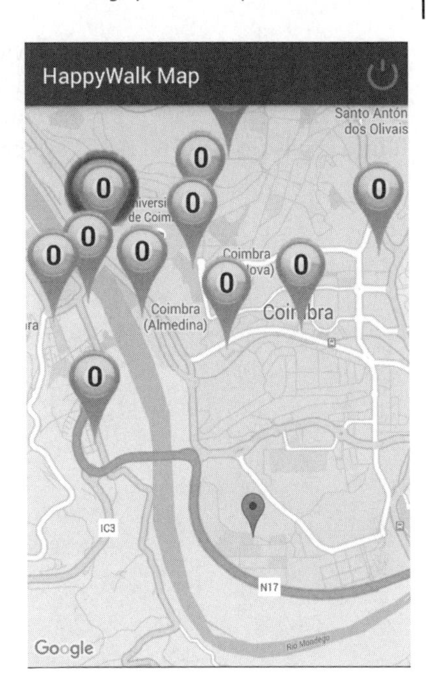

Figure 6.47 HappyWalk's map screen.

Table 6.4 Summary of the steps necessary to test the base HappyWalk system.

Step	Summarized Objective
Ensure connectivity	The server and the Android device must be able to communicate with each other, and the server needs an active Internet connection
Discover the server's IP	Used the command line / terminal together with the *ipconfig / ifconfig* command to discover the server's IP address
Set server IP in Android app	Edited the Android client's *GlobalVariables.java* class and updated the <IP_of_Server> field within the *URL* variable
Launch HappyWalk	Launched both the server and the client and made sure that the map interface appeared on the Android device, together with POI information

6.5 In Summary...

In this chapter we prepared the development environment necessary to work on HappyWalk. We went through the various steps needed for setting up AS, the IDE that we will use to work on the Android client, and Eclipse, which is used to program our server. We installed Git and used it to import the base HappyWalk client and server projects. Additionally, we obtained a *Google Maps Android API* key, a Foursquare® client ID and secret, and added them to the necessary project files.

We also installed PostgreSQL, created a database, and populated it through an SQL query. We then deployed and ran our server using Tomcat 7. Finally, we made sure that the server machine and Android device could communicate, informed the client application of the server's IP address, and tested the base HappyWalk system.

Now that HappyWalk is up and running, let us begin our work towards transforming it into a fully fledged HiTL BCI system.

7

Data Acquisition

As discussed in Chapter 5, *data acquisition* is a fundamental stage of the control loop of HiTL processes, and this is why we will dedicate the whole of the current chapter to it. With this objective in mind, we will start by presenting and describing the class that will handle most of the emotion-related tasks in our sample HappyWalk Android app, in Section 7.1. Next, in Section 7.2, we will explore the processing of sensory data.

More often than not, the burden of data acquisition and processing is too much for the application's user interface thread. Concurrently performing heavy computational tasks generally results in an application with usability issues. For example, in the HappyWalk's case, the usage of map-related functionality could suffer from stuttering if the user interface thread were not freed from heavy tasks such as data acquisition. This is, thus, the main reason why in HappyWalk we have opted for performing data acquisition in a background thread, leaving the main thread purely to user interface operation. This also has advantages from a modularity point of view, allowing to perform data acquisition as a background task through the HappyWalk Service, which is decoupled from the main application.

7.1 Creating the *EmotionTasker*

In this section, we will introduce the *EmotionTasker*. This class represents the core of our emotional inference mechanism; it will be responsible for things such as controlling when to perform emotion recognition, presenting suggestive notifications, and training the neural network. In fact, it is so important that we will continuously work on it throughout most of this tutorial.

For now, we will focus on its basic creation, setting up a constructor and some of its class variables, instantiating an object of it in our background service, and defining a method for data collection.

Right click on the *service* folder of our HappyWalk AS project, select *New* → *Java Class*, and name it *EmotionTasker*, as shown in Figure 7.1.

A Practical Introduction to Human-in-the-Loop Cyber-Physical Systems, First Edition.
David Nunes, Jorge Sá Silva and Fernando Boavida.

Companion Website URL: www.wiley.com/go/nunesloop

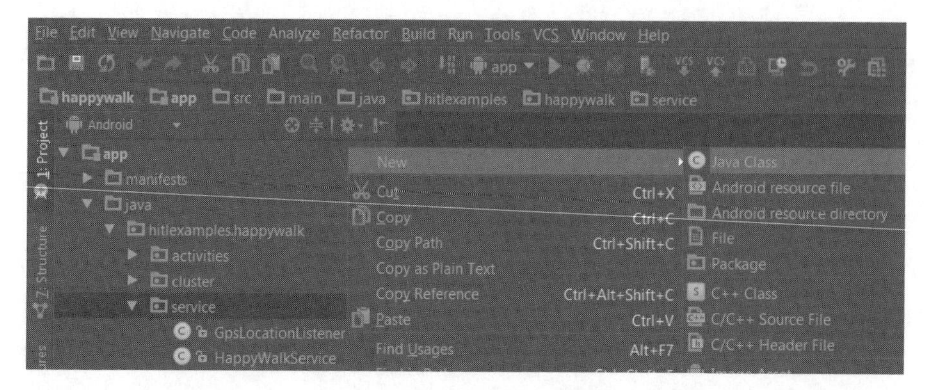

Figure 7.1 Creating a new class.

We can begin by setting up some class variables and a simple constructor. To start with, we will need a reference to our *HappyWalkService* and our *HappyWalkServiceHandler* (see lines 6 and 7 of the code below):

```
package hitlexamples.happywalk.service;

import android.os.Handler;

public class EmotionTasker {
    private HappyWalkService hWServ;
    private Handler hWServiceHandler;

    public EmotionTasker(HappyWalkService hWServ) {
        this.hWServ = hWServ;
        this.hWServiceHandler = hWServ.getHappyWalkServiceHandler();
    }
}
```

Additionally, we should also add a reference and instantiate an *EmotionTasker* object within our *HappyWalkService*. To do so, double-click the *service/HappyWalkService* file and edit it as shown below:

```
package hitlexamples.happywalk.service;

import (...)

public class HappyWalkService extends Service {

    private boolean isRunning = false;
    private final IBinder hwBinder = new HappyWalkBinder();
    private MapsActivity mapAct;
    private Thread hWServiceThread;
    private Handler hWServiceHandler;
    private HwLocationListener hwLocationListener;
    private EmotionTasker emotionTasker;

    (...)

    //Gets and Sets

    (...)
```

```
20
        public HwLocationListener getHwLocationListener() {
22          return hwLocationListener;
        }
24      public boolean isRunning() {
            return isRunning;
26      }
        public NotificationManager getNotificationManager() {
28          return mNM;
        }
30      public EmotionTasker getEmotionTasker() {
            return emotionTasker;
32      }

34    @Override
        public void onCreate() {
36
            (...)
38
            //Prepare our worker thread
40          hWServiceThread = new Thread(new Runnable() {
                public void run() {
42
            (...)
44
                    // ----------------- LOCATION ------------------------
46
                    LocationManager mlocManager = (LocationManager)
        getSystemService(Context.LOCATION_SERVICE);
48
                    List<String> providers = mlocManager.getProviders(true);
50
                    for (String provider : providers) {
52                      mlocManager.requestLocationUpdates(provider, 0, 0,
        hwLocationListener);
                    }
54
                    // Instantiate Emotion Tasker
56                  emotionTasker = new EmotionTasker(HappyWalkService.this);

58                  (...)

60                  Looper.loop();
                }
62          });
            hWServiceThread.start();
64      isRunning = true;
        }
66
        (...)
68  }
```

During this tutorial, we will often deal with classes that are very large; such as is the case of *HappyWalkService* above. Therefore, we will often abbreviate parts of code that are not relevant to the explanation at hand by using the (...) characters. This means that the line numbers shown on the left side of code snippets contained in this book are often not representative of the line numbers in the original code (which we fetched during Chapter 6). Nevertheless, we will make an effort to clearly identify the sections of the code we are working with by displaying several of the original lines around the area of interest. This should allow the reader to easily find the corresponding locations in the original code.

Returning to *HappyWalkService's* code snippet above, notice that we first added a reference to an *EmotionTasker* object, named *emotionTasker*, in line 13. We then added a *getter* function for this object in line 30. Finally, we instantiated it after a *for* loop, in line 56).

The *EmotionTasker* will make use of the ES Sensor Manager, a library for Android developed as part of the EPSRC Ubhave (Ubiquitous and Social Computing for Positive Behaviour Change) project that makes accessing and polling of smartphone sensor data an easy, highly configurable, and battery-friendly task[1] [142]. The ES Sensor Manager library is already available as part of the HappyWalk project, within the *com.ubhave.sensormanager* module. Using this library, we will write a *collectInputs()* method that will be responsible for fetching information from the Location services, the Microphone, and the Accelerometer.

```
1  package hitlexamples.happywalk.service;

3  import android.os.Handler;

5  import com.google.android.gms.maps.model.LatLng;
   import com.ubhave.sensormanager.ESException;
7  import com.ubhave.sensormanager.ESSensorManager;
   import com.ubhave.sensormanager.config.GlobalConfig;
9  import com.ubhave.sensormanager.data.pull.AccelerometerData;
   import com.ubhave.sensormanager.data.pull.MicrophoneData;
11 import com.ubhave.sensormanager.sensors.SensorUtils;

13 import hitlexamples.happywalk.utilities.GlobalVariables;

15 public class EmotionTasker {
       private HappyWalkService hWServ;
17     private Handler hWServiceHandler;
       private ESSensorManager esSensorManager;
19
       public EmotionTasker(HappyWalkService hWServ) {
21         this.hWServ = hWServ;
           this.hWServiceHandler = hWServ.getHappyWalkServiceHandler();
23         //preparing sensor manager to fetch data
           try {
25             esSensorManager = ESSensorManager.getSensorManager(hWServ);
               esSensorManager.setGlobalConfig(GlobalConfig.
       PRINT_LOG_D_MESSAGES, false);
27         } catch (ESException e) {
               e.printStackTrace();
29         }
       }
31
    /**
33     * The inputs are collected through the ubhave module
       * We use ESSensorManager's default sense window time
35     * @return - an array of doubles containing the normalized (0-1)
```

1 http://www.emotionsense.org

```
     * collected inputs. The indexes are defined in GlobalVariables
37   */
    private double[] collectInputs() throws NoCurrentPosition{
39      double[] inputs = null;
        LatLng actualPosition;
41      /*
        first , check if we have location information. This is required for
     performing emotion recognition
43      */
        if((actualPosition = hWServ.getHwLocationListener().
    getActualposition()) !=null) {
45          try {
                //normalize location data
47              double[] normalizedLocation = HwLocationProcessor.
    normalizeLatLng(actualPosition);
                //collect and process microphone and accelerometer data
49              MicrophoneData micData = (MicrophoneData) esSensorManager.
    getDataFromSensor(SensorUtils.SENSOR_TYPE_MICROPHONE);
                double averageMicValue = HwMicrophoneProcessor.
    getAverageAmplitude(micData);
51              averageMicValue = HwMicrophoneProcessor.
    normalizeAvgAmplitude(averageMicValue);

53              AccelerometerData accData = (AccelerometerData)
    esSensorManager.getDataFromSensor(SensorUtils.
    SENSOR_TYPE_ACCELEROMETER);
                double normFCTCoeffSum = HwAccelerometerProcessor.
    getNormalizedFCTCoeffSum(HwAccelerometerProcessor.getTotalAcceleration
    (accData));
55
                //insert inputs into array
57              inputs = new double[GlobalVariables.NN_INPUTS];
                inputs[GlobalVariables.NN_INPUT_ARRAY_INDEX_LATITUDE] =
    normalizedLocation[HwLocationProcessor.LATITUDE_INDEX];
59              inputs[GlobalVariables.NN_INPUT_ARRAY_INDEX_LONGITUDE] =
    normalizedLocation[HwLocationProcessor.LONGITUDE_INDEX];
                inputs[GlobalVariables.NN_INPUT_ARRAY_INDEX_NOISE] =
    averageMicValue;
61              inputs[GlobalVariables.NN_INPUT_ARRAY_INDEX_MOVEMENT] =
    normFCTCoeffSum;
            } catch (ESException e) {
63              e.printStackTrace();
            }
65      }
        else{
67          //There is no location information.
            throw new NoCurrentPosition("No current position available ,
     cannot perform emotion classification .");
69      }
        return inputs;
71  }
}
```

Notice that we need to declare an *ESSensorManager esSensorManager* class variable, at the beginning of *EmotionTasker* (line 18). This variable is initialized through the *EmotionTasker's* constructor, in line 25 (where we also deactivate its debug messages, in line 26, for a cleaner output). The *ESSensorManager* class requires us to do all of these tasks inside a *try/catch* block, to handle possible exceptions.

The *collectInputs()* method begins at line 38. It first declares two variables: a double array named *inputs*, where data will be stored, and a *LatLng actualPosition* object, which will store the user's current location in terms of latitude and longitude.

The *if* clause in line 44 makes use of the *HwLocationListener* supported by HappyWalk's background service (as the reader may remember from Figure 5.5 back on page ??). It simply checks if the *HwLocationListener* has position information and, if it does, uses a *HwLocationProcessor* to normalize it, in line 47.

It then retrieves data from the microphone (*MicrophoneData*, in line 49) and the accelerometer (*AccelerometerData*, in line 53), storing this information within the *micData* and *accData* objects. It also uses an *HwMicrophoneProcessor*, in lines 50 and 51, and a *HwAccelerometerProcessor*, in line 54, to process microphone and accelerometer data, storing the results, together with the normalized location, into *inputs* (lines 58–61). Finally, it returns *inputs*.

Notice that the size of *inputs* (line 57) and its *array indexes* (lines 58–61) are clearly defined. We have previously discussed in Section 5.3 that we would handle four types of inputs, and that information is stored on the *GlobalVariables.NN_INPUTS* variable. Knowing the array indexes where we stored the results is also important, as we will need to distinguish them later on when feeding information to our neural network. Thus, it is a good idea to define where each type of data is stored through global variables, which can be accessed from anywhere within our application. For convenience, these should be already defined within the *hitlexamples.happywalk/utilities/GlobalVariables* class:

```
//EMOTION INPUT ARRAY INDEXES
/*[0] - latitude
* [1] - longitude
* [2] - amount of noise
* [3] - amount of movement */
public static final int NN_INPUT_ARRAY_INDEX_LATITUDE = 0;
public static final int NN_INPUT_ARRAY_INDEX_LONGITUDE = 1;
public static final int NN_INPUT_ARRAY_INDEX_NOISE = 2;
public static final int NN_INPUT_ARRAY_INDEX_MOVEMENT = 3;
```

As the reader writes the *collectInputs()* method into their own copy of the project, AS will notify that it "cannot resolve" certain "symbols", highlighting certain words with a red color (see how AS cannot resolve LatLng, as the popup warns in Figure 7.2). This means that it cannot identify what those words mean and that it is necessary to add the appropriate import declarations. One easy way of doing so is by placing the writing cursor on the highlighted word and pressing the *Alt* + *Enter* keys simultaneously. A small pop-up will appear, where you can tell AS to import the appropriate class, as shown in Figure 7.3. As we delve into our project in the next sections, do not forget to add the appropriate import declarations whenever necessary.

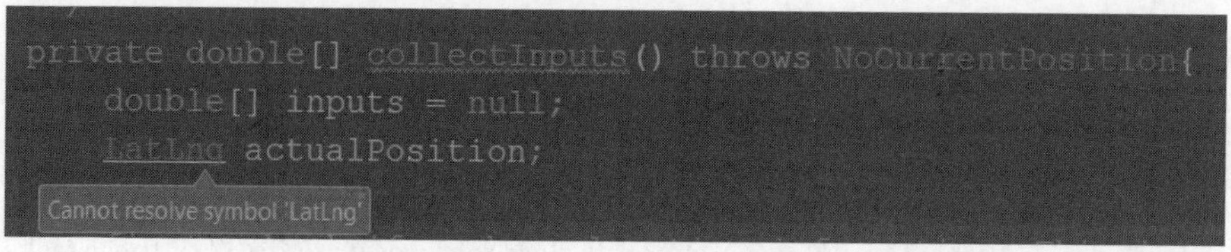

```
private double[] collectInputs() throws NoCurrentPosition{
    double[] inputs = null;
    LatLng actualPosition;
```

Cannot resolve symbol 'LatLng'

Figure 7.2 AS cannot resolve symbol issue.

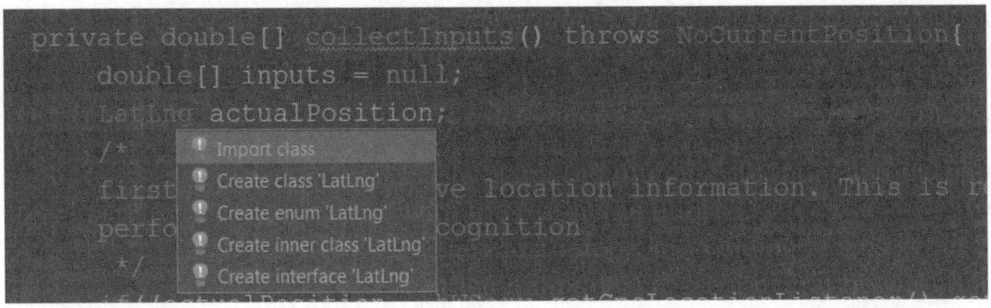

Figure 7.3 Importing the appropriate class.

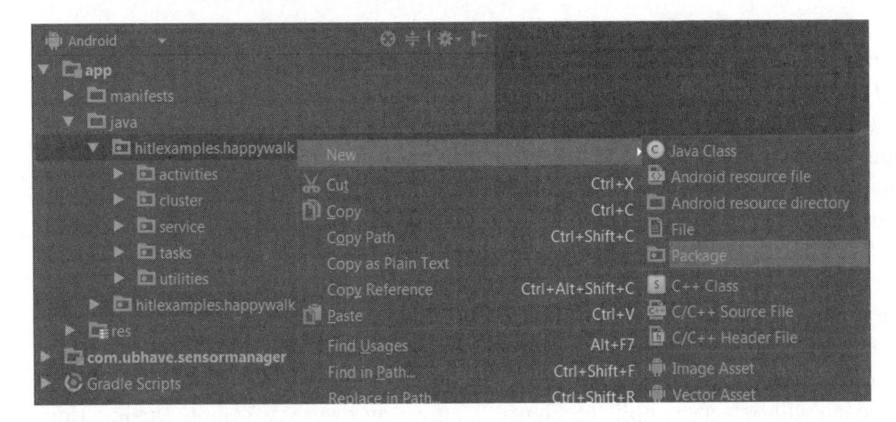

Figure 7.4 Creating a new package.

Notice that, in case there is no location information, *collectInputs()* throws a *NoCurrentPosition* exception (see lines 38 and 68 of the code snippet on page ???). This exception cannot be imported using the method above because we not implemented it yet. As we have discussed in Section 5.3.2 and showned in Figure 5.10, our neural network will require position information for its processing. Thus, it does not make sense to gather other types of data if our *HwLocationListener* cannot provide us with location information.

Let us first create a new package to hold the exception. To do so, follow the steps shown in Figure 7.4: use the project tab's *Android* view, navigate into *hitlexamples.happywalk*, *right-click* on it, and select *New → Package*. Name this new package as *exceptions*.

Now, let us right-click this new package, select *New → Java Class*, and name it *NoCurrentPosition*. The class shall extend the *Exception* Java class, as shown by the code below:

```
package hitlexamples.happywalk.exceptions;
2
public class NoCurrentPosition extends Exception {
4    public NoCurrentPosition(String detailMessage) {
        super(detailMessage);
6    }

8    public NoCurrentPosition(String detailMessage, Throwable throwable) {
        super(detailMessage, throwable);
10   }
}
```

This exception serves as a last-resort safety net. As we will see in Section 9.2, we should ensure that location information is available before attempting to use the *collectInputs()* method. Do not forget to import the new *NoCurrentPosition* class into *EmotionTasker*.

As the reader might have noticed, the *HwLocationProcessor*, the *HwMicrophoneProcessor*, and the *HwAccelerometerProcessor* classes, used to process data, also need to be implemented. Let us look into how these processors can be created.

7.2 Processing Sensory Data

In this section we shall prepare a package and several classes specifically dedicated to processing sensory data. In particular, we will focus on creating the *HwLocationProcessor*, *HwMicrophoneProcessor*, and *HwAccelerometerProcessor*, to process location, sound, and movement, respectively. These classes will normalize the inputs and implement very simple signal processing techniques, which we will also explain.

Let us first create a new package to hold emotion-related classes. Follow the steps shown in Figure 7.4: navigate into *hitlexamples.happywalk*, right-click on it, and select *New* → *Package*. Name this new package *emotion*.

Under this new package *hitlexamples.happywalk.emotion* let us create yet another package, named *processors*, where three new classes should be created: *HwLocationProcessor*, *HwAccelerometerProcessor*, and *HwMicrophoneProcessor* (see Figure 7.5).

Since signal processing is not the main focus of this book, and for the sake of brevity, our approach will be very simplistic. Figure 7.6 gives an overview of how the location, microphone, and accelerometer processors will be implemented.

When working with neural networks, it is good practice to normalize the data before using it. This is because normalization may reduce the training effort and increase the efficiency of the network. As Figure 7.6 shows, the simplest type of processing we will perform is on location. We will merely normalize latitude and longitude values on a **[0,1]**

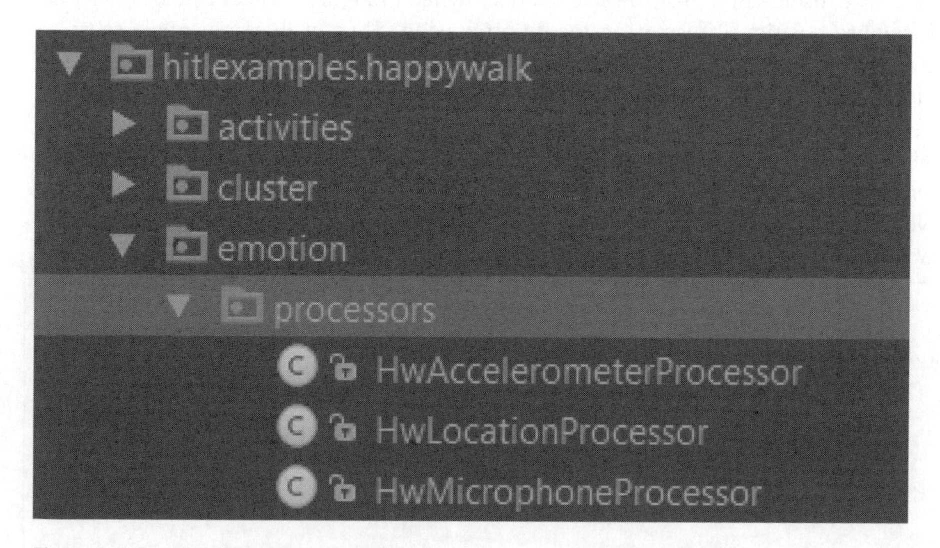

Figure 7.5 Creating the sensor processors.

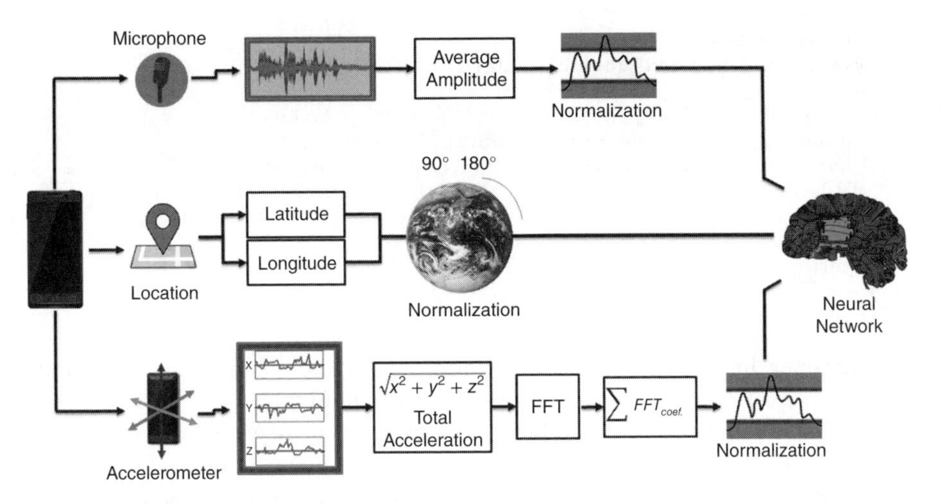

Figure 7.6 Signal processing overview.

range by considering their maximum and minimum possible values. The processing of microphone data will also be rather straightforward. We shall acquire the signal, average its amplitude, and normalize this value on a **[0,1]** scale through empirically derived minimum and a maximum thresholds. The processing of movement, however, shall be slightly more complex. We will first attempt to calculate the total acceleration (independent of direction), perform frequency analysis of the signal, and, finally, normalize the result.

Let us begin with the processing of location. Normalizing latitude and longitude values is trivial, since latitude ranges from -90 to 90 degrees, while longitude ranges from -180 to 180. As such, we simply have to employ *feature scaling*, through the formula:

$$X' = \frac{X - X_{min}}{X_{max} - X_{min}}$$

Where X' is the normalized result, X is the original value, and X_{min} and X_{max} are the minimum and maximum values of the variable, respectively. The following code implements the formula above:

```
1  package hitlexamples.happywalk.emotion.processors;
2
3  import com.google.android.gms.maps.model.LatLng;
4
5  public class HwLocationProcessor {
6      /*Latitudes range from -90 to 90.
7      Longitudes range from -180 to 180.*/
8      private static final double MIN_LAT = -90;
9      private static final double MAX_LAT = 90;
10     private static final double MIN_LNG = -180;
11     private static final double MAX_LNG = 180;
12
13     public static final int LATITUDE_INDEX = 0;
14     public static final int LONGITUDE_INDEX = 1;
15
16     /**
17      * Normalizes latitude / longitude to [0,1] range.
```

```
     * @param latLng
19   * @return normalized double - [latitude,longitude]
     */
21  public static double[] normalizeLatLng(LatLng latLng) {
        double[] normLatLng = new double[2];
23      normLatLng[LATITUDE_INDEX] = (latLng.latitude - MIN_LAT)/(MAX_LAT
    - MIN_LAT);
        normLatLng[LONGITUDE_INDEX] = (latLng.longitude - MIN_LNG)/(
    MAX_LNG - MIN_LNG);
25      return normLatLng;
    }
27 }
```

This processor implements a single method, *normalizeLatLng()* (line 21), which returns an array of doubles containing the normalized latitude and longitude values from a *LatLng* object (the standard object used by the Google Maps Android API to represent latitude/longitude coordinates). The minimum and maximum values for latitude and longitude are defined in lines 8–11. The indexes used by *HwLocationProcessor* are defined by the *LATITUDE_INDEX* (line 13) and *LONGITUDE_INDEX* (line 14) variables, which are thereafter used by *collectInputs()* as we have seen back on page ???.

Now, let us move on to the microphone signal. Here, as suggested in Figure 7.6, we will calculate the average amplitude of the signal and normalize this value within certain limits:

```
1  package hitlexamples.happywalk.emotion.processors;

3  import com.ubhave.sensormanager.data.pull.MicrophoneData;

5  import org.encog.util.arrayutil.NormalizationAction;
   import org.encog.util.arrayutil.NormalizedField;
7
   public class HwMicrophoneProcessor {
9
       private static final double MAX_AVG = 6000;
11     private static final double MIN_AVG = 150;
       private static final double NORM_HIGH = 1;
13     private static final double NORM_LOW = 0;

15     /**
        * This method returns the average value of the amplitude of a
       MicrophoneData. If the data does not contain values, it returns an
       average of zero.
17      * @param data - The microphone data
        * @return - the average amplitude
19      */
       public static double getAverageAmplitude(MicrophoneData data){
21         int[] amplitudeArray= data.getAmplitudeArray();
           double avgAmplitude = 0;
23
           if (amplitudeArray.length> 0) {
25             for (int aValue : amplitudeArray)
               {
27                 avgAmplitude += aValue;
               }
29             avgAmplitude = avgAmplitude / (double)amplitudeArray.length;
           }
31         return avgAmplitude;
```

```
    }
33
    /**
35   * This method normalizes the average value of the amplitude of a
    MicrophoneData between 0 and 1.
     * @param avgAmp - The average amplitude
37   * @return - normalized average amplitude
     */
39  public static double normalizeAvgAmplitude(double avgAmp) {
        if (avgAmp> MAX_AVG){
41          avgAmp = NORM_HIGH;
        }
43      else if (avgAmp< MIN_AVG){
            avgAmp = NORM_LOW;
45      }
        else {
47          NormalizedField normNoise = new NormalizedField(
    NormalizationAction.Normalize, "myfield", MAX_AVG, MIN_AVG, NORM_HIGH,
     NORM_LOW);
            avgAmp = normNoise.normalize(avgAmp);
49      }
        return avgAmp;
51  }
}
```

The code is rather self-explanatory, with the threshold values resting on the variables *MAX_AVG* (line 10) and *MIN_AVG* (line 11), meaning *Maximum average* and *Minimum average*, respectively. The suggested values are merely indicative. In fact, we encourage the reader to experiment empirically and determine more appropriate values.

The method *getAverageAmplitude()* (line 20) iterates over the amplitude array present in a *MicrophoneData* object and calculates its average. On the other hand, the method *normalizeAvgAmplitude()* (line 39) normalizes the average amplitude to a [*NORM_LOW*, *NORM_HIGH*] range (in this case, it is [0,1]). First, it compares the average amplitude with its thresholds. If the average amplitude is outside the [*MIN_AVG*, *MAX_AVG*] range, its value is set to *NORM_LOW* or *NORM_HIGH*, accordingly. However, if the thresholds are respected, Encog's *NormalizedField* utility is used. A new *NormalizedField* object is set up in line 47 by using Encog's *NormalizationAction.Normalized* field, as well as *MAX_AVG*, *MIN_AVG*, *NORM_HIGH*, and *NORM_LOW*, to properly define the range.

As mentioned, we will attempt something slightly more complex for the accelerometer signal. Since we simply want to have an idea of the amount of movement of the smartphone, independently of its orientation, we need to first calculate the total acceleration from all axes (x, y, z). To do so, we can define new helper methods within our *HwAccelerometerProcessor* class to calculate the square root of the sum of their squared values, as suggested by Figure 7.6. This provides us with a value representing the general acceleration of the device:

```
 /**
2 * Returns an array representing the "total" acceleration for each
     measurement, by merging the acceleration data from the x, y and z axes
     .
  * @param data - accelerometer data array
4 * @return - total acceleration array
  */
```

```
6   public static double[] getTotalAcceleration(AccelerometerData data) {
        ArrayList<float[]> readings = data.getSensorReadings();
8       double[] totalAcc = new double[readings.size()];

10      for (int i = 0; i < totalAcc.length; i++) {
            float[] sample = readings.get(i);
12          totalAcc[i] = squareRootSumSquare(sample[AcXX], sample[AcYY],
        sample[AcZZ]);
        }
14      return totalAcc;
    }
16
    private static double squareRootSumSquare(float x, float y, float z) {
18      return Math.sqrt(x * x + y * y + z * z);
    }
```

As discussed in Section 5.3.2, we will now attempt to perform a simple Fourier analysis of the transformed signal. This type of signal processing technique involves decomposing a signal into oscillatory components (frequencies). In HappyWalk, we will use a basic form of Fourier analysis. More complex approaches are outside the scope of this book. In particular, we will use the fast cosine transform (FCT), a type of Fourier transformation that only uses real values (as opposed to using complex numbers). Since we are only interested in the real component of our signal, we can use the FCT for our frequency analysis.

The Apache Commons Mathematics Library (*org.apache.commons.math3*) allows us to access methods that calculate the FCT of a signal. To use this library we will need to reference it in the *app/build.gradle* file, under *dependencies*, as indicated in line 19 of the following code snippet:

```
1   apply plugin: 'com.android.application'

3   android {
        compileSdkVersion 21
5       buildToolsVersion "21.1.2"

7       defaultConfig {
            (...)
9       }
        buildTypes {
11          (...)
        }
13  }

15  dependencies {
        compile fileTree(dir: 'libs', include: ['*.jar'])
17      compile 'com.android.support:appcompat-v7:21.0.3'
        compile 'com.google.android.gms:play-services:6.5.87'
19      compile 'org.apache.commons:commons-math3:3.0'
        compile project(':com.ubhave.sensormanager')
21  }
```

However, the present implementation of FCT requires the length of the data set to be a power of two plus one $(N = 2n + 1)$. There are several methods to handle this limitation. In this example, we will simply "zero-pad" our signal (add zero values at the end of the signal) until we achieve the necessary length. While this method is not optimal (it has some side effects in the resulting Fourier transformation), it is sufficient for our purposes. The resulting array contains a series of "FCT coefficients", which represent the frequency components contained within the original signal. These coefficients can be normalized. They are divided by a value corresponding to the original signal's length.

```
/**
 * Performs a Fast Cosine Transform (FCT) of an array of data and
 * returns the normalized Fourier coefficients
 *
 * The Fast Cosine Transform is a type of Fourier Transformation
 * that only uses real values (opposed to using complex numbers)
 * Since we are only interested in the real component of our signal,
 * we can use the FCT for our frequency analysis.
 * Also, this method only accepts array sizes that are of
 * power of 2 + 1 (e.g. 5 or 9-> 2^2 +1 = 5; 2^3 +1 = 9)
 *
 * @param data - the data
 * @return - normalized FCT coefficients
 */
private static double[] calculateFCT(double [] data) {
    double[] paddedData = zeroPadData(data);
    FastCosineTransformer transf = new FastCosineTransformer(
    DctNormalization.STANDARD_DCT_I);
    double[] dataStream_fft = transf.transform(paddedData, TransformType.
    FORWARD);
    //for normalization, we divide by the original signal's length
    for (int i = 0; i<data.length;i++) {
        dataStream_fft[i] = dataStream_fft[i]/data.length;
    }
    return dataStream_fft;
}

/***
 * This method zero pads an array of data
 * @return - padded data
 */
private static double[] zeroPadData(double[] data) {
    double[] paddedData = new double[closestUpperPowerOfTwoPlusOne(data.
    length)];
    System.arraycopy(data, 0, paddedData, 0, data.length);
    return paddedData;
}

/**
 * Finds the closest upper value of a number that is a solution for
 * 2^n + 1, with n being a real number
 * @return - the closest power of two plus one
 */
private static int closestUpperPowerOfTwoPlusOne(int number) {
    long closestPoTP1 = (long) Math.ceil(Math.log10(number) / Math.log10
    (2));
    closestPoTP1 = (long) Math.pow(2, closestPoTP1) + 1;
    return (int) closestPoTP1;
}
```

The method *calculateFCT()* (line 15) begins by calling *zeroPadData()* (line 30) which, as the name indicates, zero-pads the data. To do so, *zeroPadData()* calls, in turn, *closestUpperPowerOfTwoPlusOne()* (line 41) which, when given an integer, finds its closest upper value that is a power of two plus one. The Fourier transformation is performed through a *FastCosineTransformer* (line 17), which is part of the Apache Commons Mathematics Library[2] .

To achieve a value that represents the overall movement of the device, we sum the FCT coefficients and normalize the result between two empirically derived minimum and maximum values. Below is the full implementation of our HwAccelerometerProcessor class:

```
package hitlexamples.happywalk.emotion.processors;

import com.ubhave.sensormanager.data.pull.AccelerometerData;

import org.apache.commons.math3.transform.DctNormalization;
import org.apache.commons.math3.transform.FastCosineTransformer;
import org.apache.commons.math3.transform.TransformType;
import org.encog.util.arrayutil.NormalizationAction;
import org.encog.util.arrayutil.NormalizedField;

import java.util.ArrayList;

/**
 * In this class, we use a Fourier Transformation to evaluate the amount
    of movement being produced by the user
 */
public class HwAccelerometerProcessor {

    private static final double MAX_AVG = 400;
    private static final double MIN_AVG = 100;
    private static final double NORM_HIGH = 1;
    private static final double NORM_LOW = 0;

    private static final int AcXX =0;
    private static final int AcYY =1;
    private static final int AcZZ =2;

    /**
     * Returns an array representing the "total" acceleration for each
    measurement, by merging the acceleration data from the x, y and z axes
    .
     * @param data - accelerometer data array
     * @return - total acceleration array
     */
    public static double[] getTotalAcceleration(AccelerometerData data) {
        ArrayList<float[]> readings = data.getSensorReadings();
        double[] totalAcc = new double[readings.size()];

        for (int i = 0; i < totalAcc.length; i++) {
            float[] sample = readings.get(i);
            totalAcc[i] = squareRootSumSquare(sample[AcXX], sample[AcYY],
    sample[AcZZ]);
        }
        return totalAcc;
    }
```

2 More information can be found in the library's documentation, at http://commons.apache.org/proper/commons-math/

```
43      private static double squareRootSumSquare(float x, float y, float z) {
            return Math.sqrt(x * x + y * y + z * z);
45      }

47      /**
         * Performs a Fast Cosine Transform (FCT) of an array of data and
49       * returns the normalized Fourier coefficients
         *
51       * The Fast Cosine Transform is a type of Fourier Transformation
         * that only uses real values (opposed to using complex numbers)
53       * Since we are only interested in the real component of our signal,
         * we can use the FCT for our frequency analysis.
55       * Also, this method only accepts array sizes that are of
         * power of 2 + 1 (e.g. 5 or 9-> 2^2 +1 = 5; 2^3 +1 = 9)
57       *
         * @param data - the data
59       * @return - normalized FCT coefficients
         */
61      private static double[] calculateFCT(double [] data) {
            double[] paddedData = zeroPadData(data);
63          FastCosineTransformer transf = new FastCosineTransformer(
        DctNormalization.STANDARD_DCT_I);
            double[] dataStream_fft = transf.transform(paddedData,
        TransformType.FORWARD);
65          //for normalization, we divide by the original signal's length
            for (int i = 0; i<data.length;i++) {
67              dataStream_fft[i] = dataStream_fft[i]/data.length;
            }
69          return dataStream_fft;
        }
71
        /***
73       * This method zero pads an array of data
         * @return - padded data
75       */
        private static double[] zeroPadData(double[] data) {
77          double[] paddedData = new double[closestUpperPowerOfTwoPlusOne(
        data.length)];
            System.arraycopy(data, 0, paddedData, 0, data.length);
79          return paddedData;
        }
81
        /**
83       * Finds the closest upper value of a number that is a solution for
         * 2^n + 1, with n being a real number
85       * @return - the closest power of two plus one
         */
87      private static int closestUpperPowerOfTwoPlusOne(int number) {
            long closestPoTP1 = (long) Math.ceil(Math.log10(number) / Math.
        log10(2));
89          closestPoTP1 = (long) Math.pow(2, closestPoTP1) + 1;
            return (int) closestPoTP1;
91      }

93      /**
         * Returns a normalized sum of Fast Cosine Transform coefficients,
95       * from a totalAcceleration array
         * @param totalAccData - an array containing the total acceleration
        values
```

```
97      * @return - normalized sum of FCT coefficients
        */
99     public static double getNormalizedFCTCoeffSum(double[] totalAccData) {

           double[] fctCoeffs = calculateFCT(totalAccData);
101
           double sumFCTcoeff = 0;
103        //sum the absolute value of the coefficients
           for(double val: fctCoeffs) {
105            sumFCTcoeff += Math.abs(val);
           }
107        //perform normalization
           if (sumFCTcoeff> MAX_AVG){
109            sumFCTcoeff = NORM_HIGH;
           }
111        else if (sumFCTcoeff< MIN_AVG){
               sumFCTcoeff = NORM_LOW;
113        }
           else {
115            NormalizedField normMove = new NormalizedField(
                   NormalizationAction.Normalize, "myfield", MAX_AVG,
117                MIN_AVG, NORM_HIGH, NORM_LOW);
               sumFCTcoeff = normMove.normalize(sumFCTcoeff);
119        }
           return sumFCTcoeff;
121    }
}
```

This class is, in some ways, similar to *HwMicrophoneProcessor*. Between the lines 18 and 21 we can find the threshold values *MAX_AVG* and *MIN_AVG*. The reader is, once again, encouraged to experiment with these values to fine-tune the application. We can also find that *NORM_HIGH* and *NORM_LOW* define a [0,1] range once more. Between lines 23 and 25 we find the indexes used by *getTotalAcceleration()* to read the acceleration values in the x, y, and z axes from within an *AccelerometerData* object.

In line 99 we find the *getNormalizedFCTCoeffSum()* method. It first calculates the FCT coefficients (line 100) and then determines the sum of their absolute values (line 103). This sum is then normalized in a manner very similar to what was done for *HwMicrophoneProcessor* (line 107).

Now that they are implemented, do not forget to add the appropriate imports for the *HwLocationProcessor*, the *HwAccelerometerProcessor*, and the *HwMicrophoneProcessor* to *EmotionTasker*.

7.3 In Summary...

In this section we began our work in the *EmotionTasker* class, which represents the core of our emotional inference mechanism. In particular, we handled its creation, set up a constructor and some of its class variables, made our background service aware of its existence, and defined its first method, *collectInputs()*, for data collection.

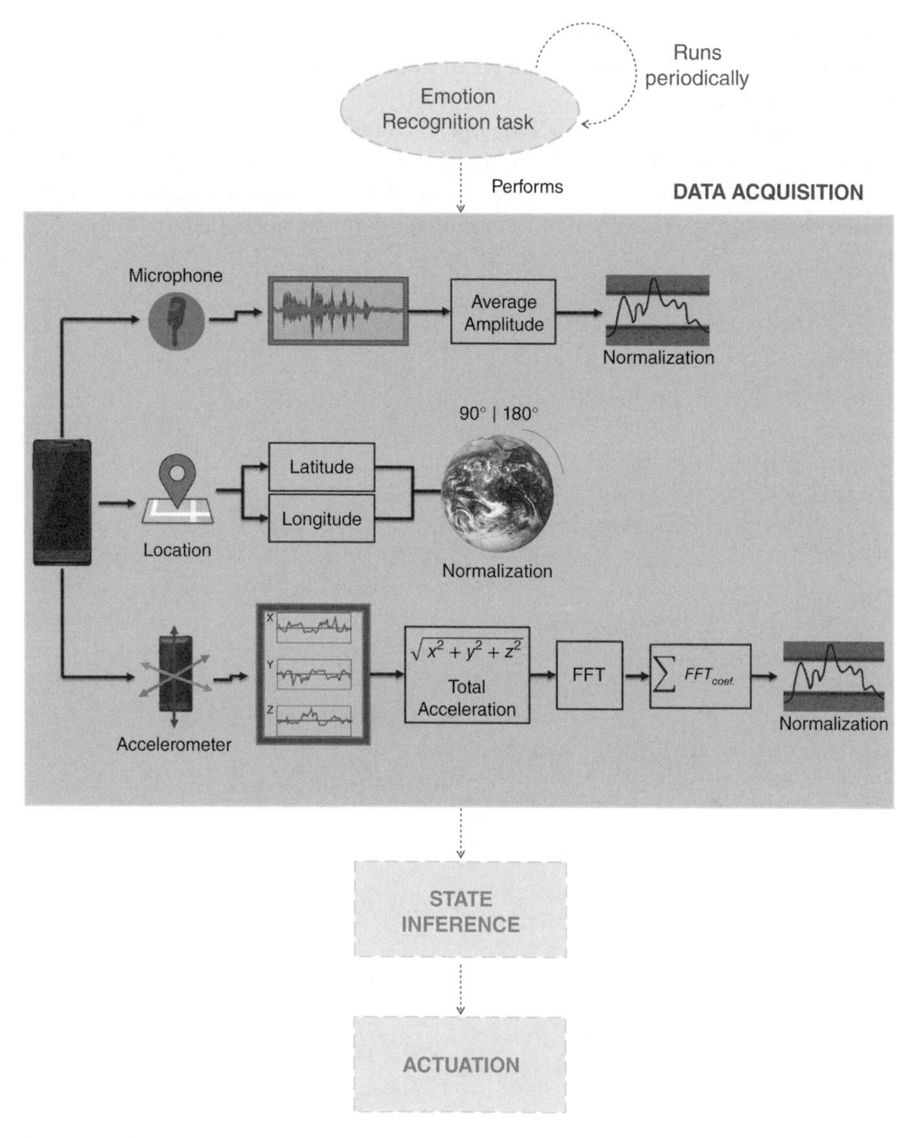

Figure 7.7 Current state of our HiTLCPS at the end of Chapter 7. (*See insert for color representation of the figure.*)

We then developed the *HwLocationProcessor*, *HwMicrophoneProcessor*, and *HwAccelerometerProcessor* classes, which are specifically dedicated to processing sensory data. We studied how to acquire and normalize the inputs, and also implemented some simple signal processing techniques for sound and movement.

With these tasks completed, we have finished the process of *Data Acquisition*. Figure 7.7 shows an overview of what we have achieved and how it fits into the larger picture of our HiTLCPS.

Several parts of Figure 7.7 are dimmed out and composed of dashed arrows and squares. These identify the tasks that we have yet to address. At the top we have an

Emotion Recognition Task, which should be responsible for periodically beginning a new emotion recognition. We also need to address the major processes of *State Inference* and *Actuation*.

In the next chapter we will focus on *State Inference* and also partially address the periodicity of emotion recognition. We shall define the core of our machine learning, which will make use of the processed inputs that we created in this chapter. Eventually, we will be one step closer to the creation of our emotional inference mechanism.

8

State Inference

In Chapter 5 we saw that, in the HiTL control loop, the stage that follows data acquisition is state inference. Subsequently, we have also justified the use of artificial neural networks (ANNs) as our mood inference tool. In this chapter, we will approach the implementation and development of our neural network, as well as the user feedback mechanisms, both of which are cornerstones of state inference in our sample HappyWalk app. In this context, our objective is to implement a learning mechanism that will receive periodic feedback from the user and learn to associate emotions with sensory inputs. The frequency of these feedback requests will be dynamically adapted to the accuracy of the network.

We will first start, in Section 8.1, by explaining how to implement the neural network that will be used as an inference tool. Next, in Section 8.2, we will deal with the problem of requesting user feedback, following which we will present the solutions for processing user feedback, in Section 8.3.

8.1 Implementing a Neural Network

HappyWalk's neural network will be implemented using the Encog Machine Learning Framework [140].[1] For this, we will perform the following steps:

- Reference the Encog library in our build dependencies.
- Declare and initialize a *BasicNetwork* object.
- Fetch input, feed it into the neural network, and collect the result.

During the first step identified above, the library must be referenced in the *app/build.gradle* file, similar to what we did for *org.apache.commons.math3* (see line 20, below):

```
apply plugin: 'com.android.application'

android {
    compileSdkVersion 21
    buildToolsVersion "21.1.2"

    defaultConfig {
```

1 http://www.heatonresearch.com/encog/

A Practical Introduction to Human-in-the-Loop Cyber-Physical Systems, First Edition.
David Nunes, Jorge Sá Silva and Fernando Boavida.

Companion Website URL: www.wiley.com/go/nunesloop

```
            (...)
        }
        buildTypes {
            (...)
        }
    }

    dependencies {
        compile fileTree(dir: 'libs', include: ['*.jar'])
        compile 'com.android.support:appcompat-v7:21.0.3'
        compile 'com.google.android.gms:play-services:6.5.87'
        compile 'org.apache.commons:commons-math3:3.0'
        compile 'org.encog:encog-core:3.0.1'
        compile project(':com.ubhave.sensormanager')
    }
```

Our next step is to declare a *BasicNetwork* object within our *EmotionTasker*, which conceptually represents our neural network (see line 11, below). We also require a method which initializes this object (lines 21–29):

```
package hitlexamples.happywalk.service;

import (...)
import org.encog.engine.network.activation.ActivationSigmoid;
import org.encog.neural.networks.BasicNetwork;
import org.encog.neural.networks.layers.BasicLayer;

public class EmotionTasker {
    private HappyWalkService hWServ;
    private Handler hWServiceHandler;
    private BasicNetwork network;
    private ESSensorManager esSensorManager;

    (...)

    /**
     *This method initializes a Neural Network with
     * two hidden layers, three neurons in the first and two
     * neurons in the second
     */
    private void initNetwork(){
        network = new BasicNetwork();
        network.addLayer(new BasicLayer(null, true, GlobalVariables.
    NN_INPUTS));
        network.addLayer(new BasicLayer(new ActivationSigmoid(), true,
    GlobalVariables.NN_HL1_NEURONS));
        network.addLayer(new BasicLayer(new ActivationSigmoid(), true,
    GlobalVariables.NN_HL2_NEURONS));
        network.addLayer(new BasicLayer(new ActivationSigmoid(), false,
    GlobalVariables.NN_OUTPUTS));
        network.getStructure().finalizeStructure();
        network.reset();
    }

    (...)

}
```

In accordance with what we discussed in Section 5.3, this initialization code creates a neural network which receives four inputs, has two hidden layers, and outputs two values.

Its first input layer will receive latitude, longitude, noise, and movement data. It is defined in line 23, where the *GlobalVariables.NN_INPUTS* variable is used (as we had seen back on page ???). As for the two hidden layers, the first one contains three neurons (line 24, determined by the *GlobalVariables.NN_HL1_NEURONS* variable) and the second two neurons (line 25, determined by the *GlobalVariables.NN_HL2_NEURONS* variable). The final output layer is defined in line 26, and the number of its neurons by *GlobalVariables.NN_OUTPUTS*.

Each layer has an *activation function*. Activation functions are mathematical functions which define the output of a neuron from its input. There are several types of activation functions typically used with neural networks. In our case, we will use *Encog's ActivationSigmoid* class, which represents an activation function with a sigmoidal shape that generates only positive numbers, similar to what is shown in Figure 8.1. As such, and since we have two neurons in our final layer, our neural network's output will be **two values, ranging from 0 to 1**.

The reader is welcome to experiment by adding more hidden layers (using the *addLayer()* method) and/or by changing the number of neurons in the hidden layers we have defined above. As we have seen in Section 5.3.2, different types of neural networks may result in different performance requirements and accuracies. For example, one could experiment and compare the implementation above with one that uses a *4-10-10-10-2* neural network configuration (three hidden layers with 10 neurons each) by changing the values of *GlobalVariables.NN_HL1_NEURONS* and

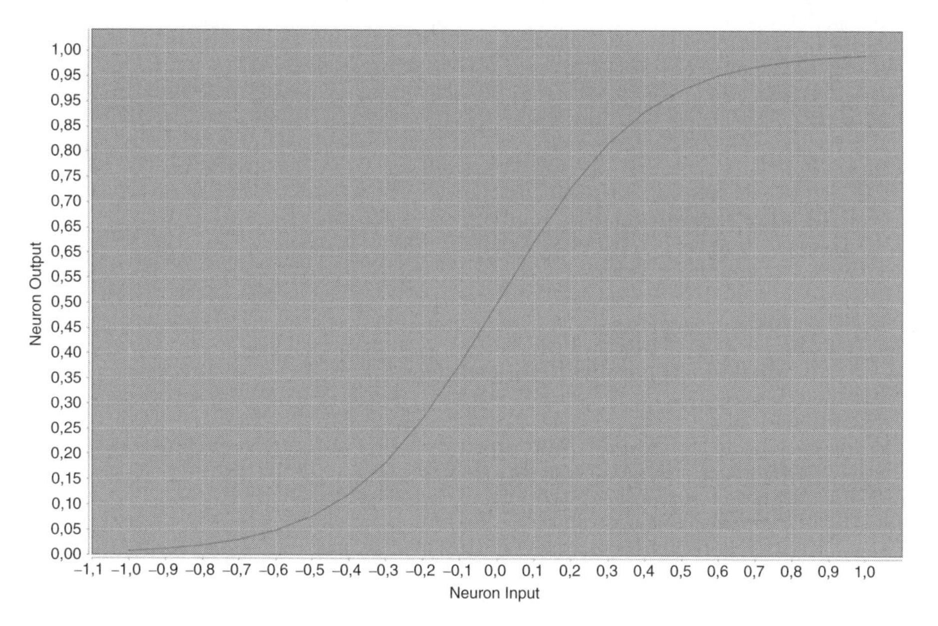

Figure 8.1 An example of a sigmoid activation function.

GlobalVariables.NN_HL2_NEURONS to *10* and by adding the following code after line 25:

network.addLayer(new BasicLayer(new ActivationSigmoid(), false, 10));

Changing the number of inputs and outputs **is not advised**, since it would conflict with the rest of the tutorial.

The most important building blocks of our *EmotionTasker* class are now in place: we can fetch inputs and use a neural network. The next logical step will be to feed these inputs to our network and collect the result. Since this is a task we will be performing periodically, we will define a new subclass within *EmotionTasker* to perform it, named *EmotionRecognitionTask* (lines 26 to 41), as well as an appropriate object declaration (line 10):

```
package hitlexamples.happywalk.service;
2
import (...)
4
public class EmotionTasker {
6     private HappyWalkService hWServ;
      private Handler hWServiceHandler;
8     private BasicNetwork network;
      private ESSensorManager esSensorManager;
10    private EmotionRecognitionTask emotionRecog;

12    public EmotionTasker(HappyWalkService hWServ) {
          (...)
14    }

16    private void initNetwork(){
          (...)
18    }

20    private double[] collectInputs() throws NoCurrentPosition{
          (...)
22    }

24    (...)

26    class EmotionRecognitionTask implements Runnable{
          private double[] outputs;
28        private double[] inputs;

30        @Override
          public void run() {
32            //Perform recognition of emotions here
          }
34
          private void fetchInputsAndCompute() throws NoCurrentPosition {
36            outputs = new double[GlobalVariables.NN_OUTPUTS];
              inputs = collectInputs();
38            //compute the emotion
              network.compute(inputs, outputs);
40        }
      }
42
}
```

The code above shows the skeleton of our new subclass. It is important to note that this class implements the *Runnable* interface. This allows us to run it on a thread other than the main one; for example, as a background task. We have already implemented an inner method named *fetchInputsAndCompute()* (lines 35-40), where we use our previous *collectInputs()* method to collect the necessary inputs and then use them on our *network* object. Notice how Encog simplifies the use of this machine learning technique: we can compute our result in a single line of code: *network.compute(inputs, outputs);* - the output is stored in the *outputs* array variable, the size of which is determined by the global variable *GlobalVariables.NN_OUTPUTS.*

8.2 Requesting User Feedback

Now that we can gather input and compute emotions, we need to present these results to the user, so that he/she may provide us with some feedback. To do so, we will create a new activity especially dedicated to requesting user feedback. This comprises the following steps:

- Create the *EmotionFeddback* activity, which will be responsible for handling the process of acquiring emotional feedback, as we showed with Figure 5.9 and discussed back on page ??.
- Implement the *EmotionSpace* view, which will represent a color space where the user can point his/her current mood.
- Finish *EmotionFeedback* by making it use *EmotionSpace* and send emotional feedback results back to *EmotionTasker.*
- Show a feedback request notification that the user may press to provide emotional feedback. We will also dynamically control the frequency of these feedback requests.

8.2.1 Creating the EmotionFeedback Activity

Right-click on the *activities* package and select *New → Activity → Basic Activity* (see Figure 8.2). Name this activity *EmotionFeedback*, as shown if Figure 8.3.

After the completion of this process, AS will have created and opened two new files, a Java class *activities/EmotionFeedback.java* and a layout file *res/layout/activity_emotion_feedback.xml*, as shown in Figure 8.4 and the code that follows it.

```
package hitlexamples.happywalk.activities;

import android.os.Bundle;
import android.support.v7.app.ActionBarActivity;

import hitlexamples.happywalk.R;

public class EmotionFeedback extends ActionBarActivity {

    @Override
    protected void onCreate(Bundle savedInstanceState) {
        super.onCreate(savedInstanceState);
        setContentView(R.layout.activity_emotion_feedback);
    }

}
```

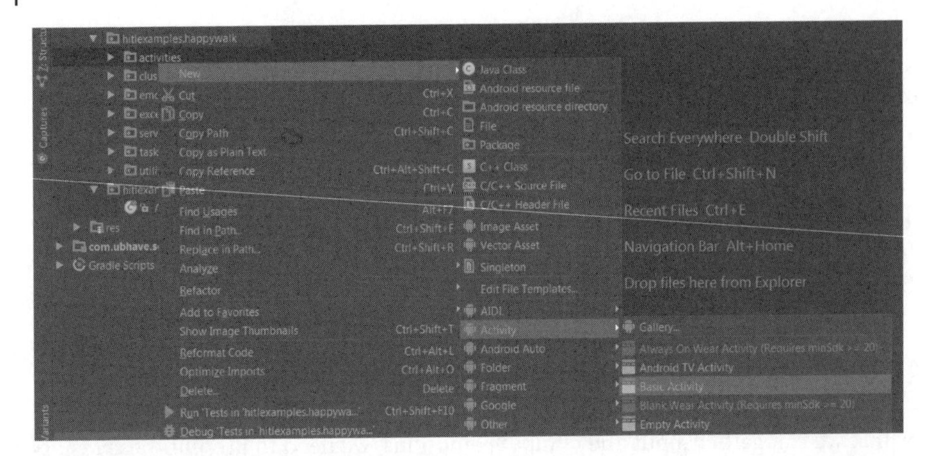

Figure 8.2 Creating a new basic activity.

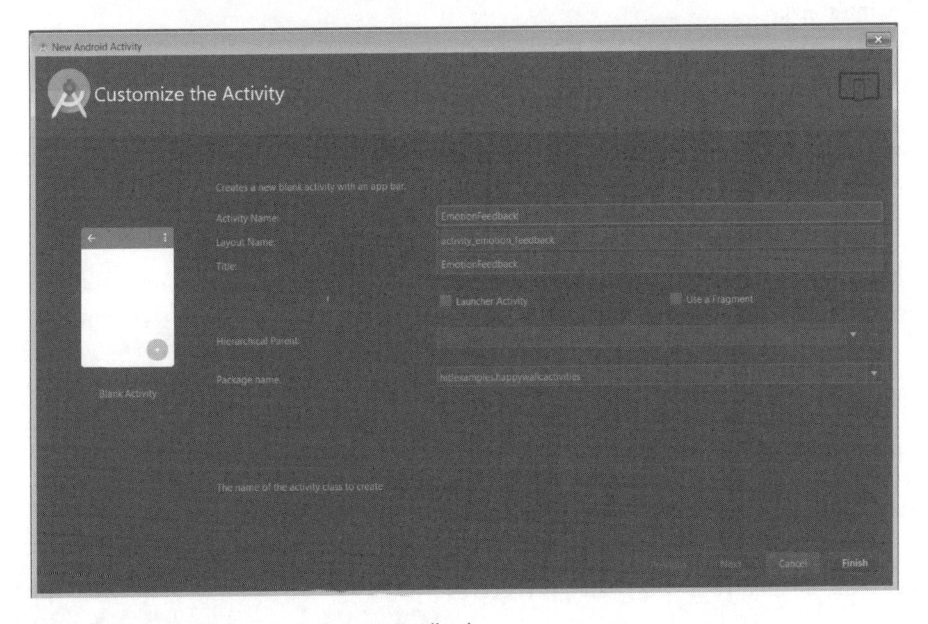

Figure 8.3 Name the activity as *EmotionFeedback*.

We want the user to provide feedback only once per emotional inference. Thus, we have to avoid the possibility of having the user return to the feedback activity after feedback has been given. We also want to avoid having multiple feedback screens opened at any one time. Therefore, we need to make sure that only a single instance of this activity is allowed. Additionally, the entire HappyWalk application is developed towards the *portrait* orientation.

Let us edit the file *happywalk/app/src/main/AndroidManifest.xml* and add the attributes *android:screenOrientation="portrait"* and *android:launchMode="single Instance"* to our new activity:

```
<?xml version="1.0" encoding="utf-8"?>
<manifest xmlns:android="http://schemas.android.com/apk/res/android"
    xmlns:tools="http://schemas.android.com/tools"
    package="hitlexamples.happywalk" >

    (...)
        <activity
            android:name=".activities.EmotionFeedback"
            android:label="@string/title_activity_emotion_feedback"
            android:launchMode="singleInstance"
            android:screenOrientation="portrait">
        </activity>
    </application>

</manifest>
```

Figure 8.4 The files that compose the *EmotionFeedback* activity.

This activity will require a means to interface with the user and receive his/her emotional feedback. In this tutorial, this will be achieved through a specialized implementation of an Android *View* named *EmotionSpace*, which will be described next.

8.2.2 Implementing the *EmotionSpace* View

The objective of *EmotionSpace* is to allow the user to point to his/her current mood. To do so, *EmotionSpace* will extend the *android.view.View* class. Our goal is to achieve an interface similar to the one shown in Figure 8.5: a view which presents the user with a color space and two circles: the yellow circle represents the output of the neural network and the green circle can be dragged by the user to provide feedback.

To achieve this view, we will go through several steps. We will begin by creating a new package and, within it, a new class that extends *android.view.View*, implementing its required constructors. We will then begin to translate the concept of an *EmotionCircle* into actual code. In doing so, we will make an effort to keep our code flexible and

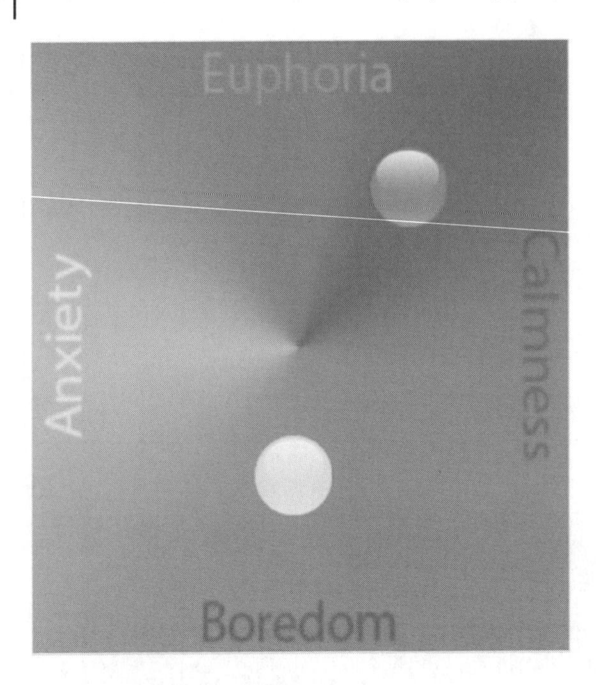

Figure 8.5 Our goal for the *EmotionSpace* view. (*See insert for color representation of the figure.*)

reusable with different kinds of graphical interfaces. Afterwards, we will focus on the task of drawing our graphical resources on screen and updating their position whenever the user interfaces with them. To do so, we will also need to consider what positioning means in the context of emotional inference and discover how to relate a certain position with an emotional output. Finally, we will handle the proper initialization of our *EmotionSpace* and the retrieval of emotional feedback. This will make our view usable by activities, including our own *EmotionFeedback*.

As mentioned above, the implementation of this *EmotionSpace* view will begin with the creation of a new class. Create a new package named *feedback* under the previous *emotion* package. Within it, we create a new class named *EmotionSpace*, as shown in Figure 8.6.

The first thing to do is make *EmotionSpace* extend *android.view.View*. To do so, simply add *extends View* in front of the class name, as shown in Figure 8.7. Do not forget to add the appropriate import declaration, as explained on page ???. However, in doing so, AS will note that *There is no default constructor available in 'android.view.View'*; we need to create one. One easy way of doing so is placing the cursor on top of the class declaration and pressing the *Alt* + *Enter* keys simultaneously. A pop-up similar to the one shown in Figure 8.7 will appear. Choose the option *Create constructor matching super* and then the option *View(context:Context, attrs:AttributeSet)* (see Figure 8.8). AS should add the necessary constructor and associated imports automatically.

Looking back at Figure 8.5, it shows us that there are several components that must be conceptually represented in *EmotionSpace*: there is a color space with the four emotions, the yellow circle for neural network output and green circle for user feedback. Let us consider the concept of *EmotionCircle*, which encompasses both the green and the yellow circles. We will represent this concept through an inner class within *EmotionSpace*:

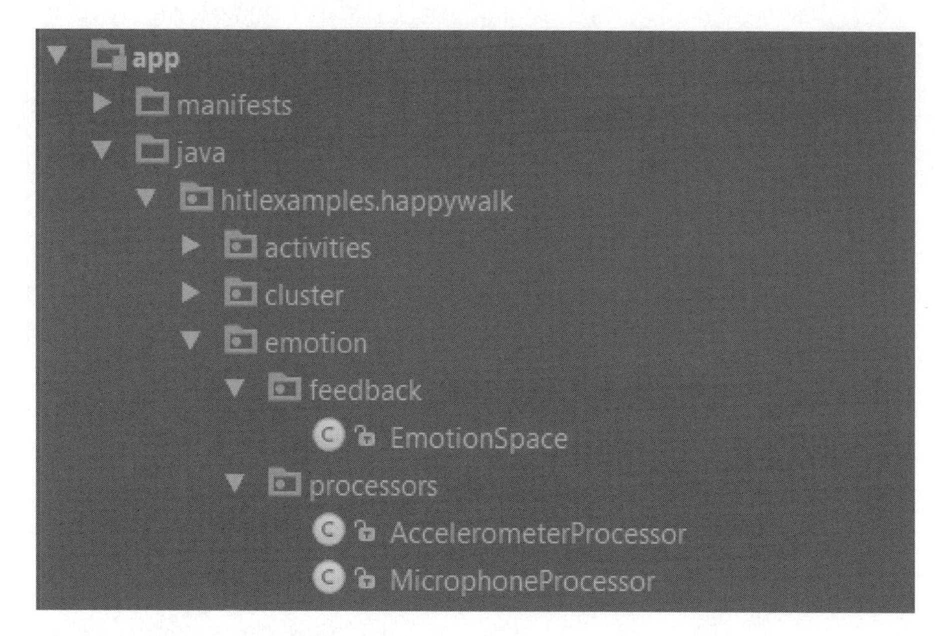

Figure 8.6 Creating the *EmotionSpace* class.

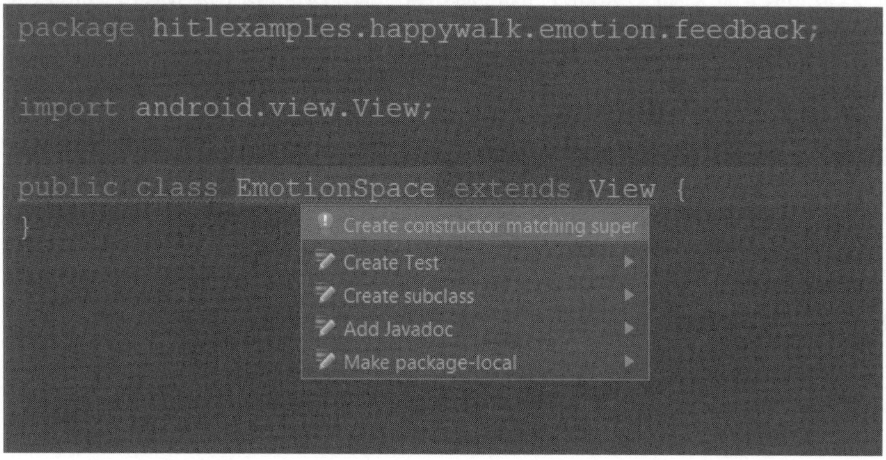

```
package hitlexamples.happywalk.emotion.feedback;

import android.view.View;

public class EmotionSpace extends View {
}
```

Create constructor matching super
Create Test
Create subclass
Add Javadoc
Make package-local

Figure 8.7 Create *EmotionSpace* constructor matching super.

```
package hitlexamples.happywalk.emotion.feedback;

import android.content.Context;
import android.graphics.Bitmap;
import android.graphics.Point;
import android.util.AttributeSet;
import android.view.View;

public class EmotionSpace extends View {
    public EmotionSpace(Context context, AttributeSet attrs) {
```

```
        super(context, attrs);
    }

    private class EmotionCircle {
        private Bitmap bitmap;
        private Point point;

        public EmotionCircle(Bitmap bitmap) {
            this.bitmap = bitmap;
            point = new Point(0,0);
        }

        public Bitmap getBitmap(){
            return bitmap;
        }

        public Point getPoint(){
            return point;
        }

        public void setPoint(Point point) {
            this.point = point;
        }
    }
}
```

The *EmotionCircle* class contains two class variables: *Bitmap*, which represents the graphics of our circle, and *Point*, which is its position within the *EmotionSpace*. We also define a constructor that takes *Bitmap* and places the circle within point *0,0* (the origin of *EmotionSpace*; we will talk about where this point is located on page ???). The developer will be able to specify the initial positions of each circle within another function we will implement later on. After the standard *Getters* and *Setters*, we also define a *setPoint* method that rewrites the internal *point* variable.

In this tutorial, the *EmotionSpace* class will only be used with the previously defined *EmotionFeedback*. However, let us imagine that, for some reason, we wanted to provide different types of *EmotionFeedback* activities, with different colors, button placements and graphics, depending on how the user is feeling. Or, in another possibility, the reader might be interested in expanding HappyWalk by adding other means of providing feedback. In these cases, it is a bad practice to "hardcode" the bitmaps used for the circles and the *EmotionSpace* background. Therefore, as an exercise, let us attempt to implement the *EmotionSpace* class in a flexible way that allows the developer to easily reuse the class and change its appearance.

A possible way of doing this is to define which bitmaps should be used within the layout of the activity that uses *EmotionSpace* (in our case, *EmotionFeedback*). Let us inspect the code behind *EmotionFeedback's* layout file, *res/layout/activity_emotion_feedback.xml*. To do so, change to the *Text* view, as shown in Figure 8.9.

```
<?xml version="1.0" encoding="utf-8"?>
<RelativeLayout xmlns:android="http://schemas.android.com/apk/res/android"
    xmlns:tools="http://schemas.android.com/tools"
    android:layout_width="match_parent"
    android:layout_height="match_parent"
    android:paddingBottom="@dimen/activity_vertical_margin"
    android:paddingLeft="@dimen/activity_horizontal_margin"
```

```
    android:paddingRight="@dimen/activity_horizontal_margin"
    android:paddingTop="@dimen/activity_vertical_margin"
    tools:context="hitlexamples.happywalk.activities.EmotionFeedback">

    <hitlexamples.happywalk.emotion.feedback.EmotionSpace
        android:layout_width="360dp"
        android:layout_height="360dp"
        android:background="@drawable/emotion_color_map"
        android:id ="@+id/emotionSpace"
        android:layout_centerHorizontal="true" />

</RelativeLayout>
```

Defining the background's bitmap this way is straightforward, since Android provides an *android:background* that can point to a bitmap of our choosing. In the example above, we use *@drawable/emotion_color_map* (line 15), which represents the bitmap

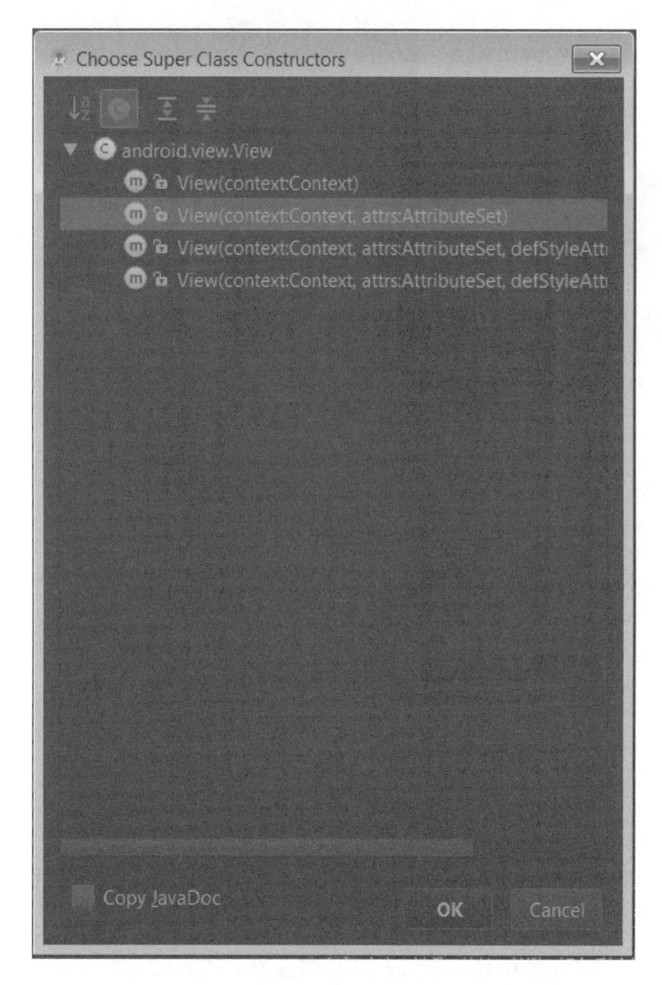

Figure 8.8 Choose *View(context:Context, attrs:AttributeSet).*

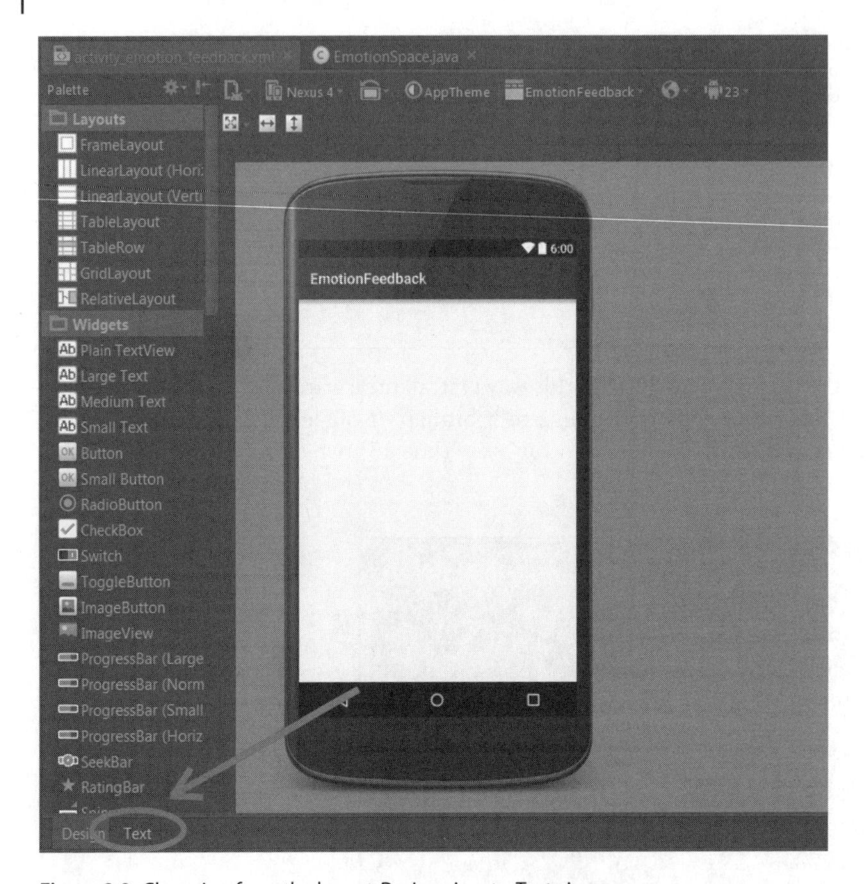

Figure 8.9 Changing from the layout *Design* view to *Text* view.

emotion_color_map.png already present in our project, placed in the folder *app/res/-drawable*. The reader is welcome to replace this background with another of his/her own choosing and to experiment with the *android:layout_width* and *android:layout_height* values.

However, what about the bitmaps for the neural network output and user feedback circles? Despite the fact that there is not a straightforward way of defining these bitmaps within the layout file, fortunately Android provides a way of defining custom layout attributes. Therefore, we can create our own attributes to define the circle bitmaps.

These attributes can be defined in a new *Values resource* file. Let us create one by right-clicking on the *app/res/values* folder and choosing *File* → *New* → *Values resource file* (see Figure 8.10). On the popup window that appears, name the file as *attrs* and leave *Directory name* as *values* (see Figure 8.11).

AS should open the new *app/res/values/attrs.xml* file, which should contain an empty *resources* root element. Our custom attributes are resources that should be declared inside a *<declare-styleable>* child element. Additionally, each individual attribute has a name and a value and can be defined within a particular *<attr>* element. As such, *attrs.xml* can be defined as shown in the code below:

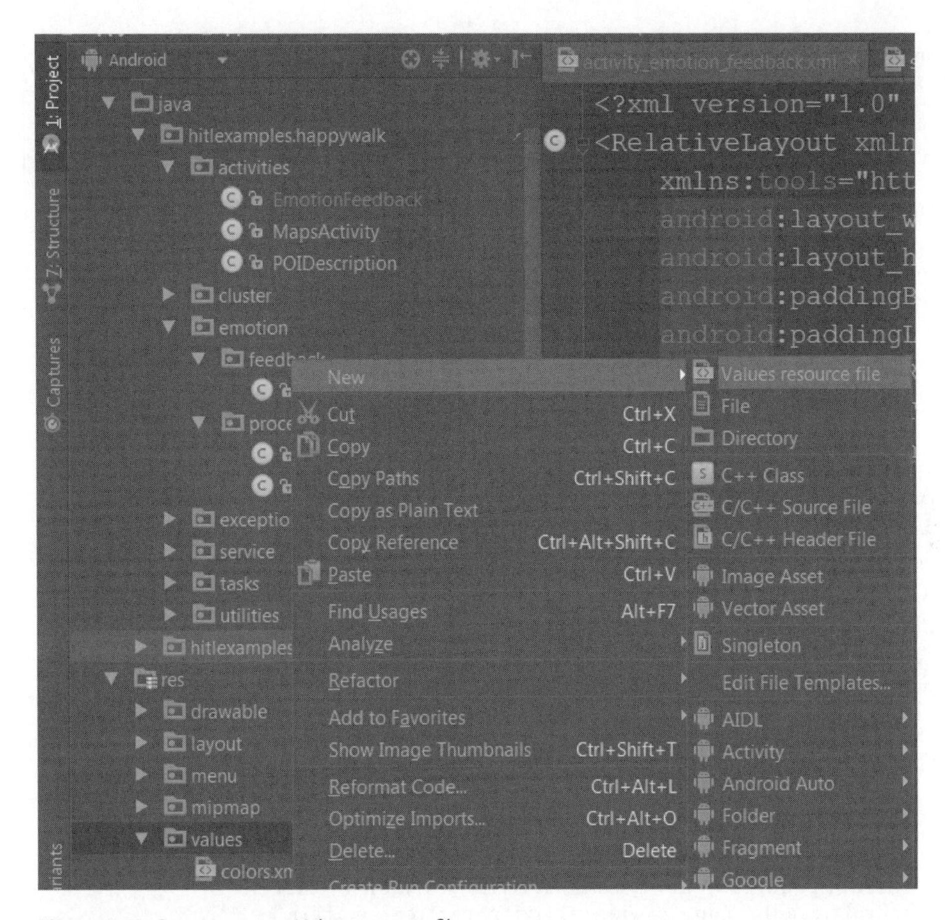

Figure 8.10 Creating a new *Values resource* file.

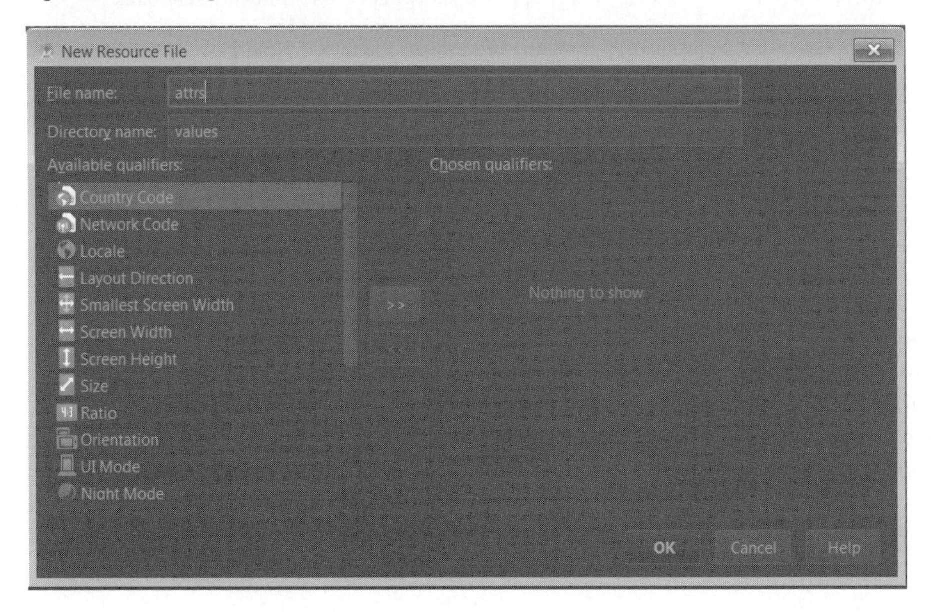

Figure 8.11 Naming the *Values resource* file.

```
1 <?xml version="1.0" encoding="utf-8"?>
  <resources>
3     <declare-styleable name="EmotionSpace">
          <attr name="neuroCircleImage" format="integer" />
5         <attr name="userCircleImage" format="integer" />
      </declare-styleable>
7 </resources>
```

Despite having a single *<declare-styleable>* element, we have named it with the same name as our *EmotionSpace* view. This allows us to easily identify which attributes belong to a view, in case we want to expand our application. As the code shows, we have named our attributes as *neuroCircleImage* (for the circle representing the neural network output) and *userCircleImage* (for the circle representing the user feedback). Readers unfamiliar with Android might be puzzled as to why the *format* of each attribute is of the type *integer*, since we are attempting to reference bitmap images. This is because resources in Android are internally referenced by a *resource ID*, an integer belonging to Android's *R* class. Therefore, our new attributes are actually intended to represent this *resource ID*, which will, in turn, point towards its respective bitmap file.

We can now use the *EmotionSpace's* new attributes within *EmotionFeedback's* layout, as illustrated below:

```
1 <?xml version="1.0" encoding="utf-8"?>
  <RelativeLayout xmlns:android="http://schemas.android.com/apk/res/android"
3     xmlns:tools="http://schemas.android.com/tools"
    xmlns:custom="http://schemas.android.com/apk/res-auto"
5     android:layout_width="match_parent"
      android:layout_height="match_parent"
7     android:paddingBottom="@dimen/activity_vertical_margin"
      android:paddingLeft="@dimen/activity_horizontal_margin"
9     android:paddingRight="@dimen/activity_horizontal_margin"
      android:paddingTop="@dimen/activity_vertical_margin"
11    tools:context="hitlexamples.happywalk.activities.EmotionFeedback">

13    <hitlexamples.happywalk.emotion.feedback.EmotionSpace
          android:layout_width="360dp"
15        android:layout_height="360dp"
          android:background="@drawable/emotion_color_map"
17        custom:neuroCircleImage="@drawable/yellow_circle"
          custom:userCircleImage="@drawable/green_circle"
19        android:id ="@+id/emotionSpace"
          android:layout_centerHorizontal="true" />
21
  </RelativeLayout>
```

Notice the additional namespace declaration at the top of the XML file, *xmlns:custom="http://schemas.android.com/apk/res-auto"* (line 4). This allows us to use our custom-defined attributes *custom:neuroCircleImage* and *custom:userCircleImage*, which represent an integer pointing towards a bitmap. In this case, we are using the *yellow_circle* and *green_circle* bitmaps already present in our project, placed in the folder *app/res/drawable*. Again, the reader is welcome to replace these bitmaps with others of his/her own choosing, although it is convenient that they still represent a circle of the same radius (we will discuss why in the next few paragraphs).

Now that the new attributes have been defined, we need to retrieve their values from within *EmotionSpace*. This can be done on the class constructor, as shown by the following code:

```
package hitlexamples.happywalk.emotion.feedback;

import android.content.Context;
import android.content.res.TypedArray;
import android.graphics.Bitmap;
import android.graphics.Point;
import android.graphics.drawable.BitmapDrawable;
import android.graphics.drawable.Drawable;
import android.util.AttributeSet;
import android.view.View;

import hitlexamples.happywalk.R;

public class EmotionSpace extends View {
    private EmotionCircle neurocircle;
    private EmotionCircle usercircle;

    public EmotionSpace(Context context, AttributeSet attrs) {
        super(context, attrs);
        /*
        This allows us to fetch EmotionSpace's custom attributes,
        as defined in values/attrs.xml. In this case, it allows for
        us to define the images to be used for the neurocircle and
    usercircle
         */
        TypedArray a = context.getTheme().obtainStyledAttributes(
                attrs,
                R.styleable.EmotionSpace,
                0, 0);

        Drawable neuroDraw = a.getDrawable(R.styleable.
    EmotionSpace_neuroCircleImage);
        Drawable userDraw = a.getDrawable(R.styleable.
    EmotionSpace_userCircleImage);

        if (!(neuroDraw instanceof BitmapDrawable) || !(userDraw
    instanceof BitmapDrawable)) {
            throw new RuntimeException("neuroCircleImage and
    userCircleImage attributes require Bitmap drawables.");
        }

        neurocircle = new EmotionCircle(
                ((BitmapDrawable)neuroDraw).getBitmap());
        usercircle = new EmotionCircle(
                ((BitmapDrawable)userDraw).getBitmap());
    }

    private class EmotionCircle {
        (...)
    }
}
```

The *TypedArray* object contains references to the previously defined attributes. From this array, we fetch their associated values and perform a check to see if these are, indeed, bitmaps that can be used and not some other form of resources (e.g. text). If they do not represent bitmaps, we throw a *RuntimeException* which will, essentially, crash our application (since the developer is providing nonsensical resources). Finally, we instantiate two class variables of type *EmotionCircle* using the provided bitmaps.

Our circles now have bitmaps and coordinates, but how and when are they displayed on the device's screen? Android abstracts the drawing of graphical information through the *onDraw* method of *android.view.View*. We can *Override* this method to tell *View* how it should update its graphical information. In our case, what exactly does *EmotionSpace* do each time it needs to update what the user is viewing? First, it needs to confirm whether it should actually be drawing anything; we should have a mechanism that avoids drawing onto the screen until we are ready to do so. In case we *can* draw, *EmotionSpace* should verify the current position of each circle and redraw their bitmaps accordingly. The following shows an implementation of *onDraw* (lines 15 –23) that follows this logic:

```
package hitlexamples.happywalk.emotion.feedback;
2
import (...)
4 import android.graphics.Canvas;

6 public class EmotionSpace extends View {
    private EmotionCircle neurocircle;
8   private EmotionCircle usercircle;
    private boolean showEmotionCircles = false;
10
    public EmotionSpace(Context context, AttributeSet attrs) {
12      (...)
    }
14
    //Called when drawing the screen
16  @Override
    protected void onDraw(Canvas canvas) {
18      if (showEmotionCircles) {
            //draw circle positions
20          canvas.drawBitmap(neurocircle.getBitmap(), neurocircle.
    getPoint().x, neurocircle.getPoint().y, null);
            canvas.drawBitmap(usercircle.getBitmap(), usercircle.getPoint
    ().x, usercircle.getPoint().y, null);
22      }
    }
24
    private class EmotionCircle {
26      (...)
    }
28 }
```

The variable *showEmotionCircles* (declared in line 9) acts as a control that impedes *onDraw* from doing anything unless we want it to (through the *if* branch in line 18). The rest of the method is straightforward: we simply tell our *Canvas* (an object that represents *what* is going to be drawn) to draw the *neurocircle's* and the *emotioncircle's* bitmaps at the correct position (lines 20 and 21).

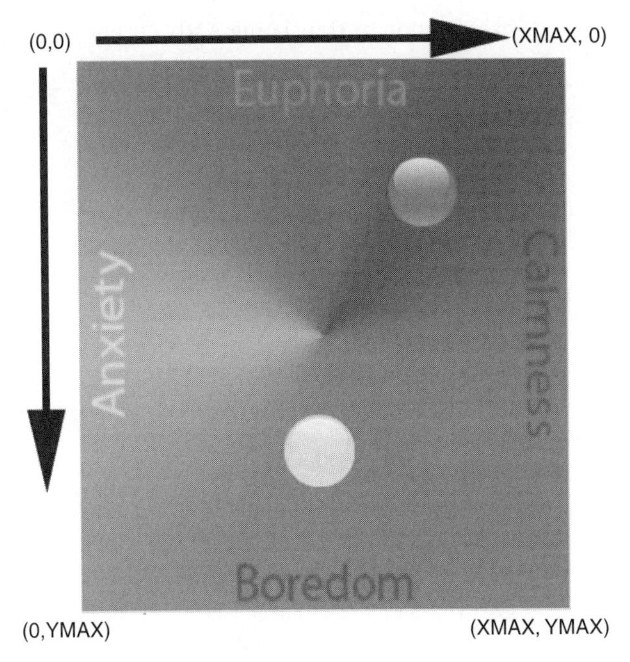

Figure 8.12 The coordinates of the *EmotionSpace* view. (*See insert for color representation of the figure.*)

Another question that we must ask ourselves is *what* exactly does this position *mean*? First, let us study how Android handles positioning coordinates within a *View* by looking at Figure 8.12.

The origin of our *View* rests on the top left corner, whereas the bottom right corner corresponds to the point where **X and Y are equivalent to its width and height, respectively**. As we have seen in Section 8.1, the output of our neural network is two decimal values, with the range **[0,1]**. As it stands, these values remain meaningless: it is left to us to give them meaning.

We will give meaning to the output of our ANN by associating each value with an axis. In this way, we can translate an emotional value into a position within *EmotionSpace*. As shown by Figure 8.12, the provided *EmotionSpace* bitmap implies that the x axis (width) corresponds to an *Anxiety-Calmness* value, whereas the y axis (height) corresponds to an *Euphoria-Boredom* value. As such, two variables are defined within *GlobalVariables* which store this associative meaning:

```
//EMOTION OUTPUT ARRAY INDEXES
/*
 * [0] - Euphoric-Bored axis
 * [1] - Anxious-Calm axis
 */
public static final int NN_OUTPUT_ARRAY_INDEX_EUPHORIC_BORED = 0;
public static final int NN_OUTPUT_ARRAY_INDEX_ANXIOUS_CALM = 1;
```

We can now simply refer to these two variables to know which position of the output array an emotional axis corresponds to.

Now, let us continue to build upon this association by translating ANN output values to *EmotionCircle* points:

```
package hitlexamples.happywalk.emotion.feedback;

import (...)

public class EmotionSpace extends View {

    (...)

    public EmotionSpace(Context context, AttributeSet attrs) {
        (...)
    }

    @Override
    protected void onDraw(Canvas canvas) {
        (...)
    }

    /**
     * Updates the position of the neurocircle, which represents the
     * result of the neural network output
     * @param euphoricBored - value of the euphoric-bored axis
     * @param anxiousCalm - value of the anxious-calm axis
     */
    private void updateNeuroCirclePosition(double euphoricBored, double
    anxiousCalm) {
        updateCirclePosition(neurocircle, euphoricBored, anxiousCalm);
    }

    private void updateUserCirclePosition(double euphoricBored, double
    anxiousCalm) {
        updateCirclePosition(usercircle, euphoricBored, anxiousCalm);
    }

    private void updateCirclePosition(EmotionCircle circle, double
    euphoricBored, double anxiousCalm) {
        circle.setPoint(calculatePointFromEmotion(euphoricBored,
    anxiousCalm));
        //redraws the screen
        invalidate();
    }

    private Point calculatePointFromEmotion(double euphoriaBoring, double
    anxietyCalm) {
        if (euphoriaBoring < 0 || euphoriaBoring > 1 ||
                anxietyCalm < 0 || anxietyCalm > 1) {
            throw new AssertionError("euphoriaBoring and/or anxietyCalm
    outside of the [0,1] range");
        }
        /* Our max value is actually the height/width of the view minus
    the height/width of the circle, since the circle should remain inside
    the view */
        int y = (int) ((getHeight()-usercircle.getBitmap().getHeight())*
    euphoriaBoring);
        int x = (int) ((getWidth()-usercircle.getBitmap().getWidth())*
    anxietyCalm);
        return new Point(x,y);
```

```
    }
48
    private class EmotionCircle {
50      (...)
    }
52 }
```

Here, we have implemented four new methods; *updateNeuroCirclePosition* (line 24) and *updateUserCirclePosition* (line 28) are minor helper methods which simply use *updateCirclePosition* (line 32) to update and redraw the position of the *neurocircle* and the *emotioncircle*, respectively.

How does the redrawing of the position work? In Android, the developer does not have direct control over when the screen is updated. In fact, *onDraw* (discussed on page ???) is called when *View* is initially drawn and whenever the system feels it is necessary. As developers, the only thing we can do is *request* a current *View* to be redrawn, through a method named *invalidate()*. As such, the method *updateCirclePosition* updates an *EmotionCircle's Point* variable (line 33) and calls *invalidate()* (line 35) to request Android for a redrawing of the *View*.

To update the *Point* variable, *updateCirclePosition* uses another method named *calculatePointFromEmotion* (defined between lines 38 and 47) which receives two variables of type double, *euphoricBored* and *anxiousCalm*, that *should* correspond to outputs from the neural network (line 32). In any case, we implemented a simple assertion (between lines 39 and 42) to ensure that these values are at least within the [0,1] range; otherwise, our application would probably crash. This is because we use these values to determine the corresponding position of the *EmotionCircle* within the *View*; for values greater than 1 or less than 0, we would most likely return negative positions or ones that are outside of the *View*'s limits.

At first glance, one could be tempted to simply multiply the emotion outputs by their corresponding axis (*euphoriaBoring* by the *View*'s height and *anxietyCalm* by its width). However, it is important to notice that the *EmotionCircles'* bitmaps also have their own height and width. For the sake of usability, we will now impose that these bitmaps **always remain *fully inside* the *View*.** Since the coordinates of bitmaps follow the same convention as those of *Views* (shown in Figure 8.12), the **maximum value of an emotional output corresponds to the size of the view minus the size of the circle's bitmap**. This is reflected in the proposed implementation of the *calculatePointFromEmotion* method.

Since the idea behind *EmotionSpace* is for the user to drag and drop the *usercircle*, marking how he/she is feeling, we need some way of distinguishing whether or not the circle is being touched. There are several ways to accomplish this. One way of doing so is to assume that the provisioned bitmap is, in fact, a circle, with a diameter equivalent to the bitmap's width. We can then take advantage of this fact to compute the circle's radius and, therefore, determine whether the point where the user touched is within this radius. This is illustrated below.

```
  package hitlexamples.happywalk.emotion.feedback;
2
  public class EmotionSpace extends View {
4     (...)
      private boolean isTouchingUserCircle = false;
6     private final double TOUCHING_DEVIATION = 1.35;

8     public EmotionSpace(Context context, AttributeSet attrs) {
```

```
            (...)
10      }

12      (...)

14      private boolean isTouchingUserCircle(Point touchPoint) {
            boolean isTouching = false;
16          Point usercircleCenter = usercircle.getCenterPosition();

18          double distanceFromCenter = Math.sqrt((double) (usercircleCenter.x
        - touchPoint.x)*(usercircleCenter.x - touchPoint.x) + (
        usercircleCenter.y - touchPoint.y)*(usercircleCenter.y - touchPoint.y)
        );
            /*
20          if the distance is lesser than the radius, the user
            is touching inside the circle
22           */
            if(distanceFromCenter<usercircle.radius*TOUCHING_DEVIATION) {
24              isTouching = true;
            }
26          return isTouching;
        }
28
        /**
30       * Represents an emotional circle.
         * It is assumed that the provisioned bitmap
32       * has identical width and height and represents a circle
         * with a diameter equal to the bitmap's size.
34       */
        private class EmotionCircle {
36          private Bitmap bitmap;
            private Point point;
38          private int radius;

40          public EmotionCircle(Bitmap bitmap) {
                this.bitmap = bitmap;
42              point = new Point(0,0);
                //calculate radius:
44              radius = bitmap.getWidth()/2;
            }
46
            public Bitmap getBitmap(){
48              return bitmap;
            }
50
            public Point getPoint(){
52              return point;
            }
54
            public void setPoint(Point point) {
56              this.point = point;
            }
58
            public Point getCenterPosition() {
60              return new Point(point.x + radius, point.y + radius);
            }
62      }
    }
```

Let us begin by focusing on the code that follows line 29. To avoid constantly calculating the circle's radius each time the user touches the *View*, we have extended the *EmotionCircle* class. After adding a small documentation comment where we express our assumptions regarding the provided bitmap, we added a new inner variable named *radius* (line 38), which is calculated during the construction of the *EmotionCircle* (line 44). This variable is used within a new method named *getCenterPosition()*, which returns the current position of circle's center (lines 59–61).

These new functionalities are used by the method *isTouchingUserCircle(Point touchPoint)* (lines 14–27) which implements the logic proposed above. Considering the circle radius as a strict frontier makes sense in theory (the user is either touching *inside or outside* the circle), but in practice we have found this to be too restrictive. As such, a new final variable named *TOUCHING_DEVIATION* was created (line 6), which increases (or decreases, if its value is set to less than 1) the effective touching radius (see line 23). The reader is free to adjust this value to his/her own liking. Finally, we have also defined a new control Boolean, named *isTouchingUserCircle* (line 5), which will be used by *EmotionSpace* to evaluate the current touching state.

As previously mentioned, the *usercircle* should remain fully inside *View*. This is important, since otherwise the user could fully or partially drag the circle to the outside of *View* and not be able to drag it back in. Let us write a simple method that verifies and corrects the usercircle's positioning:

```
1  package hitlexamples.happywalk.emotion.feedback;

3  import (...)

5  public class EmotionSpace extends View {

7      (...)

9      public EmotionSpace(Context context, AttributeSet attrs) {
           (...)
11     }

13     private void correctUserCirclePosition() {
           //If the usercircle is beyond the View's borders, bring it back in
       .
15         if (usercircle.getPoint().x + usercircle.getBitmap().getWidth() >
       getWidth()) {
               usercircle.getPoint().x = getWidth() - usercircle.getBitmap().
       getWidth();
17         }
           else if(usercircle.getPoint().x < 0) {
19             usercircle.getPoint().x = 0;
           }
21         if (usercircle.getPoint().y + usercircle.getBitmap().getHeight()
       > getHeight()) {
               usercircle.getPoint().y = getHeight() - usercircle.getBitmap()
       .getHeight();
23         }
           else if(usercircle.getPoint().y < 0) {
25             usercircle.getPoint().y = 0;
           }
27         invalidate();
       }
```

```
29
        (...)
31  }
```

Remember that the lone *getWidth()* and *getHeight()* methods are implemented by the parent class. Since *EmotionSpace* extends *android.view.View*, they return *EmotionSpace's* width and height, respectively. The method is rather self-explanatory: if its bitmap is extending beyond the *View*'s limits (lines 15, 18, 21, and 24), we set the *usercircle's* coordinates to the outermost value that is still inside the *View* (lines 16, 19, 22, and 25). Afterwards, we use *invalidate()* to redraw *EmotionSpace* (line 27).

We now have all the necessary methods to handle the user's touch gestures. Each time the user touches our *EmotionSpace* view, Android calls upon the *View*'s method *onTouchEvent(MotionEvent event)*. There are three types of touch events which we need to consider:

- ***ACTION_DOWN***: The user has just touched the screen. Here, we need to check if the user is touching the *usercircle*.
- ***ACTION_MOVE***: The user is still touching the screen and moving around. If the user was previously touching the *usercircle*, we shall move it according to the input gesture.
- ***ACTION_UP***: The user has lifted his/her finger from the screen. Now is the time to check if the *usercircle* has been left inside the *View* and, if not, correct its position.

Let us translate this logic into Java code:

```
1   package hitlexamples.happywalk.emotion.feedback;

3   import (...)
    import android.view.MotionEvent;
5
    public class EmotionSpace extends View {
7       private EmotionCircle neurocircle;
        private EmotionCircle usercircle;
9       private boolean isTouchingUserCircle = false;
        private boolean showEmotionCircles = false;
11
        //These values were experimentally derived to improve usercircle
        placement.
13      private final double TOUCHING_DEVIATION = 1.35;
        private final double MOVING_DEVIATION = 1.15;
15
        public EmotionSpace(Context context, AttributeSet attrs) {
17          (...)
        }
19
        (...)
21
        //This method handles touch events
23      public boolean onTouchEvent(MotionEvent event) {
            //first, check if we can move the emotionCircle
25          if (showEmotionCircles) {
                //figure out the type of event
27              switch (event.getAction()) {
                    case MotionEvent.ACTION_DOWN: //the user has touched the
        screen
29                      Point touchPoint = new Point((int) event.getX(), (int)
         event.getY());
```

```
                    isTouchingUserCircle = isTouchingUserCircle(touchPoint
    );
                    break;
                case MotionEvent.ACTION_MOVE:
                    //if the user is touching the userCircle, we move it.
                    if (isTouchingUserCircle) {
                        //set the center of the circle to the new touching
     point
                        usercircle.setPoint(new Point((int) ((event.getX()
     - usercircle.getBitmap().getWidth()) * MOVING_DEVIATION), (int) ((
    event.getY() - usercircle.getBitmap().getHeight()) * MOVING_DEVIATION)
    ));
                    }
                    //redraw new usercircle position
                    invalidate();
                    break;
                case MotionEvent.ACTION_UP:
                    //fixes the cases where the user places the circle
    beyond the View boundaries.
                    correctUserCirclePosition();
                    break;
            }
        }
        return true;
    }

    private boolean isTouchingUserCircle(Point touchPoint) {
        (...)
    }

    private void correctUserCirclePosition() {
        (...)
    }

    private class EmotionCircle {
        (...)
    }
}
```

Again, after some experimentation, we found that slightly adjusting the placement of the *usercircle* resulted in a more intuitive manipulation. As such, we defined a new double variable *MOVING_DEVIATION*, which influences said placement (line 14). The reader is free to adjust and fine-tune this value.

To effectively use *EmotionSpace* we still need to allow the developer to somehow initialize it. There has to be some way to define the initial positions of the *neurocircle* and the *usercircle*. After such initialization has taken place, we can then set the *showEmotionCircles* variable to *true*, which will trigger the drawing and manipulation capabilities of *View*. However, this process is not completely straightforward. Before we allow drawing and dragging, we need to make sure that the parent activity (in our case, *EmotionFeedback*) has fully drawn *EmotionSpace*. This is because, if *EmotionSpace* has yet to be drawn, the view's coordinate system is nonexistent and we cannot properly place the bitmap circles. Therefore, we need to somehow know when our view is ready to be initialized. This can be achieved through a *ViewTreeObserver* object, as shown by the code below:

```
1  package hitlexamples.happywalk.emotion.feedback;

3  import (...)
   import android.view.ViewTreeObserver;
5
   public class EmotionSpace extends View {
7      (...)

9      private boolean showEmotionCircles = false;

11     (...)

13     public EmotionSpace(Context context, AttributeSet attrs) {
           (...)
15     }

17     (...)

19     /**
        * Initializes this EmotionSpace through the neural network outputs.
21      * Triggers the display of the circle markers and user interaction.
        * @param euphoricBored
23      * @param anxiousCalm
        */
25     public void initialize(final double euphoricBored, final double
       anxiousCalm)
       {
27         /*
           * Now, we need to wait until the emotionSpace has been draw,
29         * so we can place our user and neuro circles. To do this, we use a
           * GlobalLayoutListener
31         */
           ViewTreeObserver vto = getViewTreeObserver();
33         vto.addOnGlobalLayoutListener(new ViewTreeObserver.
       OnGlobalLayoutListener() {
               /*
35             On the code below, we need to remove the layout observer we
       added above. The correct method to do this is
       removeOnGlobalLayoutListener(). However, we are targeting Android API
       level 10, where this method is not supported. Thus, we are forced to
       use the older and deprecated removeGlobalOnLayoutListener(), which
       works for our purposes.
               */
37             @SuppressWarnings("deprecation")
               @Override
39             public void onGlobalLayout() {
                   getViewTreeObserver().removeGlobalOnLayoutListener(this);
41                 //update circle positions based on ANN outputs;
                   updateNeuroCirclePosition(euphoricBored,
43                         anxiousCalm);
                   updateUserCirclePosition(euphoricBored,
45                         anxiousCalm);
                   showEmotionCircles = true;
47             }
           });
49     }

51     (...)
```

```
53      private class EmotionCircle {
            (...)
55      }
    }
```

Note that, as mentioned in the comments (lines 34–36), we are using a deprecated function. This shouldn't cause any issues. However, if for some reason the above code does not work in future Android versions, the reader is encouraged to try to use *removeOnGlobalLayoutListener()* instead of *removeGlobalOnLayoutListener()*. When *onGlobalLayout()* is called, we know that the global layout is fully drawn and, as such, so is our *EmotionSpace* view. Therefore, we then use the emotion axes provided by the developer (which should reflect the output of the ANN) to place our emotion circles (both start at the same position). Finally, we set *showEmotionCircles* to *true*.

Now that the user can see, move, and place his/her feedback intent, we need a way to retrieve it. In our particular example, we want a way for the *EmotionFeedback* activity to *get* the emotion values corresponding to the last position of the *usercircle*:

```
    package hitlexamples.happywalk.emotion.feedback;
2
    import (...)
4   import hitlexamples.happywalk.exceptions.InitializationException;
    import hitlexamples.happywalk.utilities.GlobalVariables;
6
    public class EmotionSpace extends View {
8
       (...)
10
        public EmotionSpace(Context context, AttributeSet attrs) {
12          (...)
        }
14
        (...)
16
        public double[] getEmotionFeedback() throws InitializationException {
18          if (!showEmotionCircles) {
                throw new InitializationException("EmotionSpace has not been
        initialized.");
20          }
            else {
22              double[] emotionFeedback = new double[GlobalVariables.
        NN_OUTPUTS];
                 /*
24              The value of the user feedback is given by position of the
        circle, divided by the maximum value (height/width view minus the
        height/width of the circle)
                 */
26              if (getHeight()-usercircle.getBitmap().getHeight()== 0 ||
                        getWidth()-usercircle.getBitmap().getWidth() == 0) {
28                  throw new AssertionError("EmotionSpace view - bitmap (
        height/width) = 0");
                }
30              emotionFeedback[GlobalVariables.
        NN_OUTPUT_ARRAY_INDEX_EUPHORIC_BORED] =
                        ((double) usercircle.getPoint().y) / (getHeight()-
        usercircle.getBitmap().getHeight());
```

```
32              emotionFeedback[GlobalVariables.
        NN_OUTPUT_ARRAY_INDEX_ANXIOUS_CALM] =
                        ((double) usercircle.getPoint().x) / (getWidth()-
        usercircle.getBitmap().getWidth());
34              return emotionFeedback;
            }
36      }

38      (...)

40      private class EmotionCircle {
            (...)
42      }
    }
```

The public method *getEmotionFeedback*, as shown above, handles this task (line 17). First, we need to check if *EmotionSpace* has been initialized. We can do this through the *showEmotionCircles* variable (line 18). If it was not, it is not possible to return meaningful feedback. Therefore, an exception is thrown, warning the developer of his/her mistake (line 19). We encourage the reader to create a new class under the package *exceptions* named *InitializationException*, similarly to what was done for the *NoCurrentPosition* exception back on page ???. Below, a possible implementation of *InitializationException* is presented:

```
1   package hitlexamples.happywalk.exceptions;

3   public class InitializationException extends Exception {
        public InitializationException(String detailMessage) {
5           super(detailMessage);
        }
7
        public InitializationException(String detailMessage, Throwable
        throwable) {
9           super(detailMessage, throwable);
        }
11  }
```

Let us return our focus to the *getEmotionFeedback()* code. If, on the other hand, *EmotionSpace* has been properly initialized the *else* branch in line 21 is run. In this case, all we need to do is fetch the position of the *usercircle.* Since emotion values are restricted to the [0,1] range, the value of the user feedback is given by the position of the circle divided by the maximum possible value, which is the size of the view minus the size of the bitmap (as discussed on page ???). This reasoning is applied in lines 30 and 32. To avoid divisions by zero, we also verify if the bitmap does not fill the entire view (which could happen, if the developer is not careful in picking a bitmap which is sufficiently small), and throw an *AssertionError* if necessary (line 28).

Our feedback acquisition view should finally be complete! The full code is presented below:

```
1   package hitlexamples.happywalk.emotion.feedback;

3   import android.content.Context;
    import android.content.res.TypedArray;
5   import android.graphics.Bitmap;
```

```
import android.graphics.Canvas;
import android.graphics.Point;
import android.graphics.drawable.BitmapDrawable;
import android.graphics.drawable.Drawable;
import android.util.AttributeSet;
import android.view.MotionEvent;
import android.view.View;
import android.view.ViewTreeObserver;

import hitlexamples.happywalk.R;
import hitlexamples.happywalk.exceptions.InitializationException;
import hitlexamples.happywalk.utilities.GlobalVariables;

public class EmotionSpace extends View {
    private EmotionCircle neurocircle;
    private EmotionCircle usercircle;
    private boolean isTouchingUserCircle = false;
    private boolean showEmotionCircles = false;

    //These values were experimentally derived to improve usercircle
    placement.
    private final double TOUCHING_DEVIATION = 1.35;
    private final double MOVING_DEVIATION = 1.15;

    public EmotionSpace(Context context, AttributeSet attrs) {
        super(context, attrs);
        /*
        This allows us to fetch EmotionSpace's custom attributes,
        as defined in values/attrs.xml. In this case, it allows for
        us to define the images to be used for the neurocircle and
    usercircle
         */
        TypedArray a = context.getTheme().obtainStyledAttributes(
                attrs,
                R.styleable.EmotionSpace,
                0, 0);

        Drawable neuroDraw = a.getDrawable(R.styleable.
    EmotionSpace_neuroCircleImage);
        Drawable userDraw = a.getDrawable(R.styleable.
    EmotionSpace_userCircleImage);

        if (!(neuroDraw instanceof BitmapDrawable) || !(userDraw
    instanceof BitmapDrawable)) {
            throw new RuntimeException("neuroCircleImage and
    userCircleImage attributes require Bitmap drawables.");
        }

        neurocircle = new EmotionCircle(
                ((BitmapDrawable)neuroDraw).getBitmap());
        usercircle = new EmotionCircle(
                ((BitmapDrawable)userDraw).getBitmap());
    }

    /**
     * Initializes this EmotionSpace through the neural network outputs.
     * Triggers the display of the circle markers and user interaction.
     * @param euphoricBored
     * @param anxiousCalm
```

```
59       */
        public void initialize(final double euphoricBored, final double
        anxiousCalm)
61      {
            /*
63          * Now, we need to wait until the emotionSpace has been draw,
            * so we can place our user and neuro circles. To do this, we use a
65          * GlobalLayoutListener
            */
67          ViewTreeObserver vto = getViewTreeObserver();
            vto.addOnGlobalLayoutListener(new ViewTreeObserver.
        OnGlobalLayoutListener() {
69              /*
                On the code below, we need to remove the layout observer we
        added above. The correct method to do this is
        removeOnGlobalLayoutListener(). However, we are targeting Android API
        level 10, where this method is not supported. Thus, we are forced to
        use the older and deprecated removeGlobalOnLayoutListener(), which
        works for our purposes.
71              */
                @SuppressWarnings("deprecation")
73              @Override
                public void onGlobalLayout() {
75                  getViewTreeObserver().removeGlobalOnLayoutListener(this);
                    //update circle positions based on ANN outputs;
77                  updateNeuroCirclePosition(euphoricBored,
                            anxiousCalm);
79                  updateUserCirclePosition(euphoricBored,
                            anxiousCalm);
81                  showEmotionCircles = true;
                }
83          });
        }
85
        //Called when drawing the screen
87      @Override
        protected void onDraw(Canvas canvas) {
89          if (showEmotionCircles) {
                //draw circle positions
91              canvas.drawBitmap(neurocircle.getBitmap(), neurocircle.
        getPoint().x, neurocircle.getPoint().y, null);
                canvas.drawBitmap(usercircle.getBitmap(), usercircle.getPoint
        ().x, usercircle.getPoint().y, null);
93          }
        }
95
        /**
97       * Updates the position of the neurocircle, which represents the
         * result of the neural network output
99       * @param euphoricBored - value of the euphoric-bored axis
         * @param anxiousCalm - value of the anxious-calm axis
101      */
        private void updateNeuroCirclePosition(double euphoricBored, double
        anxiousCalm) {
103         updateCirclePosition(neurocircle, euphoricBored, anxiousCalm);
        }
105
        private void updateUserCirclePosition(double euphoricBored, double
        anxiousCalm) {
```

```
107         updateCirclePosition(usercircle, euphoricBored, anxiousCalm);
        }
109
        private void updateCirclePosition(EmotionCircle circle, double
        euphoricBored, double anxiousCalm) {
111         circle.setPoint(calculatePointFromEmotion(euphoricBored,
        anxiousCalm));
            //redraws the screen
113         invalidate();
        }
115
        private Point calculatePointFromEmotion(double euphoriaBoring, double
        anxietyCalm) {
117         if (euphoriaBoring < 0 || euphoriaBoring > 1 ||
                    anxietyCalm < 0 || anxietyCalm > 1) {
119             throw new AssertionError("euphoriaBoring and/or anxietyCalm
        outside of the [0,1] range");
            }
121         /* Our max value is actually the height/width of the view minus
        the height/width of the circle, since the circle should remain inside
        the view */
            int y = (int) ((getHeight()-usercircle.getBitmap().getHeight())*
        euphoriaBoring);
123         int x = (int) ((getWidth()-usercircle.getBitmap().getWidth())*
        anxietyCalm);
            return new Point(x,y);
125     }

127     public double[] getEmotionFeedback() throws InitializationException {
            if (!showEmotionCircles) {
129             throw new InitializationException("EmotionSpace has not been
        initialized.");
            }
131         else {
                double[] emotionFeedback = new double[GlobalVariables.
        NN_OUTPUTS];
133             /*
                The value of the user feedback is given by position of the
        circle, divided by the maximum value (height/width view minus the
        height/width of the circle)
135             */
                if (getHeight()-usercircle.getBitmap().getHeight()== 0 ||
137                     getWidth()-usercircle.getBitmap().getWidth() == 0) {
                    throw new AssertionError("EmotionSpace view - bitmap (
        height/width) = 0");
139             }
                emotionFeedback[GlobalVariables.
        NN_OUTPUT_ARRAY_INDEX_EUPHORIC_BORED] =
141                     ((double) usercircle.getPoint().y) / (getHeight()-
        usercircle.getBitmap().getHeight());
                emotionFeedback[GlobalVariables.
        NN_OUTPUT_ARRAY_INDEX_ANXIOUS_CALM] =
143                     ((double) usercircle.getPoint().x) / (getWidth()-
        usercircle.getBitmap().getWidth());
                return emotionFeedback;
145         }
        }
147
        //This method handles touch events
```

```
    public boolean onTouchEvent(MotionEvent event) {
        //first, check if we can move the emotionCircle
        if (showEmotionCircles) {
            //figure out the type of event
            switch (event.getAction()) {
                case MotionEvent.ACTION_DOWN: //the user has touched the
    screen
                    Point touchPoint = new Point((int) event.getX(), (int)
     event.getY());
                    isTouchingUserCircle = isTouchingUserCircle(touchPoint
    );
                    break;
                case MotionEvent.ACTION_MOVE:
                    //if the user is touching the userCircle, we move it.
                    if (isTouchingUserCircle) {
                        //set the center of the circle to the new touching
     point
                        usercircle.setPoint(new Point((int) ((event.getX()
     - usercircle.getBitmap().getWidth()) * MOVING_DEVIATION), (int) ((
    event.getY() - usercircle.getBitmap().getHeight()) * MOVING_DEVIATION)
    ));
                    }
                    //redraw new usercircle position
                    invalidate();
                    break;
                case MotionEvent.ACTION_UP:
                    //fixes the cases where the user places the circle
    beyond the View boundaries.
                    correctUserCirclePosition();
                    break;
            }
        }
        return true;
    }

    private boolean isTouchingUserCircle(Point touchPoint) {
        boolean isTouching = false;
        Point usercircleCenter = usercircle.getCenterPosition();

        double distanceFromCenter = Math.sqrt((double) (usercircleCenter.x
     - touchPoint.x)*(usercircleCenter.x - touchPoint.x) + (
    usercircleCenter.y - touchPoint.y)*(usercircleCenter.y - touchPoint.y)
    );
        /*
        if the distance is lesser than the radius, the user
        is touching inside the circle
         */
        if(distanceFromCenter<usercircle.radius*TOUCHING_DEVIATION) {
            isTouching = true;
        }
        return isTouching;
    }

    private void correctUserCirclePosition() {
        //If the usercircle is beyond the View's borders, bring it back in
    .
        if (usercircle.getPoint().x + usercircle.getBitmap().getWidth() >
    getWidth()) {
```

```
            usercircle.getPoint().x = getWidth() - usercircle.getBitmap().
    getWidth();
        }
        else if(usercircle.getPoint().x < 0) {
            usercircle.getPoint().x = 0;
        }
        if (usercircle.getPoint().y + usercircle.getBitmap().getHeight()
    > getHeight()) {
            usercircle.getPoint().y = getHeight() - usercircle.getBitmap()
    .getHeight();
        }
        else if(usercircle.getPoint().y < 0) {
            usercircle.getPoint().y = 0;
        }
        invalidate();
    }

    /**
     * Represents an emotional circle.
     * It is assumed that the provisioned bitmap
     * has identical width and height and represents a circle
     * with a diameter equal to the bitmap's size.
     */
    private class EmotionCircle {
        private Bitmap bitmap;
        private Point point;
        private int radius;

        public EmotionCircle(Bitmap bitmap) {
            this.bitmap = bitmap;
            point = new Point(0,0);
            //calculate radius:
            radius = bitmap.getWidth()/2;
        }

        public Bitmap getBitmap(){
            return bitmap;
        }

        public Point getPoint(){
            return point;
        }

        public void setPoint(Point point) {
            this.point = point;
        }

        public Point getCenterPosition() {
            return new Point(point.x + radius, point.y + radius);
        }
    }
}
```

8.2.3 Finishing *EmotionFeedback*

In this section we will finish our *EmotionFeedback* activity by connecting it with our background *HappyWalkService*. We will then create a "provide feedback" button and make use of *EmotionSpace*. Finally, we will handle sending emotional feedback results back to *EmotionTasker*.

Returning to our *EmotionFeedback* activity, after we get the user's feedback we will need to make use of it by sending it back to *EmotionTasker* somehow. As such, one of our priorities should be to connect and get a reference to our background *HappyWalkService*:

```
package hitlexamples.happywalk.activities;
2
import android.content.ComponentName;
4 import android.content.Context;
import android.content.Intent;
6 import android.content.ServiceConnection;
import android.os.Bundle;
8 import android.os.IBinder;
import android.support.v7.app.ActionBarActivity;
10
import hitlexamples.happywalk.R;
12 import hitlexamples.happywalk.service.HappyWalkService;

14 public class EmotionFeedback extends ActionBarActivity {

16     private HappyWalkService hWService;

18     @Override
    protected void onCreate(Bundle savedInstanceState) {
20         super.onCreate(savedInstanceState);
        setContentView(R.layout.activity_emotion_feedback);
22     }

24     /**
     * Here we have to bind to our HappyWalk service
26      */
    @Override
28     protected void onResume() {
        bindService(new Intent(this, HappyWalkService.class), hwConnection
    , Context.BIND_AUTO_CREATE);
30         super.onResume();
    }
32
    private ServiceConnection hwConnection = new ServiceConnection() {
34         public void onServiceConnected(ComponentName className, IBinder
    service) {
            /* This is called when the connection with the service has
    been established, giving us a service object we can use to interact
    with the service. Because we have bound to a explicit service that we
     know is running in our own process, we can cast its IBinder to a
    concrete class and directly access it. */
36             hWService = ((HappyWalkService.HappyWalkBinder)service).
    getService();
            /*
38              * now that our service is connected, we can finally
             * set up the emotion space
40              */
            setupEmotionSpace();
42         }
        public void onServiceDisconnected(ComponentName className) {
```

```
44              /* This is called when the connection with the service has
        been unexpectedly disconnected -- that is, its process crashed.
        Because it is running in our same process, we should never see this
        happen. */
                hWService = null;
46          }
        };
48
        /**
50       * Here we unbind the service
         */
52      @Override
        protected void onPause() {
54          unbindService(hwConnection);
            hWService = null;
56          super.onPause();
        }
58  }
```

The above code shows our proposed implementation. As we may remember from Section 5.2.1, the method *onResume()* is run after *onCreate()* and every time the user returns to the activity from the foreground (see Figure 5.3). Therefore, *onResume()* (lines 27–31) is a prime candidate for attempting to connect and get a reference to our background service, by calling *bindService()* (available to Activities) to connect to *HappyWalkService* (line 29). While doing so, we provide an appropriate Intent and *Context.BIND_AUTO_CREATE* flag (which automatically creates the service as long as the binding exists). We also provide a reference to an *hwConnection* object, which is responsible for handling the connection of the service.

As the reader may see, *hwConnection* is an object of type *ServiceConnection* (line 33) which implements the *onServiceConnected()* method (line 34), called right after the connection to *HappyWalkService* has been established, and where a reference to the *HappyWalkService* object is acquired (line 36). After we have our reference we may set up the *EmotionSpace* through the *setupEmotionSpace()* method (in line 41, which we shall implement in the next paragraphs). Also, whenever the activity is paused, we unbind the service (line 54).

Now, we need a way for the user to indicate that he/she is ready to provide feedback. We can do this through a simple button. First, let us define the text that appears inside this new button. In Android, it is good practice to define all strings in a proper XML resource file, located at *src/main/res/values/strings.xml*:

```
    <resources>
2       <string name="app_name">HappyWalk</string>
        <string name="title_activity_maps">HappyWalk Map</string>
4       <string name="title_activity_emotion_feedback">EmotionFeedback</string>

6       <!-- GPS states -->
        <string name="locationUnavailable">HappyWalk requires location
        information to work properly!</string>
8
      (...)
10
        <!--Emotion Feedback -->
12      <string name="emotFeedButton">This is how I feel</string>
    </resources>
```

For the sake of readability, we define a new separator comment *"<!–Emotion Feedback –>"*. We then define a new string resource *emotFeedButton* whose value is *"This is how I feel"* (the reader is welcome to change this value to one of his/her own preference).

Now, we can easily add our button to *EmotionFeedback* through its layout *activity_emotion_feedback.xml*:

```
<RelativeLayout xmlns:android="http://schemas.android.com/apk/res/android"
    xmlns:tools="http://schemas.android.com/tools"
    xmlns:custom="http://schemas.android.com/apk/res-auto"
  android:layout_width="match_parent"
    android:layout_height="match_parent"
  android:paddingLeft="@dimen/activity_horizontal_margin"
    android:paddingRight="@dimen/activity_horizontal_margin"
    android:paddingTop="@dimen/activity_vertical_margin"
    android:paddingBottom="@dimen/activity_vertical_margin"
    tools:context="hitlexamples.happywalk.activities.EmotionFeedback">

    <hitlexamples.happywalk.emotion.feedback.EmotionSpace
        android:layout_width="360dp"
        android:layout_height="360dp"
        android:background="@drawable/emotion_color_map"
        custom:neuroCircleImage="@drawable/yellow_circle"
        custom:userCircleImage="@drawable/green_circle"
        android:id ="@+id/emotionSpace"
        android:layout_centerHorizontal="true" />

    <Button
        android:layout_width="wrap_content"
        android:layout_height="wrap_content"
        android:text="@string/emotFeedButton"
        android:id="@+id/emotionFeedbackButton"
        android:layout_centerHorizontal="true"
        android:layout_alignParentBottom="true"
        android:layout_marginBottom="50dp"
        android:clickable="false"
        android:onClick="emotionFeedback"/>

</RelativeLayout>
```

As the reader might remember from page ???, *EmotionSpace* should already be set up within *app/res/layout/activity_emotion_feedback.xml*. The id of *EmotionFeedback's EmotionSpace* is defined by the *android:id* attribute. Here, we use *"@+id/emotionSpace"* (line 18) to tell Android to store a new id named *emotionSpace*, which represents our *EmotionSpace* view.

The code presented above also shows that the button's text is defined by the *android:text* attribute. This attribute is pointing to the string resource *emotFeedButton* we just defined. The id of the button is defined to be *emotionFeedbackButton* (line 25). Finally, the *android:onClick="emotionFeedback"* attribute (line 30) indicates that whenever the user presses this button the *emotionFeedback()* method should be called.

Now that our layout elements are properly defined and identified, we can use them in *EmotionFeedback's* code:

```
package hitlexamples.happywalk.activities;

import (...)
import android.os.Bundle;
```

```
import android.view.View;
import android.widget.Button;

import hitlexamples.happywalk.emotion.feedback.EmotionSpace;
import hitlexamples.happywalk.exceptions.InitializationException;
import hitlexamples.happywalk.utilities.GlobalVariables;

public class EmotionFeedback extends ActionBarActivity {
    private EmotionSpace emotionspace;
    private HappyWalkService hWService;
    private double[] nnOutput;

    (...)

    private void setupEmotionSpace() {
        emotionspace = (EmotionSpace) findViewById(R.id.emotionSpace);
        /* The bundle provides us an array of doubles containing the
    output (0-1) of the neural network. The indexes are defined in
    GlobalVariables. */
        nnOutput = getIntent().getExtras().getDoubleArray(
                GlobalVariables.BND_EXTRA_EMOTION_OUTPUT_ARRAY_KEY);
        emotionspace.initialize(nnOutput[GlobalVariables.
    NN_OUTPUT_ARRAY_INDEX_EUPHORIC_BORED],
                nnOutput[GlobalVariables.
    NN_OUTPUT_ARRAY_INDEX_ANXIOUS_CALM]);
        //user can now press the feedback button
        Button feedbackButton = (Button) findViewById(R.id.
    emotionFeedbackButton);
        feedbackButton.setClickable(true);
    }

    /**
     * Run when the user presses the feedback button
     * @param view
     */
    public void emotionFeedback(View view) {
        if (emotionspace != null) {
            try {
                double[] userEmotionFeedback = emotionspace.
    getEmotionFeedback();
                double[] inputs = getIntent().getExtras().getDoubleArray(
    GlobalVariables.BND_EXTRA_EMOTION_INPUT_ARRAY_KEY);
                long timestamp = getIntent().getExtras().getLong(
    GlobalVariables.BND_EXTRA_EMOTION_TIMESTAMP_KEY);
                hWService.getEmotionTasker().processUserFeedback(inputs,
    nnOutput, userEmotionFeedback, timestamp);
                //show the map
                Intent intent = new Intent(this, MapsActivity.class);
                startActivity(intent);
            } catch (InitializationException e) {
                e.printStackTrace();
            }
            //finish this activity
            this.finish();
        }
    }
}
```

The method *setupEmotionSpace()* (lines 19–29) is called right after we get a reference to *HappyWalkService*. First, it fetches a reference to *EmotionSpace* through its id (line 20). Then, we expect that whoever starts *EmotionFeedback* also provides an array of doubles containing the output of the neural network. This can be done through the ***extras* of the *Intent* that started *EmotionFeedback***. An *Intent's* extras is a bundle that can be used to pass any additional information. The keys used to fetch this array from the *Intent's extras* should be already defined within the *GlobalVariables* class (*BND_EXTRA_EMOTION_OUTPUT_ARRAY_KEY*, in line 22).

The output of the neural network is stored in a class private variable *nnOutput* (declared in line 15). We then use its values to initialize *EmotionSpace* (line 24). Now that the activity is ready, we enable the feedback button (lines 27 and 28).

The method *emotionFeedback()* (lines 35–51) is called after the button has been pressed. It is responsible for storing the position of the green circle within a double array (line 38) and passing this information, together with the corresponding neural network inputs and outputs, and a timestamp, to the *EmotionTasker* (line 41). Notice that the information is passed through the *processUserFeedback()* method, which we will implement later on.

It is equally important to note that we expect from whoever starts *EmotionFeedback* to also provide the inputs and timestamp, through the *Intent's extras*. The global variables *BND_EXTRA_EMOTION_INPUT_ARRAY_KEY* and *BND_EXTRA_EMOTION_TIMESTAMP_KEY* hold the keys needed to fetch these values (lines 39 and 40). Afterwards, the method shows the map (lines 43 and 44) and closes the EmotionFeedback activity (line 49).

We have now finished our *EmotionFeedback* activity! Below, the reader can find its complete Java code:

```
package hitlexamples.happywalk.activities;
2
import android.content.ComponentName;
4 import android.content.Context;
import android.content.Intent;
6 import android.content.ServiceConnection;
import android.os.IBinder;
8 import android.support.v7.app.ActionBarActivity;
import android.os.Bundle;
10 import android.view.View;
import android.widget.Button;
12
import hitlexamples.happywalk.R;
14 import hitlexamples.happywalk.emotion.feedback.EmotionSpace;
import hitlexamples.happywalk.exceptions.InitializationException;
16 import hitlexamples.happywalk.service.HappyWalkService;
import hitlexamples.happywalk.utilities.GlobalVariables;
18
public class EmotionFeedback extends ActionBarActivity {
20     private EmotionSpace emotionspace;
    private HappyWalkService hWService;
22     private double[] nnOutput;

24     @Override
    protected void onCreate(Bundle savedInstanceState) {
26         super.onCreate(savedInstanceState);
        setContentView(R.layout.activity_emotion_feedback);
```

```
28      }

30      /**
         * Here we have to bind to our HappyWalk service
32       */
        @Override
34      protected void onResume() {
            bindService(new Intent(this, HappyWalkService.class), hwConnection
        , Context.BIND_AUTO_CREATE);
36          super.onResume();
        }
38
        private ServiceConnection hwConnection = new ServiceConnection() {
40          public void onServiceConnected(ComponentName className, IBinder
        service) {
                /* This is called when the connection with the service has
        been established, giving us a service object we can use to interact
        with the service. Because we have bound to a explicit service that we
         know is running in our own process, we can cast its IBinder to a
        concrete class and directly access it. */
42              hWService = ((HappyWalkService.HappyWalkBinder)service).
        getService();
                /*
44               * now that our service is connected, we can finally
                 * set up the emotion space
46               */
                setupEmotionSpace();
48          }
            public void onServiceDisconnected(ComponentName className) {
50              /* This is called when the connection with the service has
        been unexpectedly disconnected -- that is, its process crashed.
        Because it is running in our same process, we should never see this
        happen. */
                hWService = null;
52          }
        };
54
        /**
56       * Here we unbind the service
         */
58      @Override
        protected void onPause() {
60          unbindService(hwConnection);
            hWService = null;
62          super.onPause();
        }
64
        private void setupEmotionSpace() {
66          emotionspace = (EmotionSpace) findViewById(R.id.emotionSpace);
            /* The bundle provides us an array of doubles containing the
        output (0-1) of the neural network. The indexes are defined in
        GlobalVariables. */
68          nnOutput = getIntent().getExtras().getDoubleArray(
                    GlobalVariables.BND_EXTRA_EMOTION_OUTPUT_ARRAY_KEY);
70          emotionspace.initialize(nnOutput[GlobalVariables.
        NN_OUTPUT_ARRAY_INDEX_EUPHORIC_BORED],
                    nnOutput[GlobalVariables.
        NN_OUTPUT_ARRAY_INDEX_ANXIOUS_CALM]);
72          //user can now press the feedback button
```

```
        Button feedbackButton = (Button) findViewById(R.id.
    emotionFeedbackButton);
74      feedbackButton.setClickable(true);
    }
76
    /**
78   * Run when the user presses the feedback button
     * @param view
80   */
    public void emotionFeedback(View view) {
82      if (emotionspace != null) {
            try {
84              double[] userEmotionFeedback = emotionspace.
    getEmotionFeedback();
                double[] inputs = getIntent().getExtras().getDoubleArray(
    GlobalVariables.BND_EXTRA_EMOTION_INPUT_ARRAY_KEY);
86              long timestamp = getIntent().getExtras().getLong(
    GlobalVariables.BND_EXTRA_EMOTION_TIMESTAMP_KEY);
                hWService.getEmotionTasker().processUserFeedback(inputs,
    nnOutput, userEmotionFeedback, timestamp);
88              //show the map
                Intent intent = new Intent(this, MapsActivity.class);
90              startActivity(intent);
            } catch (InitializationException e) {
92              e.printStackTrace();
            }
94          //finish this activity
            this.finish();
96      }
    }
98 }
```

8.2.4 Showing a Feedback Request Notification

Before processing the feedback, we still need to request it first. As discussed on page ??, a possibility is to use a notification which will prompt the user to teach our neural network mechanism. In this section we will implement a way of showing a feedback request notification to the user, which may be pressed to trigger the emotional feedback process. To do so, we will begin by extending our *EmotionTasker* with a new method that can create and remove this notification. We will then focus on creating a dynamic mechanism that adapts the frequency of feedback requests to the accuracy of our neural network.

Let us create a method within *EmotionTasker* that will present a notification to the user which, when pressed, will call the *EmotionFeedback* activity. We should also assume that our emotion inference has an expiration date: it does not make sense for the user to provide feedback to an inference task that was performed a long time ago. Therefore, we need to consider a mechanism that will revert notifications and filter feedback which has "expired". This can be achieved using a simple timestamp:

```
  package hitlexamples.happywalk.service;
2
  import (...)
4 import android.content.Intent;
  import android.os.Bundle;
6 import android.media.RingtoneManager;
  import android.app.PendingIntent;
```

```
8  import android.support.v4.app.NotificationCompat;
   import android.util.Log;
10
   import hitlexamples.happywalk.R;
12 import hitlexamples.happywalk.activities.EmotionFeedback;

14 public class EmotionTasker {

16   private HappyWalkService hWServ;
       private Handler hWServiceHandler;
18     private BasicNetwork network;
       private ESSensorManager esSensorManager;
20     private EmotionRecognitionTask emotionRecog;

22     /*this variable keeps track of the time when we fired our last
       emotion notification */
24     private long lastEmotionNotifMillis = 0;
       private NotificationRemovalTask currentNotifRemovTask;
26
     (...)
28
     private void showEmotionFeedbackNotification(double[] inputs, double[]
       outputs) {
30     //First, cancel previous notification removal tasks
           hWServiceHandler.removeCallbacks(currentNotifRemovTask);
32         //Now, prepare a Bundle with the information to be passed to
       EmotionFeedback
           Bundle bnd = new Bundle();
34         bnd.putDoubleArray(GlobalVariables.
       BND_EXTRA_EMOTION_INPUT_ARRAY_KEY,
                   inputs);
36         bnd.putDoubleArray(GlobalVariables.
       BND_EXTRA_EMOTION_OUTPUT_ARRAY_KEY,
                   outputs);
38         /* put a timestamp on this bundle to avoid expired feedback. We
       store this same value within lastEmotionNotifMillis long, to keep
       track of when the last emotion feedback notification was sent.*/
           lastEmotionNotifMillis = System.currentTimeMillis();
40         bnd.putLong(GlobalVariables.BND_EXTRA_EMOTION_TIMESTAMP_KEY,
       lastEmotionNotifMillis);
           bnd.putInt(GlobalVariables.BND_EXTRA_REQ_CODE_KEY,
42                 GlobalVariables.AREQ_EMOTION_FEEDBACK_NOTIF);

44         Intent intent = new Intent(hWServ, EmotionFeedback.class);
           intent.putExtras(bnd);
46
           PendingIntent resultPendingIntent =
48                 PendingIntent.getActivity(
                           hWServ,
50                         0,
                           intent,
52                         PendingIntent.FLAG_UPDATE_CURRENT
                   );
54
           NotificationCompat.Builder mNotifyBuilder = new NotificationCompat
       .Builder(hWServ)
56                 .setTicker(hWServ.getResources().getString(R.string.
       app_name) + " " + hWServ.getResources().getString(R.string.
       emotionFeedbackNotifContent))
```

```
                .setContentTitle(hWServ.getResources().getString(R.string.
    app_name))
58              .setContentText(hWServ.getResources().getString(R.string.
    emotionFeedbackNotifContent))
                .setSmallIcon(R.drawable.emot_notif_icon)
60              .setContentIntent(resultPendingIntent)
                .setOngoing(true)
62              .setSound(RingtoneManager.getDefaultUri(RingtoneManager.
    TYPE_NOTIFICATION));

64        hWServ.getNotificationManager().notify(
                hWServ.hHNotificNum,
66              mNotifyBuilder.build());
    }
68
    (...)
70
    class EmotionRecognitionTask implements Runnable{
72      private double[] outputs;
        private double[] inputs;
74
        @Override
76      public void run() {
          //Perform recognition of emotions here
78      }

80      private void postNotificationRemovalTask() {
                /* post a notificationRemovalTask, which will revert the
    notification in case the user takes too long to provide input.
82              It runs a little after the expected expiration time.
                */
84              currentNotifRemovTask = new NotificationRemovalTask();
                hWServiceHandler.postDelayed(currentNotifRemovTask, (long)
     (GlobalVariables.EXPIRE_EMOTION_MILLIS *1.05));
86          }

88      private void fetchInputsAndCompute() throws NoCurrentPosition {
          (...)
90      }
    }
92
    /** This runnable task reverts our notification to its default
94    * state in case the inferred emotion has already expired.
    */
96  class NotificationRemovalTask implements Runnable {
      @Override
98    public void run() {
        if (System.currentTimeMillis() - lastEmotionNotifMillis >
    GlobalVariables.EXPIRE_EMOTION_MILLIS) {
100       //revert the notification
          Log.d("EMOTION NOTIFICATION","lastEmotioNotifMillis shows that
    current notification has expired. Reverting...");
102       hWServ.showNotification(false);
        }
104   }
    }
106 }
```

EmotionTasker's new *showEmotionFeedbackNotification()* method is responsible for creating a notification that asks the user for feedback (lines 29–67). First, it cancels previous notification removal tasks (those that revert to expired notifications, which we will discuss below) since a new one will be created (line 31). As the reader may remember from page ???, if we intend to use the *EmotionFeedback* activity, we need to provide additional data in an *extras* Bundle. Therefore, our method also includes the sensory input, the neural network output, as well as a timestamp, into a Bundle to pass to *EmotionFeedback*. Notice that we use the correct global variable key for each data element (lines 34, 36, and 40). The timestamp is also stored in the *lastEmotionNotifMillis* variable. This will be used to check if the user feedback is still valid.

In line 41, we add an integer to our *extras* Bundle with the key *BND_EXTRA_REQ_CODE_KEY*. This integer represents a "request code key", which we will use later on in this tutorial. Its purpose is to help our *MapsActivity* to initialize differently, depending on who called it. In this case, the value *GlobalVariables.AREQ_EMOTION_FEEDBACK_NOTIF* is used to clearly identify that redirections to *MapsActivity* come from emotion feedback notifications.

Afterwards, *showEmotionFeedbackNotification()* builds an *Intent* describing the desire to call *EmotionFeedback* (line 44). It also builds a *PendingIntent* from this original *Intent* using the *getActivity()* method (line 47). A *PendingIntent* is a type of *Intent* that can be handed to other elements so that they can perform the action at a later time. In our case, this means that our notification will only call *EmotionFeedback* when pressed. Notice that we pass *FLAG_UPDATE_CURRENT* during our *PendingIntent's* construction. This way, if the described *PendingIntent* already exists it will be kept, but its extra data shall be replaced with the one from the new *Intent*.

The method then uses a *NotificationCompat.Builder*, several resource strings, and a bitmap icon to construct our notification message (lines 55–62). A working bitmap should already be present in *src/main/res/drawable/emot_notif_icon.png* (represented by the id *R.drawable.emot_notif_icon*). The reader is free to use his/her own image. As for the strings, we need to define *emotionFeedbackNotifContent* within *src/main/res/-values/strings.xml*), as shown below (line 8):

```
<resources>
    <string name="app_name">HappyWalk</string>

  (...)

  <!-- Notifications -->
    <string name="serviceNotifContent">is walking with you</string>
    <string name="emotionFeedbackNotifContent">Can you tell me how you
    feel?</string>

    <!--Emotion Feedback -->
    <string name="emotFeedButton">This is how I feel</string>
</resources>
```

The final result should be a notification that says, "Can you tell me how you feel?" (or any other sentence, in case the reader decides to personalize the *emotionFeedbackNotifContent* resource), as shown in Figure 8.13.

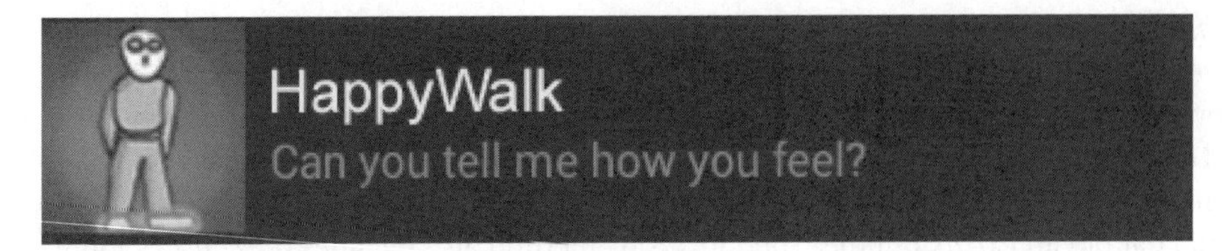

Figure 8.13 The emotion feedback notification.

Going back to our *EmotionTasker* code on page ???, we have also implemented yet another method within *EmotionRecognitionTask* named *postNotificationRemovalTask()* (line 80). It makes use of a *NotificationRemovalTask* inner class, which also implements the *Runnable* interface (much like *EmotionTasker*, as we saw on page ???). The *NotificationRemovalTask* uses the *lastEmotionNotifMillis* and the *GlobalVariables.EXPIRE_EMOTION_MILLIS* variables to evaluate if an emotion notification has expired (see lines 96–105). This mechanism serves as a failsafe, in case a newer notification is fired before the previous *NotificationRemovalTask* has been run. This could happen when the period of emotion recognition tasks (dictated by the variable *GlobalVariables.RECOG_MIN_HOURS_WITHOUT_FEEDBACK*) is lower than the time an emotion is considered valid (*GlobalVariables.EXPIRE_EMOTION_MILLIS*). However, since we cancel previous notification removal tasks before creating a new notification, this situation should never occur under normal circumstances.

Now that we have our feedback request notifications in place, we can begin thinking about when they should be fired. If we repeatedly request user feedback, there is no point to our learning mechanism: the main focus of our HiTL app should be to become as unobtrusive as possible. Therefore, we are going to devise a simple mechanism that will determine when user feedback requests are necessary. Since we want to avoid overly complex solutions, let us consider the following pseudocode:

```
void function UpdateEmotionAccuracy (neuralNetworkOutput, UserFeedback) {
2  currentEuclidean = calculate the euclidean distance between
      neuralNetworkOutput and UserFeedback;
   weightedMeanEuclidean = use the currentEuclidean to update a weighted
      mean value;
4  newFeedbackTime = directLinearVariation (weightedMeanEuclidean);

6  /* the base time to the next emotion feedback request is calculated from
   a weighted arithmetic mean that includes this newFeedbackTime and also a
      bit of randomization */
8  baseTimeToNextEmoFdbckReq = weightedMean(newFeedbackTime) + randomFactor
      ();

10 //keep track of when we last performed this feedback request
   lastEmoFeedbackReq = currentTime();
12 }
```

The previous approach simply associates the time it takes until the next feedback request with the distance between the result of the neural network and the user feedback (line 2). To smooth this change, we perform a weighted mean (line 3); that is, newer emotional feedback with very low accuracy won't punish a previously well-performing neural network too harshly. A *newFeedbackTime* value is calculated based on a direct linear variation of the weighted mean of the Euclidean distance (line 4). This value is then used to calculate the base time to the next emotion feedback request through another weighted mean. We also introduce some randomness to the equation: this lessens the possibility of users starting to associate their feedback with specific feedback request intervals (line 8). Finally, we store a timestamp to keep track of when the last feedback request was performed (line 11). Below follows this reasoning in concrete Java code:

```
package hitlexamples.happywalk.service;

import (...)
import java.util.Random;

public class EmotionTasker {

  (...)

  /*
    we keep the current base time, the last time and the
    previous euclidean distance for emotion feedback requests
    in memory, to avoid constantly accessing the disk
     */
    private long baseTimeToNextEmoFdbckReq = 0;
    private long lastEmoFeedbackReq = 0;
    private float wMeanEuclideanDistance = 0;

  (...)

  /**
   * This method calculates the euclidean distance between the user
    feedback and the neural network output
   */
  private double computeEuclideanDistance (double[] output, double[]
    idealOutput) {
    double euclideanDistance = 0;
    for (int i = 0; i<output.length; i++) {
      euclideanDistance += Math.pow((output[i]-idealOutput[i]),2);
    }
    euclideanDistance = Math.sqrt(euclideanDistance);
    return euclideanDistance;
  }

  (...)

  /**
   * This class is responsible for updating the emotion accuracy feedback
    times and previous euclidean distance values
   */
  class UpdateEmotionAccuracyTask implements Runnable {
    private double[] output;
    private double[] idealOutput;

    public UpdateEmotionAccuracyTask(double[] output, double[] idealOutput
    ) {
      if (output.length != idealOutput.length) {
        throw new AssertionError("output and idealOutput are of different
    sizes");
      }
      else {
```

```
            this.output = output;
48          this.idealOutput = idealOutput;
        }
50      }

52      @Override
        public void run() {
54        /*calculate the euclidean distance
          This gives us an estimate on how accurate our last inference was. */
56        double euclideanDistance = computeEuclideanDistance(output,
        idealOutput);
          /*
58        This weighted mean will be used to check if our neural network
          is performing well enough to trigger notifications and send
        information to the server.
60        */
          wMeanEuclideanDistance = (float) (euclideanDistance *
        GlobalVariables.WEIGHT_OF_NEW_EUCLIDEAN_DISTANCE +
62            wMeanEuclideanDistance * (1-GlobalVariables.
        WEIGHT_OF_NEW_EUCLIDEAN_DISTANCE));

64        float maxEuclideanDistance = (float) Math.sqrt(GlobalVariables.
        NN_OUTPUTS);
          /*
66        we compute a new feedback time through a direct linear variation
          based on the weighted mean of the euclidean distance
68
          number of milliseconds in an hour = 3600000
70         */
          long newFeedbackTime = (long) ((GlobalVariables.
        RECOG_MAX_HOURS_WITHOUT_FEEDBACK -
72            ((GlobalVariables.RECOG_MAX_HOURS_WITHOUT_FEEDBACK -
        GlobalVariables.RECOG_MIN_HOURS_WITHOUT_FEEDBACK)*
                wMeanEuclideanDistance/maxEuclideanDistance))*3600000);
74
          /* update the feedback time through a weighted arithmetic mean, with
         a bit of randomization */
76        baseTimeToNextEmoFdbckReq = (long) (newFeedbackTime *
        GlobalVariables.WEIGHT_OF_NEW_EMOTION_FEEDBACK_TIME +
              baseTimeToNextEmoFdbckReq * (1-GlobalVariables.
        WEIGHT_OF_NEW_EMOTION_FEEDBACK_TIME));
78        Random rand = new Random();
          long margin =  (long) (baseTimeToNextEmoFdbckReq*GlobalVariables.
        MARGIN_PERCNT_RANDOM_EMO_FDBCK_TIME);
80        //the final value will oscillate between baseValue -/+ (margin/2)
          baseTimeToNextEmoFdbckReq = (baseTimeToNextEmoFdbckReq-(margin/2)) +

82            ((long) (rand.nextDouble()*margin));

84        /* update the last emotion feedback timestamp */
          lastEmoFeedbackReq = System.currentTimeMillis();
86      }
      }
88  }
```

We define and use several important new class variables on the code above:

- *wMeanEuclideanDistance* (line 17): is a weighted mean of the Euclidean distance, which relates to the history of the performance of our neural network.

- *baseTimeToNextEmoFdbckReq* (line 15): represents the time, in milliseconds, until our next emotion feedback request.
- *lastEmoFeedbackReq* (line 16): is a timestamp of the last emotion feedback request.

The new inner class *UpdateEmotionAccuracyTask* (lines 38–87) allows us to easily compute and update each of these values. As the reader may see from line 38, this class implements the *Runnable* interface and, as such, can be used as a background task. We define a simple constructor (lines 42–50) which receives two arrays of doubles: *output* (the output of the neural network) and *idealOutput* (the feedback from the user). The constructor also serves to verify if the two arrays are of equal size.

The actual workload of *UpdateEmotionAccuracyTask* rests on its *run()* method (lines 53–86). It begins by calculating the Euclidean distance between *output* and *idealOutput* (line 56). To do so, it uses the *computeEuclideanDistance()* method, which is defined between lines 24 and 31.

The value of *wMeanEuclideanDistance* is calculated in line 61. It is a simple weighted mean where new Euclidean distance values have a weight of *WEIGHT_OF_NEW_EUCLIDEAN_DISTANCE* (which, by default, is 0.4).

We then calculate the maximum possible Euclidean distance (line 64). This is used to compute *newFeedbackTime* (line 71) through a simple direct linear variation based on the weighted mean of the Euclidean distance. Here, *newFeedbackTime* represents a temporary value that will be used to compute *baseTimeToNextEmoFdbckReq*, the time until the next emotion feedback request.

As we can see in line 76, *baseTimeToNextEmoFdbckReq* is also calculated through a simple weighted mean, where *newFeedbackTime* has a weight equivalent to *WEIGHT_OF_NEW_EMOTION_FEEDBACK_TIME* (which, by default, is also 0.4). We then shuffle this value by adding a randomized margin, calculated from a percentage of *baseTimeToNextEmoFdbckReq* in line 79, and applied in line 81. This shuffling is a preventive measure against user habituation (e.g. the user expects a notification at certain intervals), which may happen when the user repeatably provides similar feedback.

The variables *wMeanEuclideanDistance, baseTimeToNextEmoFdbckReq, lastEmoFeedbackReq,* and *lastEmotionNotifMillis* remain important even when the user stops using our app, since they provide us with the necessary history that allows us to comprehend how well our neural network has been performing and when the user should be consulted. Thus, they should be made persistent. To do this, we will be using Android's *SharedPreferences APIs.*[2] These allow us to easily store a relatively small collection of key-value pairs. As such, let us write some helper methods that store and retrieve these values from the app's preferences:

```
package hitlexamples.happywalk.service;
2
import (...)
4 import android.content.SharedPreferences;
import android.preference.PreferenceManager;
6
public class EmotionTasker {
8
    (...)
```

2 https://developer.android.com/training/basics/data-storage/shared-preferences.html

```
10
    private void restoreEmoFeedbackValsFromPreferences() {
12      SharedPreferences pref = PreferenceManager.
    getDefaultSharedPreferences(hWServ);

14      /* if, for some reason, we cannot find this value, we use
    getDefaultBaseEmoFeedbackTime() */
        baseTimeToNextEmoFdbckReq = pref.getLong(GlobalVariables.
    PREF_EMOTION_BASE_FEEDBACK_MILLIS_KEY,
16              getDefaultBaseEmoFeedbackTime());

18      /* if, for some reason, we cannot find this value, set it to zero
    to ensure a feedback request in the next emotion detection task */
        lastEmoFeedbackReq = pref.getLong(GlobalVariables.
    PREF_LAST_EMOTION_FEEDBACK_REQ_KEY, 0);
20
        // same thing for lastEmotionNotifMillis
22      lastEmotionNotifMillis = pref.getLong(GlobalVariables.
    PREF_LAST_EMOTION_NOTIF_KEY, 0);

24      /* by default, we define half of largest distance possible to
    ensure that new neural nets will require some training before having
    their results taken into account
         */
26      float maxEuclideanDistance = (float) Math.sqrt(GlobalVariables.
    NN_OUTPUTS);
        wMeanEuclideanDistance = pref.getFloat(GlobalVariables.
    PREF_NEURALNETWORK_EUCLDIST_WGT_AVG_KEY,
28              maxEuclideanDistance/2);

30  }

32  /**
     * Stores the current value for the base and last emotion feedback
    times into the shared preferences
34   */
    public void saveEmoFeedbackValsToSharedPreferences() {
36      SharedPreferences pref = PreferenceManager.
    getDefaultSharedPreferences(hWServ);
        SharedPreferences.Editor editor = pref.edit();
38      editor.putLong(GlobalVariables.
    PREF_EMOTION_BASE_FEEDBACK_MILLIS_KEY, baseTimeToNextEmoFdbckReq);
        editor.putLong(GlobalVariables.PREF_LAST_EMOTION_FEEDBACK_REQ_KEY,
     lastEmoFeedbackReq);
40      editor.putLong(GlobalVariables.PREF_LAST_EMOTION_NOTIF_KEY,
    lastEmotionNotifMillis);
        editor.putFloat(GlobalVariables.
    PREF_NEURALNETWORK_EUCLDIST_WGT_AVG_KEY, wMeanEuclideanDistance);
42      editor.commit();
    }
44
    /**
46   * Returns the default baseTimeToNextEmoFdbckReq value
     * by averaging between the maximum and minimum values.
48   */
    private long getDefaultBaseEmoFeedbackTime() {
50      //number of milliseconds in an hour = 3600000
        return (long) ((GlobalVariables.RECOG_MAX_HOURS_WITHOUT_FEEDBACK +
     GlobalVariables.RECOG_MIN_HOURS_WITHOUT_FEEDBACK) / 2
```

```
52                  * 3600000);
        }
54
        (...)
56  }
```

Let us begin with the *saveEmoFeedbackValsToSharedPreferences()* method (lines 35–43). As the reader can see, storing values on *SharedPreferences* is a relatively simple task. We begin by getting a reference to the default *SharedPreferences*, in line 36. This allows us to acquire a *SharedPreferences.Editor* (line 37), which we use to store the desired values together with their corresponding global variable keys (lines 38–41). Finally, we commit the changes using the editor's *commit()* method (line 42).

We also wrote a *restoreEmoFeedbackValsFromPreferences()* method to load the values from storage into memory (lines 11–30). We begin fetching a reference to *SharedPreferences* in line 12. We then attempt to restore *baseTimeToNextEmoFdbckReq* (line 15). One property of *SharedPreferences's getter* methods is that they allow us to define a *default* value, in case the provided key does not exist. We use this to our advantage by defining a *getDefaultBaseEmoFeedbackTime()* method (lines 49–53). This method simply averages the global variables *RECOG_MAX_HOURS_WITHOUT_FEEDBACK* and *RECOG_MIN_HOURS_WITHOUT_FEEDBACK*, which define the maximum and minimum periods for feedback requests (in hours), and returns the result in milliseconds.

We use a similar strategy for restoring *lastEmoFeedbackReq* (line 19) and *lastEmotionNotifMillis* (line 22), except we provide the default value of 0 for both. For the *maxEuclideanDistance* variable, we define its default value as half of the maximum Euclidean distance. As we will see in Section 9.2, these default values make sense, since they will ensure that we perform feedback requests if there is no data stored and that newly created neural networks perform a certain amount of training before having their results taken into account.

Lest we forget, let us actually use the *saveEmoFeedbackValsToSharedPreferences()*. Since *EmotionTasker* is intimately related to our service, we should save its values when the service is about to be destroyed, as can be seen below:

```
    public class HappyWalkService extends Service {
2
        (...)
4
        @Override
6       public void onDestroy() {
            isRunning = false;
8           if (hWServiceHandler != null) {
            /*
10          * This uses the Handler to send a "quit" message to the Looper,
            * thus terminating our background Thread.
12          */
                hWServiceHandler.getLooper().quit();
14          }
            //save last known position to shared preferences
16          hwLocationListener.storeActualPositionInSharedPrefs();
            //save current base emotion feedback time to shared preferences
18          emotionTasker.saveEmoFeedbackValsToSharedPreferences();
```

```
            //remove listener
20          LocationManager mlocManager = (LocationManager) getSystemService(
        Context.LOCATION_SERVICE);
            mlocManager.removeUpdates(hwLocationListener);
22
            //remove notification
24          mNM.cancel(hHNotificNum);
            super.onDestroy();
26      }

28      (...)
    }
```

As we have seen in Section 5.2.1, the method *onDestroy()* is called at the end of the service's lifecycle (see Figure 5.3). In order to save *EmotionFeedback's* values, we changed *HappyWalkService's onDestroy()* method by adding a call to *saveEmoFeedbackValsToSharedPreferences()*, in line 18.

Now that we know when to request and how to retrieve user feedback, it is time to start thinking about how to process this information.

8.3 Processing User Feedback

In this section we will handle what happens after the user presses the *emotionFeedback* button in the *EmotionFeedback* activity. This triggers the processing of user feedback, which, as mentioned on page ???, is passed on to the *EmotionTasker* through the *processUserFeedback()* method. We will begin by implementing this method, where we restore the HappyWalk notification to its default status and verify if the provided emotional data is still valid. If it is valid, we shall use it to train the neural network and then send it to the server. As such, let us detail each of these tasks in its respective subsection.

8.3.1 Processing Feedback on the *EmotionTasker*

As mentioned above, the *EmotionTasker* shall process feedback within a *processUserFeedback()* method, which we will now implement:

```
1   package hitlexamples.happywalk.service;

3   import (...)

5   public class EmotionTasker {

7     (...)

9       /**
         * Handles user emotion feedback. Training of the neural network only
        takes place if the emotion notification has not expired.
11       * This avoids the possible issue of users providing feedback long
        after our notification has been shown.
         */
13      public void processUserFeedback(double[] inputs, double[] outputs,
        double[] idealOutput, long emotionTimestamp) {
            /*
```

```
15      We request a timestamp instead of relying on our internal
    lastEmotionNotifMillis because there is a possibility that a new
    emotion was inferred while the user was still using the feedback
    screen. Thus, while the lastEmotionNotifMillis works for notifications
    , there are no guarantees that it is associated with the emotion we
    are about to process right now. Hence, timestamps are necessary.

17      If no new notifications were created in the meantime, we can
    cancel the notification removal tasks and revert the notification to
    its normal state.
         */
19      if (emotionTimestamp == lastEmotionNotifMillis) {
            //cancel notification removal task
21          hWServiceHandler.removeCallbacks(currentNotifRemovTask);
            //revert notification
23          hWServ.showNotification(false);
        }
25      //perform training and send to server, if emotion has not expired
        if (System.currentTimeMillis() - emotionTimestamp <
    GlobalVariables.EXPIRE_EMOTION_MILLIS) {
27          hWServiceHandler.post(new NeuralNetworkTrainingTask(inputs,
    idealOutput));
            sendEmotionToServer(idealOutput);
29          //update emotion feedback frequency
            hWServiceHandler.post(new UpdateEmotionAccuracyTask(outputs,
    idealOutput));
31      }
        else {
33          Log.d("NEURALNETWORK TRAINING", "Expired emotion, discarding
    feedback...");
        }
35  }

37  (...)

39 }
```

This method achieves several objectives. First, it checks if the last fired notification corresponds to the emotion currently being processed (line 19). If it does, we can safely remove the pending notification removal task (line 21) and revert the notification to its default status ("HappyWalk is walking with you") by passing a *false* Boolean value to *HappyWalkService's showNotification()* method (line 23). If there are newer notifications, we shall do nothing that interferes with their lifecycle.

Afterwards, the method uses the timestamp to check if the emotion we are about to process is still valid (line 26). As noted in the comments (lines 14–18), it is possible that the user opened the *EmotionFeedback* activity and forgot to provide feedback until much later. If it is valid, we use the feedback to train the neural network (through a *NeuralNetworkTrainingTask* class, in line 27), send this information to the server (through a *sendEmotionToServer()* method, in line 28), and update the frequency of user feedback requests, through the *UpdateEmotionAccuracyTask* that we implemented in the previous section (line 30).

As the reader might have noticed, we still need to implement the *NeuralNetworkTrainingTask* class and the *sendEmotionToServer()* method. Let us consider each of these issues separately.

8.3.2 Training the Neural Network

In this section we will implement our neural network's training process. We will first create a task specifically tailored for this purpose, which will use the Encog library to train the network. We will then use the *SharedPreferences* API to make our neural network's state persistent.

As we have seen in the previous section, the *processUserFeedback()* method references a *NeuralNetworkTrainingTask*, which is responsible for training the neural network based on the provided feedback. Thankfully, the Encog library makes this step a rather painless process:

```
1  package hitlexamples.happywalk.service;

3  import (...)
   import org.encog.ml.data.basic.BasicMLDataSet;
5  import org.encog.neural.networks.training.propagation.resilient.
       ResilientPropagation;

7
   public class EmotionTasker {
9
     (...)
11
       /**
13      * Trains the neural network based on the Resilient Propagation
       heuristic
        */
15     class NeuralNetworkTrainingTask implements Runnable {
           private double[] inputs;
17         private double[] idealOutput;

19         public NeuralNetworkTrainingTask(double[] inputs, double[]
       idealOutput) {
               this.inputs = inputs;
21             this.idealOutput = idealOutput;
           }
23
           @Override
25         public void run() {
               double[][] trainingInput = {inputs};
27             double[][] trainingIdealOutput = {idealOutput};
               BasicMLDataSet trainingSet = new BasicMLDataSet(trainingInput,
        trainingIdealOutput);
29             ResilientPropagation rProp = new ResilientPropagation(network,
       trainingSet);
               //train the network
31             do {
                   rProp.iteration();
33             } while(rProp.getError() >= GlobalVariables.
       NN_MAX_TRAINING_ERROR);

35             //save the new weights
               saveNNWeightsToPreferences(network.dumpWeights());
37         }
       }
39
     (...)
41 }
```

The *NeuralNetworkTrainingTask's* code is rather self-explanatory. Similar to what we previously did for *UpdateEmotionAccuracyTask* (back on page ???), this task also implements the *Runnable* interface (line 15) and defines a simple constructor that takes two arrays of doubles: *inputs* and *idealOutput* (lines 19–22). The *inputs* array represents the original sensory input, which should still be accompanying the current emotion (the result of *collectInputs()*, discussed back on page ???, which is then passed to *showEmotionFeedbackNotification()*, *EmotionFeedback* and, finally, *processUserFeedback()*). The *idealOutput*, on the other hand, represents the user feedback that was provided by *EmotionFeedback*.

In the class *run()* method these two arrays are used in conjunction to create a *BasicMLDataSet* object, which is Encog's representation of a set of training data (line 28). Notice that the *BasicMLDataSet's* constructor takes two double[][] objects and we only have a single emotion to process. As such, we first translate *inputs* and *idealOutput* into two double[][] arrays (lines 26 and 27).

Afterwards, we use a technique named "Resilient Propagation" (line 29) to train the network until its error is below a certain customizable threshold. The training is performed inside a *do while* loop (lines 31–33), and the threshold is defined by the *NN_MAX_TRAINING_ERROR* global variable (line 33). Ideally, this training process should use more than one training value; however, since this task will be performed each time the user provides feedback, we should still expect the neural network to adapt over a considerable period.

The result of the neural network training is another element that has to be made permanent. Each time the app is started, the previous state of the neural network must be loaded from the device's storage, so that HappyWalk may progressively become more accurate and personalized. Thus, we reference a method *saveNNWeightsToPreferences()* to save this information. Let us implement this and another helper method (*restoreNNWeightsFromPreferences()*) to save and load the neural network's state information from the app's preferences:

```
1  package hitlexamples.happywalk.service;

3  import (...)

5  public class EmotionTasker {

7    (...)

9      private BasicNetwork network;

11     (...)

13   private void restoreNNWeightsFromPreferences() {
           //Try to get network from preferences first
15         SharedPreferences pref = PreferenceManager.
       getDefaultSharedPreferences(hWServ);
           if (pref.contains(GlobalVariables.PREF_NEURALNETWORK_WEIGHT_KEY))
       {
17             String weights = pref.getString(GlobalVariables.
       PREF_NEURALNETWORK_WEIGHT_KEY, null);
               if(weights!=null) {
19                 String[] weights_string_array = weights.split(",");
                   double[] weights_array = new double[weights_string_array.
       length];
```

```
21                  int i = 0;
                    for (String value : weights_string_array) {
23                      weights_array[i++] = Double.parseDouble(value);
                    }
25                  network.decodeFromArray(weights_array);
                }
27          } else {
                Log.d("NEURAL NETWORK", "No weights were found; neural network
            has been reset");
29          }
        }
31
        private void saveNNWeightsToPreferences(String weights) {
33          SharedPreferences pref = PreferenceManager.
        getDefaultSharedPreferences(hWServ);
            SharedPreferences.Editor editor = pref.edit();
35          editor.putString(GlobalVariables.PREF_NEURALNETWORK_WEIGHT_KEY,
        weights);
            editor.commit();
37      }

39      (...)

41  }
```

The *saveNNWeightsToPreferences()* method (lines 32–37) is somewhat similar to *saveEmoFeedbackValsToSharedPreferences()*, which we wrote back on page ???. We begin by getting a reference to the default *SharedPreferences* (line 33) and then to its *Editor* (line 34). The major difference is that we only have a single key-value pair to store: a *weights* string and its key (stored in the *REF_NEURALNETWORK_WEIGHT_KEY* global variable). These shall contain our neural network's weights (line 35). Once again, we commit the changes using the editor's *commit()* method (line 36).

The *restoreNNWeightsFromPreferences()* method (lines 13–30) also begins by acquiring a reference to the default *SharedPreferences* (line 15). We then verify if *PREF_NEURALNETWORK_WEIGHT_KEY* exists in the *SharedPreferences* (line 16). If it does, we attempt to store its corresponding string within a variable called *weights* (line 17). If the *weights* string does exist (line 18), we split it by the "," character, and store the corresponding substrings within a *weights_string_array* (line 19). These substrings are then converted into an array of doubles (lines 20–24). Finally, we use the *BasicNetwork.decodeFromArray()* method to restore our neural network (line 25). If we cannot find weights stored within *SharedPreferences* our neural network will reset, and we log this occurrence (line 28).

We can make use of *restoreNNWeightsFromPreferences()* and the other helper methods we have written so far to further initialize our *EmotionTasker*:

```
1   package hitlexamples.happywalk.service;

3   import (...)

5   public class EmotionTasker {

7       (...)

9       public EmotionTasker(HappyWalkService hWServ) {
```

```
        this.hWServ = hWServ;
        this.hWServiceHandler = hWServ.getHappyWalkServiceHandler();
        //preparing sensor manager to fetch data
        try {
            esSensorManager = ESSensorManager.getSensorManager(hWServ);
            esSensorManager.setGlobalConfig(GlobalConfig.
    PRINT_LOG_D_MESSAGES, false);
        } catch (ESException e) {
            e.printStackTrace();
        }
        //fetch the base emotion feedback time interval
        restoreEmoFeedbackValsFromPreferences();
        //initialize Neural Network
        initNetwork();
        //restore neural network weights from our preferences, if we have
    them
        restoreNNWeightsFromPreferences();
    }

    (...)

}
```

As the code above shows, we have edited our *EmotionTasker's* constructor to better initialize the class. We now use *restoreEmoFeedbackValsFromPreferences()* to restore our emotion feedback values (line 20), initialize the neural network using the *initNetwork()* method we previously implemented on page ??? (line 22), and restore the neural network's state using *restoreNNWeightsFromPreferences()* (line 24). This way, the constructor now handles the initialization of the sensing library, fetches the base values of emotional feedback requests, initializes the ANN, and restores its weights. Let us move on to the task of sending emotional information to the server.

8.3.3 Sending Emotional Information to the Server

As previously discussed back in Section 5.3, one of the objectives of HappyWalk is to display a near real-time average of the mood at each POI, through heatmaps with different colors, so that users may pick walking destinations with the moods they desire. To do this, the server needs to collectively acquire the current moods of HappyWalk users, and it is up to the Android app to send this information. As such, what kind of data do we need to send?

First of all we need to obviously send emotional data, so that the server may aggregate it. We also need to send location data, since we need to know where the user is before we can associate his/her emotions with a certain POI. Finally, to avoid duplicate data, we also need to somehow identify individual user records.

Here, it is important that we comply with the requirement of anonymization, since we are dealing with possibly sensitive information. To do so, HappyWalk generates a pseudo-random universally unique identifier (UUID), a "practically unique" 128-bit value. This identifier will serve as the only means of "tracking" each user's data, thus avoiding the storage of any type of personal information. *HappyWalkService* makes use of Java's *java.util.UUID* utility class to handle the generation of these identifiers.

Therefore, our implementation should send this anonymous UUID of the user, his/her current position and his/her emotional feedback. This information is going to

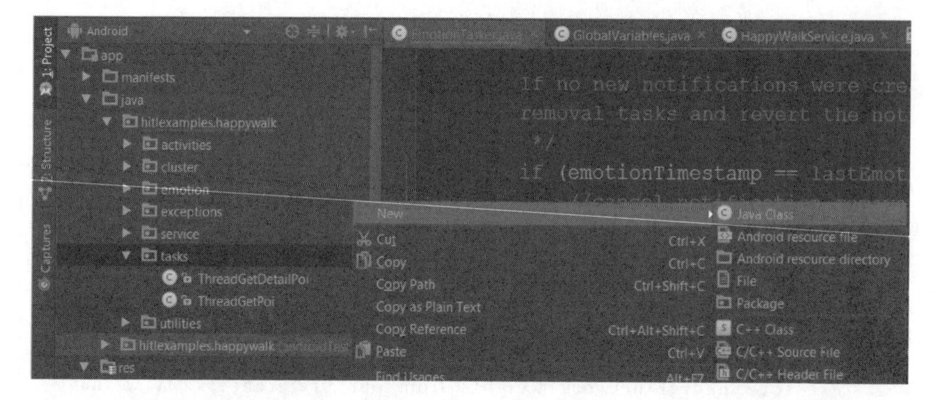

Figure 8.14 Creating *TaskSendEmotion*.

be aggregated by the server in order to be shared with the community, so that users may know the average mood at a certain POI as well as the number of people in its surroundings.

Let us write a new class that sends this information to the server. Create a new class under *hitlexamples.happywalk/tasks*, as shown in Figure 8.14.

Name this new class *TaskSendEmotion*. Below, the reader can find a suggested implementation:

```
package hitlexamples.happywalk.tasks;

import android.util.Log;
import org.json.JSONObject;
import hitlexamples.happywalk.utilities.CommunicationClass;

public class TaskSendEmotion implements Runnable {

    private String uuid;
    private double userEuphoriaBored;
    private double userAnxietyCalm;
    private double latitude;
    private double longitude;

    public TaskSendEmotion(String uuid, double userEuphoriaBored, double
    userAnxietyCalm, double latitude, double longitude) {
        this.uuid = uuid;
        this.userEuphoriaBored = userEuphoriaBored;
        this.userAnxietyCalm = userAnxietyCalm;
        this.latitude=latitude;
        this.longitude=longitude;
    }

    public void run() {
        Log.d("requests", "TaskSendEmotion");
        JSONObject ob = requestSendEmotion(uuid, userEuphoriaBored,
                userAnxietyCalm, latitude, longitude);

        if (ob != null) {
            CommunicationClass.sendPostMessage(ob, "setEmotion/");
        }
    }
```

```
32
    /**
34   * This function build one object json with the request to server.
     *
36   * @return Json object with username password and location. If an
    error occurs, it returns null
     */
38  public JSONObject requestSendEmotion(String uuid, double
    userEuphoriaBored, double userAnxietyCalm, double latitude, double
    longitude) {

40      JSONObject json = new JSONObject();

42      try {
            json.put("uuid", uuid);
44          json.put("userEuphoriaBored", userEuphoriaBored);
            json.put("userAnxietyCalm", userAnxietyCalm);
46          json.put("latitude", latitude);
            json.put("longitude", longitude);
48
        } catch (Exception e) {
50          e.printStackTrace();
        }
52      Log.i("->setEmotion", json.toString());
        return json;
54  }
}
```

The *TaskSendEmotion* class implements the *Runnable* interface; as such, it can be fed as a task to be run in the background through our *HappyWalkServiceHandler* (line 7). Its constructor takes a *uuid* string, two double values corresponding to the *Anxiety–Calmness* and *Euphoria–Boredom* axes, as well as the user's latitude and longitude (lines 15–21). The *requestSendEmotion()* method creates a *JSON* object containing the necessary information (lines 38–54). Finally, the *run()* method uses *CommunicationClass* (a utility class that contains several communication methods) to send our *JSON* object to a service hosted by the server named *setEmotion/* (line 29). We will consider the implementation of this service later on, in Section 9.1.

Now that we have *TaskSendEmotion*, we can use it to easily implement *EmotionTasker's sendEmotionToServer()* method, previously discussed on page ???:

```
1   package hitlexamples.happywalk.service;

3   import (...)
    import hitlexamples.happywalk.tasks.TaskSendEmotion;
5
    public class EmotionTasker {
7
        (...)
9
        /**
11       * Sends output emotion to the server
         * @param outputs - the emotion output to be sent
13       */
        private void sendEmotionToServer(double[] outputs) {
15          LatLng currentPos = hWServ.getHwLocationListener().
        getActualposition();
```

```
            hWServiceHandler.post(new TaskSendEmotion(
17                  GlobalVariables.UUID,
                    outputs[GlobalVariables.
        NN_OUTPUT_ARRAY_INDEX_EUPHORIC_BORED],
19                  outputs[GlobalVariables.NN_OUTPUT_ARRAY_INDEX_ANXIOUS_CALM
        ],
                    currentPos.latitude,
21                  currentPos.longitude));
        }
23
        (...)
25
    }
```

The *sendEmotionToServer()* method simply fetches the device's current position (line 15) and sends it to the server, together with the user's UUID and emotional information (lines 16–21).

With its last piece in place, our *processUserFeedback()* method should now be complete. With this, we have completed the app's ability to acquire and use feedback to train the neural network.

8.4 In Summary...

In this chapter we have handled the state inference process of our HiTLCPS. Figure 8.15 shows a summary of the current state of our HiTLCPS. It illustrates many of the tasks we

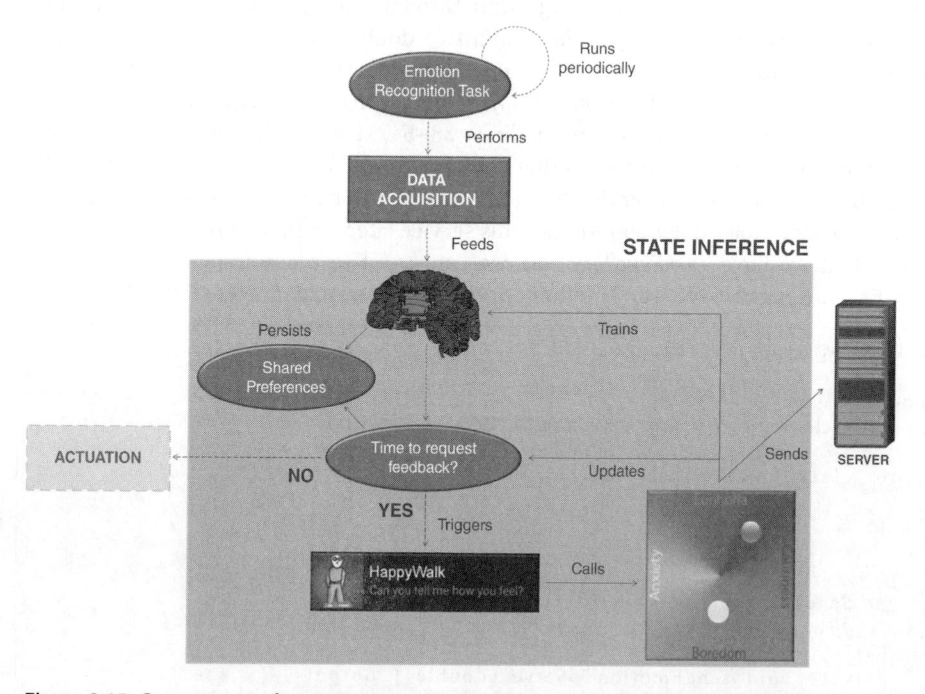

Figure 8.15 Current state of our HiTLCPS at the end of Chapter 8. (*See insert for color representation of the figure.*)

have performed during this chapter. We began by implementing a neural network and creating an activity for acquiring user feedback. This activity is triggered by an emotion feedback notification, the frequency of which is controlled by a dynamic mechanism based on the performance of the neural network. We also handled persistence by saving and restoring the values pertaining to this dynamic notification frequency and to the state of our neural network's weights. Finally, the processing of user feedback was covered, where we performed the training of our neural network and sent relevant data to the server.

However, as the reader may have noticed, there are several parts of Figure 8.15 that we have yet to address, identified by dashed arrows and squares. At the top we have *EmotionRecognitionTask*, which is still incomplete. As the reader may remember, its *run()* method (line 77 of the code on page ???) still needs to be implemented. This task should run periodically (once or twice an hour), as we decided back on page ??. It should also trigger the beginning of the *data acquisition* process, which we handled in Chapter 7. We still need to feed the acquired data to our neural network. This has been partially handled by the *fetchInputsAndCompute()* method (lines 35–40 of the code on page ???), but the method has yet to be used within *EmotionRecognitionTask*.

Additionally, the dynamic feedback mechanism we implemented in Section 8.2.4 only computes the neural network's performance and the feedback notification period. These values still need to be put to use, to decide if we should trust the neural network (i.e. if its performance is good enough), and to determine whether the application should show a feedback request or actuate.

In what concerns feedback requests, we covered most of the necessary work in this chapter. However, within our HiTL application we still need to be able to use the neural network to effectively produce a positive effect on the user. This challenge lies within the *Actuation* part of the app. In the next chapter, we will discuss how to use the ANN results to present suggestions to the user through the map interface.

9

Actuation

After tackling data acquisition and state inference in the previous two chapters, in this chapter we will address the final step of our HiTLCPS sample app: actuation. To do so, in Section 9.1 we will first handle emotions on the server side. This implies the need to implement intelligence that saves and updates emotional information and associates it with certain POIs. We will also need to prune the database for outdated emotions; that is, each emotion has limited validity, after which it should be deleted from the database. Afterwards, in Section 9.2, we will keep working on *EmotionTasker*, defining when the results of our neural network should be considered, as well as what to do with them. Last but not least, in Section 9.3, we will also discuss how to provide positive reinforcement to the user and how to represent emotional information on the map.

9.1 Handling Emotions on the Server

As previously mentioned in Section 5.2.2, this book does not cover how to implement and handle the database. Instead, it focuses on the intelligence associated with the handling of emotions in the server. In particular, we will detail how to save and update emotional information and how to prune outdated emotions.

In this section we will be implementing the *setEmotion* web service referenced back on page ???. Before we begin, we suggest a revision of the server's class structure, previously discussed on page ??. We will need to create new classes in the *Model*, *Web*, and *Com* packages. As such, we also need to familiarize ourselves with HappyWalk's database schema, in order to make proper use of the necessary *DAOs* and *HibernateMaps*. Figure 9.1 shows an overview of its conceptual schema.

HappyWalk's database has only two entities: *Pointofinterest* and *Emotion*. These entities are related by a "many-to-many" cardinality, that is a *Pointofinterest* can have several *Emotions* and an *Emotion* can affect several *Pointofinterests*.

This is because **each emotion will affect every POI within a certain range of the user**. This design decision stems from the fact that the objective of HappyWalk is to provide general information about areas through heatmaps, not about specific places. It also accounts for areas with several nearby POIs: considering that users only send emotional updates periodically, their location may change between updates. Additionally, not associating an emotion with a unique POI further contributes towards our anonymization requirement.

A Practical Introduction to Human-in-the-Loop Cyber-Physical Systems, First Edition.
David Nunes, Jorge Sá Silva and Fernando Boavida.

Companion Website URL: www.wiley.com/go/nunesloop

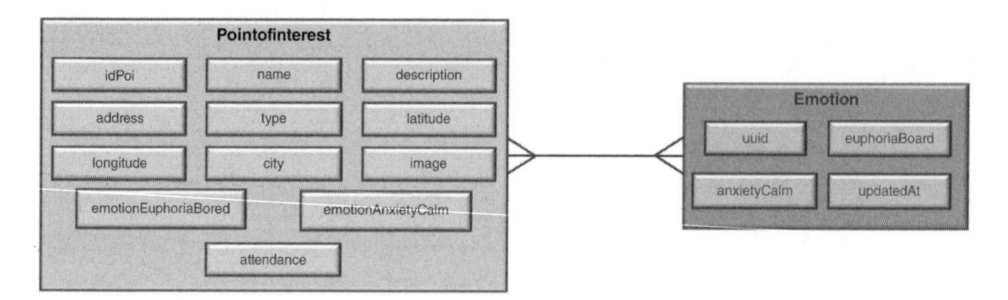

Figure 9.1 HappyWalk's database conceptual schema.

The *Pointofinterest* entity contains various attributes:

- ***idPoi***: an alphanumeric string which uniquely identifies the POI.
- ***name***: a string containing the name of the POI.
- ***description***: a string containing a description of the POI.
- ***address***: a string with the POI's address.
- ***type***: a string with the POI's type (plaza, garden, park, etc.).
- ***latitude/longitude***: doubles containing the POI's coordinates.
- ***city***: the name of the POI's location.
- ***image***: a binary representation of an image associated with the POI.
- ***emotionEuphoriaBored/emotionAnxietyCalm***: doubles containing the average values of each emotional axis calculated from all the users at the POI.
- ***attendance***: an integer representing the number of users at a POI.

The *Emotion* entity contains only four attributes:

- ***uuid***: a string with the UUID of the user to which this emotion belongs.
- ***euphoriaBored/anxietyCalm***: doubles representing the emotional axes of this emotion.
- ***updatedAt***: a long representing a timestamp of the last time this emotion was updated.

In the next few sections, we will be making use of these entities through their associated *DAOs* and *HibernateMaps*.

- We will begin with the modeling of *setEmotion* requests. This will allow us to easily parse subsequent JSON messages and their contents (Section 9.1.1).
- We will then create an interface to act as a "point of entry" for our *setEmotion* web service (Section 9.1.2).
- Afterwards, we will delve more deeply into our server and create a background thread to perform emotion-related tasks (Section 9.1.3).
- Using our background thread, we will process incoming emotions by updating the respective POI information (Section 9.1.4).
- Finally, our background thread will also be used to periodically prune outdated emotions from the database (Section 9.1.5).

9.1.1 Parsing JSON Requests

To begin, we shall create a *model* of incoming *setEmotion* requests. This will be, essentially, a basic Java class whose attributes will represent the data elements being sent by

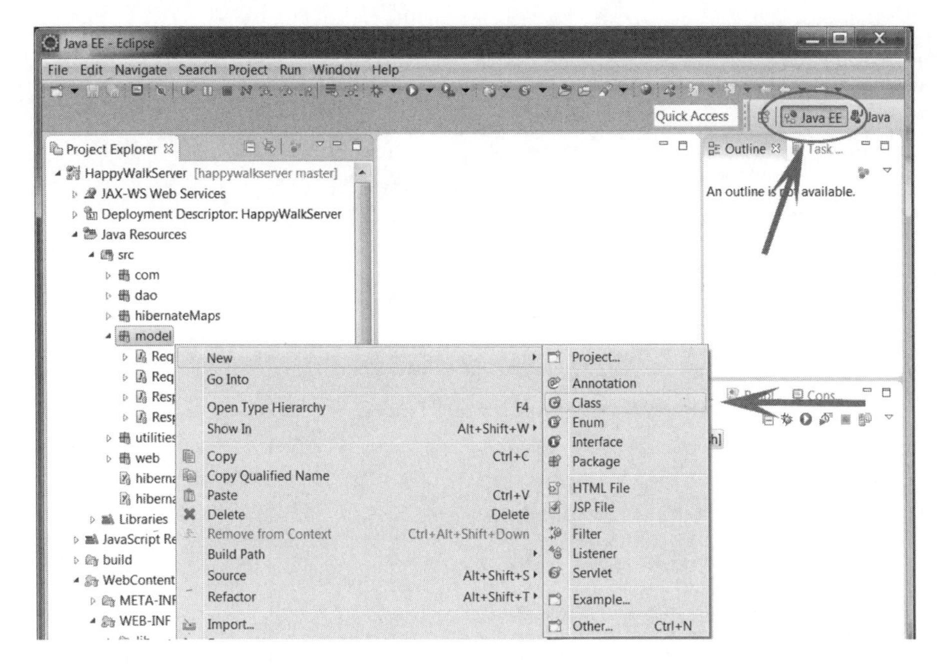

Figure 9.2 Creating a new class in Eclipse.

the Android client. Constructing this type of *model* allows us to easily use the Jersey library to parse JSON data.

Open the server's project in Eclipse, the one previously cloned in Section 6.3.1. Make sure Eclipse is on its "Java EE" perspective and create a new class within the *model* package, as shown in Figure 9.2. Name this new class *RequestSetEmotion* (see Figure 9.3).

Eclipse should generate an empty class within the *model* package. This *RequestSetEmotion* will be used by Jersey to parse incoming JSON messages belonging to our *setEmotion*.

As the user may remember from our *TaskSendEmotion* Android class, back on page ???, the *requestSendEmotion()* method created a *JSON* containing a *uuid* string, two double values corresponding to the *Anxiety-Calmness* and *Euphoria-Boredom* axes, as well as the user's latitude and longitude. As such, our server's *RequestSetEmotion* class has to reflect all of this incoming information accurately, so that no data is missed. Let us begin by creating class variables for each of these elements:

```
1  package model;
2
3  public class RequestSetEmotion {
4
5    private String uuid;
6    private Double userEuphoriaBored;
7    private Double userAnxietyCalm;
8    private Double latitude;
9    private Double longitude;
10
11 }
```

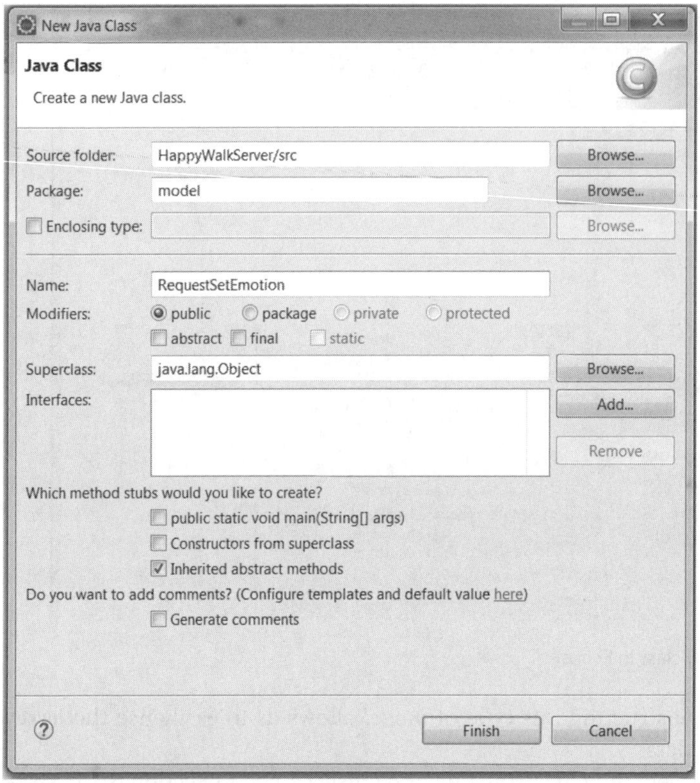

Figure 9.3 Naming *RequestSetEmotion*.

The above implementation shows that *RequestSetEmotion* is just a regular Java object. However, in order for Jersey to properly parse incoming JSONs, there are some important particularities that need to be taken into consideration.

As the reader may have noticed, *RequestSetEmotion's* class variables' names match the corresponding JSON attributes defined by *TaskSendEmotion* (see page ???). This is relevant, since we need to generate some boilerplate code based on these names.

We need to provide two class constructors: Jersey requires a non-argument default constructor (based on Java's *Object* superclass) and a constructor using all the fields we have just defined. We also need to provide the variables' *Getters* and *Setters*. Fortunately, Eclipse makes this task rather effortless: as shown in Figure 9.4, simply right-click on *RequestSetEmotion's* editor window and hover your mouse cursor over *Source*. Here, one can access four useful code-generation options: *Generate Getters and Setters...*, *Generate toString()...*, *Generate Constructor using Fields...*, and *Generate Constructor from Superclass...*.

Let us first generate a constructor using fields. After clicking on the associated option, a window similar to Figure 9.5 should appear. Select all of our previously defined variables. As for the other options, follow Figure 9.5 and click "Ok". Now, click on *Generate Constructor from Superclass...*, follow the options shown in Figure 9.6 and click "Ok".

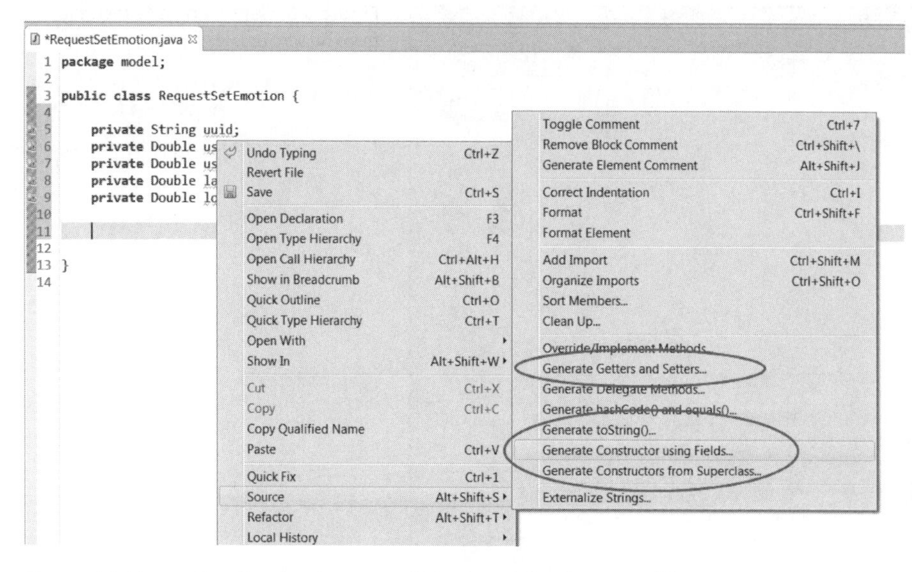

Figure 9.4 Generating the *Constructors, toString()*, and the *Getters and Setters*.

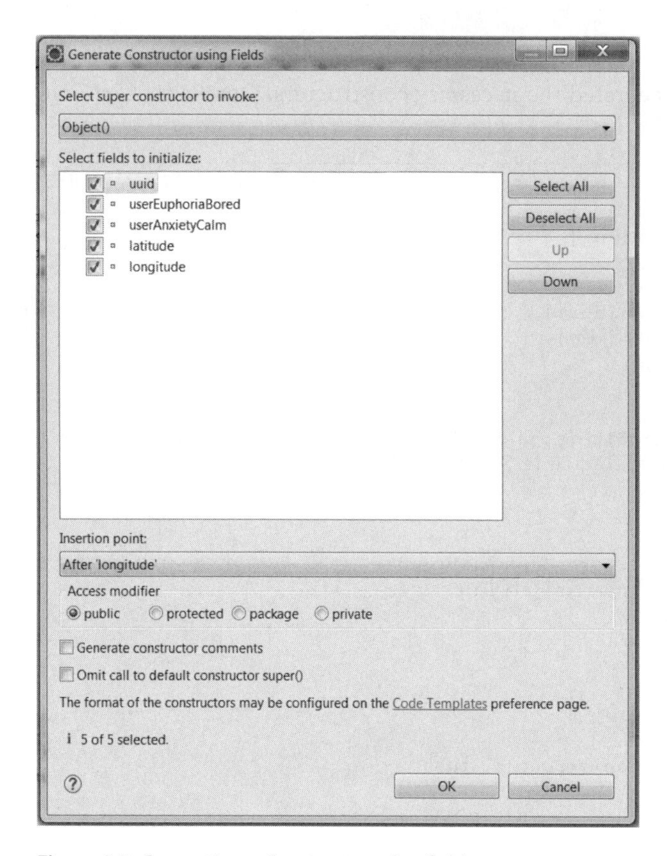

Figure 9.5 Generating a *Constructor* using fields.

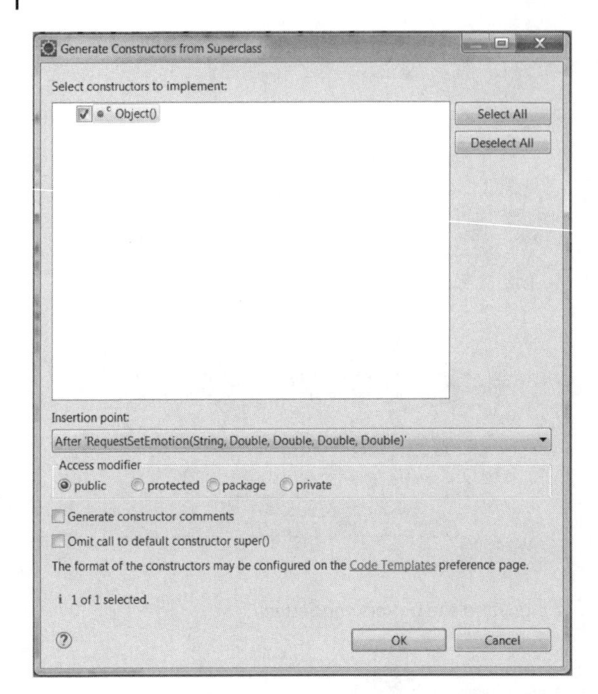

Figure 9.6 Generating a *Constructor* from *Superclass*.

These steps should have generated the necessary constructors. Our *RequestSetEmotion* class should now look like this:

```
package model;

public class RequestSetEmotion {

  private String uuid;
  private Double userEuphoriaBored;
  private Double userAnxietyCalm;
  private Double latitude;
  private Double longitude;

  public RequestSetEmotion(String uuid, Double userEuphoriaBored, Double
    userAnxietyCalm, Double latitude,
      Double longitude) {
    super();
    this.uuid = uuid;
    this.userEuphoriaBored = userEuphoriaBored;
    this.userAnxietyCalm = userAnxietyCalm;
    this.latitude = latitude;
    this.longitude = longitude;
  }

  public RequestSetEmotion() {
    super();
    // TODO Auto-generated constructor stub
  }

}
```

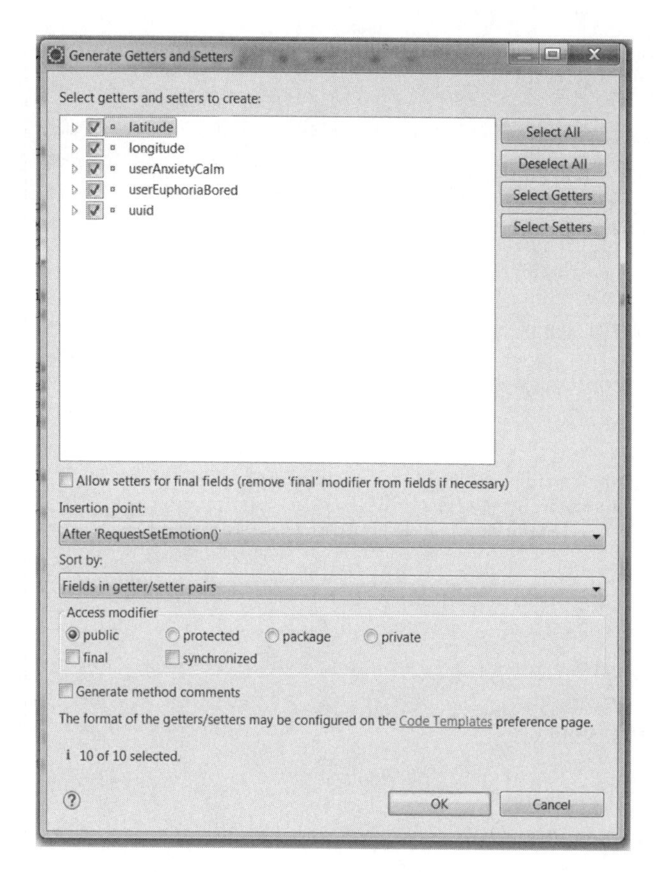

Figure 9.7 Generating the *Getters and Setters*.

Next, we shall generate the necessary *Getters* and *Setters*. Use the code-generation option *Generate Getters and Setters…* shown in Figure 9.4 and enter the settings shown in Figure 9.7 in the popup that appears.

Several new methods should have been generated: a getter and a setter for each of our class variables. Finally, we should also override the class default *toString()* method in order to more easily visualize and debug incoming JSON requests. Simply use the *Generate toString()…* option and follow the settings shown in Figure 9.8.

Now that all of the necessary code generators have been used, *RequestSetEmotion* should look similar to the following:

```
package model;

public class RequestSetEmotion {

  private String uuid;
  private Double userEuphoriaBored;
  private Double userAnxietyCalm;
  private Double latitude;
  private Double longitude;

```

```
    public RequestSetEmotion(String uuid, Double userEuphoriaBored, Double
      userAnxietyCalm, Double latitude,
12      Double longitude) {
      super();
14    this.uuid = uuid;
      this.userEuphoriaBored = userEuphoriaBored;
16    this.userAnxietyCalm = userAnxietyCalm;
      this.latitude = latitude;
18    this.longitude = longitude;
    }
20
    public RequestSetEmotion() {
22    super();
    }
24
    @Override
26  public String toString() {
      return "RequestSetEmotion [uuid=" + uuid + ", userEuphoriaBored=" +
      userEuphoriaBored + ", userAnxietyCalm="
28        + userAnxietyCalm + ", latitude=" + latitude + ", longitude=" +
      longitude + "]";
    }
30
    public String getUuid() {
32    return uuid;
    }
34
    public void setUuid(String uuid) {
36    this.uuid = uuid;
    }
38
    public Double getUserEuphoriaBored() {
40    return userEuphoriaBored;
    }
42
    public void setUserEuphoriaBored(Double userEuphoriaBored) {
44    this.userEuphoriaBored = userEuphoriaBored;
    }
46
    public Double getUserAnxietyCalm() {
48    return userAnxietyCalm;
    }
50
    public void setUserAnxietyCalm(Double userAnxietyCalm) {
52    this.userAnxietyCalm = userAnxietyCalm;
    }
54
    public Double getLatitude() {
56    return latitude;
    }
58
    public void setLatitude(Double latitude) {
60    this.latitude = latitude;
    }
62
    public Double getLongitude() {
64    return longitude;
    }
66
```

```
    public void setLongitude(Double longitude) {
68      this.longitude = longitude;
    }
70  }
```

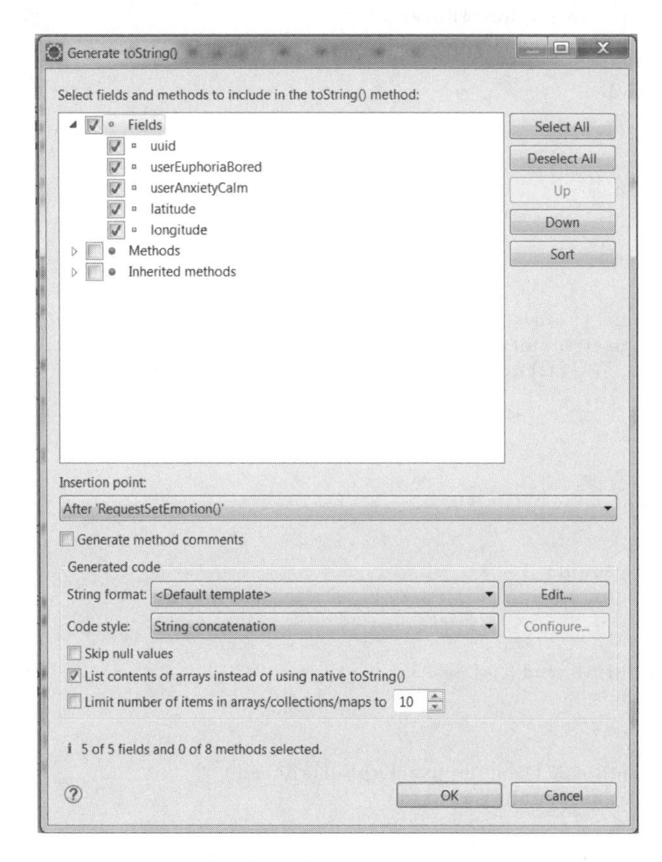

Figure 9.8 Overriding the default *toString()* method.

Our *RequestSetEmotion* is nearly complete. We only need to inform Jersey that this class represents the *root* element of incoming JSONs. The following code represents *RequestSetEmotion's* final form:

```
    package model;
2
    import javax.xml.bind.annotation.XmlRootElement;
4
    @XmlRootElement
6   public class RequestSetEmotion {

8     private String uuid;
      private Double userEuphoriaBored;
10    private Double userAnxietyCalm;
      private Double latitude;
12    private Double longitude;
```

```
14    public RequestSetEmotion(String uuid, Double userEuphoriaBored, Double
        userAnxietyCalm, Double latitude,
          Double longitude) {
16      super();
        this.uuid = uuid;
18      this.userEuphoriaBored = userEuphoriaBored;
        this.userAnxietyCalm = userAnxietyCalm;
20      this.latitude = latitude;
        this.longitude = longitude;
22    }

24    public RequestSetEmotion() {
        super();
26    }

28    @Override
      public String toString() {
30      return "RequestSetEmotion [uuid=" + uuid + ", userEuphoriaBored=" +
        userEuphoriaBored + ", userAnxietyCalm="
            + userAnxietyCalm + ", latitude=" + latitude + ", longitude=" +
        longitude + "]";
32    }

34    public String getUuid() {
        return uuid;
36    }

38    public void setUuid(String uuid) {
        this.uuid = uuid;
40    }

42    public Double getUserEuphoriaBored() {
        return userEuphoriaBored;
44    }

46    public void setUserEuphoriaBored(Double userEuphoriaBored) {
        this.userEuphoriaBored = userEuphoriaBored;
48    }

50    public Double getUserAnxietyCalm() {
        return userAnxietyCalm;
52    }

54    public void setUserAnxietyCalm(Double userAnxietyCalm) {
        this.userAnxietyCalm = userAnxietyCalm;
56    }

58    public Double getLatitude() {
        return latitude;
60    }

62    public void setLatitude(Double latitude) {
        this.latitude = latitude;
64    }

66    public Double getLongitude() {
        return longitude;
68    }
```

```
70    public void setLongitude(Double longitude) {
        this.longitude = longitude;
72    }
   }
```

Despite its somewhat misleading naming, notice that we only need to import and use the *@XmlRootElement* annotation, placing it before the beginning of the class. Jersey will understand that the class does, in fact, represent a JSON (and not an XML) when we use it later on our sevice's web interface.

9.1.2 Creating the Web Interface

Now that we have a way of parsing JSON objects into a Java class, we can define a web interface that serves as a "point of entry" for our *setEmotion* web service. Similarly to what we did for *RequestSetEmotion* in Figure 9.2, create a new class, this time in the **web** package, and name it *SetEmotion*. Below follows a possible implementation of our new web interface:

```
1  package web;

3  import javax.ws.rs.Consumes;
   import javax.ws.rs.POST;
5  import javax.ws.rs.Path;
   import javax.ws.rs.core.MediaType;
7
   import model.RequestSetEmotion;
9
   import com.ComEmotions;
11
   @Path("/setEmotion")
13 public class SetEmotion {
       @POST
15     @Consumes(MediaType.APPLICATION_JSON)
       public void setEmotion(RequestSetEmotion requestSetEmotion){
17     System.out.println("->Setting emotion:"+requestSetEmotion.toString());
       ComEmotions comEmotions= new ComEmotions();
19     comEmotions.setEmotionFromRequest(requestSetEmotion);
       }
21 }
```

We use Jersey's *@Path* annotation to define the path to our service. Notice that it matches the *"/setEmotion"* path we have previously used in *TaskSendEmotion* (page ???). Next, we define that the *setEmotion()* method shall implement the POST operation of our RESTful web service (by using the *@POST* annotation). We also define that this operation consumes a JSON (see the *@Consumes(MediaType.APPLICATION_JSON)* annotation). Note that *setEmotion()* receives a *RequestSetEmotion* object. Jersey will interpret this as meaning that *RequestSetEmotion* represents the incoming JSON and will, thus, attempt to parse it accordingly.

Within our *setEmotion()* method, we print a logging message which uses *RequestSetEmotion's toString()*. This allows us to confirm that the JSON information is being properly parsed, as well as easily study the contents of incoming *setEmotion* requests. Finally, we refer to a non-implemented *com* class, *ComEmotions*, which will be responsible for processing the request. We will begin implementing this class in Section 9.1.4.

9.1.3 Creating the Server's Background Thread

In order to provide near real-time information about the attendance and the emotional status of each POI, we need to make sure that each emotional input has a limited lifespan. As such, there are two major tasks that we have to perform when dealing with incoming emotions. First, we need to **process incoming emotions**, creating or updating them as necessary. Second, we need to **periodically prune outdated emotions from the database**.

Processing incoming emotions is a task that will be performed whenever a new *setEmotion* request arrives. Pruning outdated emotions is a periodic task that shall be run on a background thread on the server, similar to what happens with the Android app's background service. However, since we cannot predict when a user will send us emotional input, we have to be careful. This is because **these tasks may interfere with each other**. Imagine, for example, that we are about to delete a certain emotional input that has just expired. What if, at that exact moment, the corresponding user sends us a request with updated data? The server may become confused, since the emotion we were supposed to update suddenly got deleted by our pruning task! Therefore, we need to make sure that emotional pruning and updates occur *sequentially* and never at the same time.

One way of dealing with this requirement is to create a single background thread that accepts tasks in a "first come, first served" basis. We can then feed this background thread with both emotion processing and pruning tasks, and they will never overlap. This is essentially the same approach we have been using on the Android app whenever we passed tasks to the *HappyWalkServiceHandler*. However, the *android.os.Handler* class belongs to the Android OS. Therefore, we will need to find an alternative way of implementing our background thread. An appropriate class that we may use to achieve this goal is the *java.util.concurrent.ScheduledExecutorService*. In particular, we can instantiate a *SingleThreadScheduledExecutor* that never executes anything in parallel, which is exactly what we need. But how and when can we tell our server to use this executor?

As the reader may remember from Section 5.2.2, our server is implemented through the Java Servlet API. This API offers an interface for receiving notification events about the server's lifecycle changes. In particular, we are interested in the *contextInitialized* and *contextDestroyed* events which represent the initialization and the finalization of our server, respectively. As such, we can instantiate an executor when the server starts and shut it down when the server stops. We can then use this executor to perform background tasks whenever we see fit.

With this objective in mind, create a new class on the *utilities* package and name it *EmotionListener* (in a manner similar to what we did for *RequestSetEmotion*, back on page ???). This class shall implement a *ServletContextListener* that will handle the server's initialization and destruction events:

```
1 package utilities;
2 
3 import java.util.concurrent.Executors;
4 import java.util.concurrent.ScheduledExecutorService;
5 
6 import javax.servlet.ServletContextEvent;
7 import javax.servlet.ServletContextListener;
8 
9 public class EmotionListener implements ServletContextListener {
```

```
    /*
     * public scheduler that will execute all tasks related
     * with Emotions.
     */
    public static ScheduledExecutorService scheduler;

    @Override
    public void contextInitialized(ServletContextEvent event) {
      scheduler = Executors.newSingleThreadScheduledExecutor();
    }

    @Override
    public void contextDestroyed(ServletContextEvent event) {
      scheduler.shutdownNow();
    }
}
```

The public variable *scheduler* of type *ScheduledExecutorService* is initialized as a *SingleThreadScheduledExecutor* when the server starts and shut down when the server stops.

Finally, we need to inform our server of the existence of *EmotionListener*. This can be done within the *HappyWalkServer/WebContent/WEB-INF/web.xml* file (see Figure 9.9).

Here, we will need a new *<listener>* element within the *<web-app>* root element. Within it, we need to add yet another *<listener-class>* sub-element, declaring our *utilities.EmotionListener*.

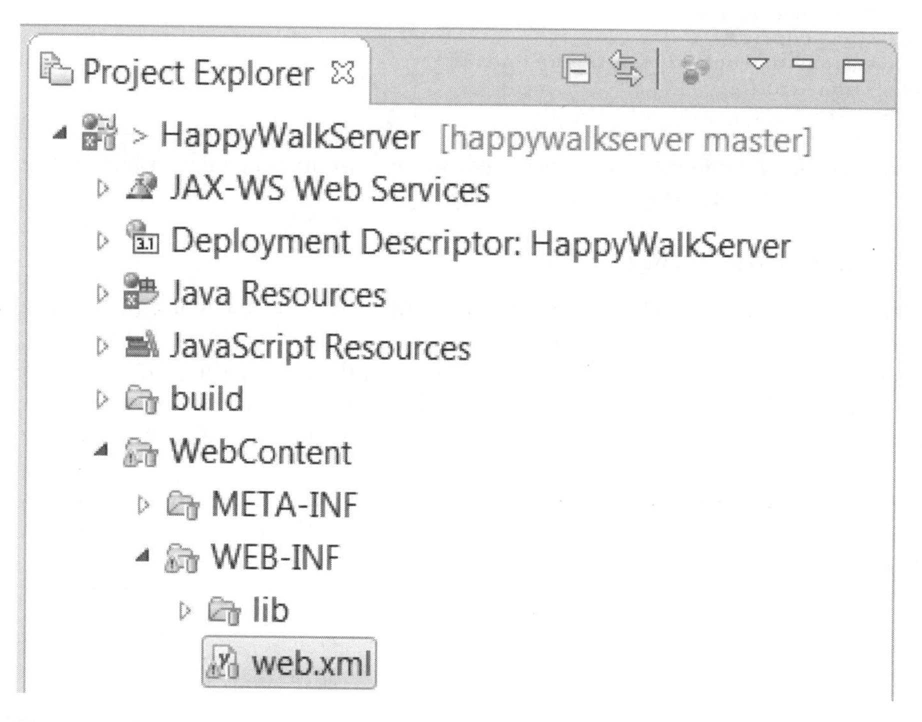

Figure 9.9 The location of the *HappyWalkServer's* web.xml.

The following example shows a modified *web.xml* where the *<listener>* element was added between the *<servlet>* and the *<servlet-mapping>* elements (lines 18–20):

```
<?xml version="1.0" encoding="UTF-8"?>
<web-app xmlns="http://java.sun.com/xml/ns/javaee" xmlns:web="http://java.
    sun.com/xml/ns/javaee/web-app_2_5.xsd">
  <servlet>
    <servlet-name>HappyWalkServlet</servlet-name>
    <servlet-class>
      com.sun.jersey.spi.container.servlet.ServletContainer
    </servlet-class>
    <init-param>
      <param-name>com.sun.jersey.config.property.packages</param-name>
      <param-value>web</param-value>
    </init-param>
    <init-param>
      <param-name>com.sun.jersey.spi.container.ContainerRequestFilters</
    param-name>
      <param-value>com.sun.jersey.api.container.filter.LoggingFilter</
    param-value>
    </init-param>
    <load-on-startup>1</load-on-startup>
  </servlet>
  <listener>
    <listener-class>utilities.EmotionListener</listener-class>
  </listener>
  <servlet-mapping>
    <servlet-name>HappyWalkServlet</servlet-name>
    <url-pattern>/rest/*</url-pattern>
  </servlet-mapping>
  <session-config>
    <session-timeout>120</session-timeout>
  </session-config>
</web-app>
```

Now that our *EmotionListener* is declared, we will use its scheduler to handle any emotion-related tasks during the next sections.

9.1.4 Processing Incoming Emotions

In this section we will tackle the requirement of processing incoming *setEmotion* requests. Emotional requests may pertain to new emotional records or they may be updates to existing ones. As the user moves around, the surrounding POIs affected by his/her emotional information also change. Therefore, we will need to write methods able to add, remove, or update the influence of emotional information to the corresponding sets of POIs.

Before we continue developing code, let us take a step back and think on what needs to be done whenever a new request arrives at the server. This is actually a somewhat complex issue that can be illustrated by the following pseudocode:

```
void function processIncomingEmotion (request) {
  emotion = searchDatabaseForEmotion(request.uuid);

  //If we do not have emotional information from that uuid...
  if (emotion == null) {
    //create emotion and provide a timestamp
```

```
    emotion = createNewEmotion(request.uuid,
8       request.euphoriaBored, request.anxietyCalm,
        System.currentTime);
10  //search for POIs within a certain range of the user's location
    listPOIs = searchForPOIsWithinRange(request.latitude,
12    request.longitude);
    for each (poi : listPOIs) {
14    //associate each POI with this new emotion
      poi.associate(emotion);
16    //increment its attendance
      poi.attendance += 1;
18    //update emotional mean
      poi.updateEmotionMean(emotion);
20  }
  }
22  //If this is updated information about an emotion...
  else {
24  //update values
    emotion.anxietyCalm(request.anxietyCalm);
26  emotion.euphoriaBored(request.euphoriaBored);
    emotion.updatedAt(System.currentTime);
28
    /*consider the new location to update the list of POIs associated with
30  this emotion and their values */
    listPreviousPOIs = emotion.getAssociatedPOIs();
32  listNewPOIs = searchForPOIsWithinRange(request.latitude,
      request.longitude);
34
    for each (newpoi : listNewPOIs) {
36    for each (oldpoi : listPreviousPOIs) {
        /* this boolean keeps track of whether
38      the newpoi was previously associated */
        boolean isACurrentPOI = false;
40      if (oldpoi == newpoi) {
          /* This newpoi was previously associated
42        let us remove it from the previous list */
          listPreviousPOIs.remove(oldpoi);
44        isACurrentPOI = true;
        }
46      if (isACurrentPOI) {
          //just need to update the emotional mean
48        newpoi.updateEmotionMean(emotion);
        }
50      else {
          //associate the POI with emotion
52        poi.associate(emotion);
          //increment its attendance
54        poi.attendance += 1;
          //update emotional mean
56        poi.updateEmotionMean(emotion);
        }
58    }
    }
60
    /* Now that we have iterated through all the newPOIs,
62  those remaining on listPreviousPOIs are older
    associations that need to be removed and have
64  their means and attendance updated accordingly */
```

```
66      for each (oldpoi : listPreviousPOIs) {
          oldpoi.removeEmotionFromMean(emotion);
68        poi.disassociate(emotion);
        }
70    }
  }
```

As the above code suggests, we have a lot of ground to cover. First and foremost, whenever we begin to process a new request, we need to check if we already have information from its UUID on the database (line 2). Since a UUID should uniquely identify a user, it makes no sense to have multiple records of different emotions belonging to the same person. Therefore, by checking if a certain UUID is already present, we can determine whether we are dealing with a new emotion or an update to an existing value.

In case no emotion with the request's UUID is found (line 5), we create a new emotional record in the database by providing the request's UUID and emotional axes (lines 7 and 8). We also need to register a timestamp that will be used by our pruning task (to simplify, this can be the time of registration within the server) (line 9). As discussed on page ???, in order for the new emotion to have an impact, it must be associated with nearby POIs. Therefore, we need to search for every POI within a certain range of the location of the request (line 11). For each of these, we perform an association with the new emotion, increment the attendance count, and, finally, calculate the new emotional average (lines 13–20).

When an emotion with the same UUID as the incoming request already exists in the database (line 23), we first need to update its data and timestamp, which is rather straightforward; a simple database request is enough (lines 25–27). However, updating its POI associations is a bit more complex. Since the user may have moved his/her physical location between requests, the previous POI associations may be outdated. Therefore, we need to first retrieve a list of previously associated POIs (line 31), which we will compare with a list of POIs located nearby the new location (line 32). For each of the new POIs, we check if it was previously associated with the request's UUID. If it was, we remove it from the previous POI list and update its emotional mean. If it was not, we associate it with the emotion, increment its attendance count, and update its mean (lines 35–59). After iterating through all of the new POIs, the previous POI list is left with the outdated associations. As such, we iterate over this list one last time to remove the emotion's stale associations and their effect on the old POIs' means (lines 66–69).

Now that we have an idea of what we need to accomplish, we can focus on writing Java code. As we had previously referenced when impl ementing *SetEmotion* back on page ???, we shall implement the above logic in a new ***com*** package class. Create this new class by right-clicking on the *com* package → *New* → *Class*, and naming it *ComEmotions*.

Let us first identify the necessary building blocks to support emotional processing. Considering the tasks associated with both the pruning of old emotions and the processing of new ones, our *ComEmotions* shall implement the following helper methods:

- ***addEmotionValToPOI***: adds the influence of an emotion to a POI's attendance and emotional average.
- ***removeEmotionValFromPOI***: removes the influence of an emotion from a POI's attendance and emotional average.
- ***updatePOIsOfAddedEmotion***: finds, associates, and updates the appropriate POIs with a new emotion.

- ***updatePOIsOfUpdatedEmotion***: finds which new POIs should be associated with an emotion, from its updated location, and performs the necessary associations and value updates. It also removes outdated associations and influences.
- ***updatePOIsOfRemovedEmotion***: finds the POIs currently associated with an emotion and removes the corresponding influences.

Since *addEmotionValToPOI* and *removeEmotionValFromPOI* are more fundamental methods, let us implement them first:

```
package com;

import hibernateMaps.Emotion;
import hibernateMaps.Pointofinterest;

public class ComEmotions {

    /**
     * This method adds the influence of an emotion to a poi's
     * attendance, EmotionAnxietyCalm and EmotionEuphoriaBored values.
     *
     * @param poi - POI to be updated
     * @param emo - Emotion to be added
     *
     * @return updated POI
     */
    private static Pointofinterest addEmotionValToPOI(Pointofinterest poi,
     Emotion emo){
        Double newEmotionAnxietyCalm;
        Double newEmotionEuphoriaBored;

        //increase attendance by 1
        int newAttendance = poi.getAttendance()+1;

        //avoid nulls and check if emotion values are greater than 0 to
    avoid unnecessary computation
        if (poi.getEmotionAnxietyCalm() != null && poi.
    getEmotionAnxietyCalm().compareTo(0d)>0) {
            newEmotionAnxietyCalm = (poi.getEmotionAnxietyCalm()*poi.
    getAttendance() + emo.getAnxietyCalm().doubleValue())/newAttendance;
        }
        else {
            newEmotionAnxietyCalm = emo.getAnxietyCalm().doubleValue();
        }

        if (poi.getEmotionEuphoriaBored() != null && poi.
    getEmotionEuphoriaBored().compareTo(0d)>0) {
            newEmotionEuphoriaBored = (poi.getEmotionEuphoriaBored()*poi.
    getAttendance() + emo.getEuphoriaBored().doubleValue())/newAttendance;

        }
        else{
            newEmotionEuphoriaBored = emo.getEuphoriaBored().doubleValue()
    ;
        }

        poi.setEmotionAnxietyCalm(newEmotionAnxietyCalm);
        poi.setEmotionEuphoriaBored(newEmotionEuphoriaBored);
        poi.setAttendance(newAttendance);
```

```
43          return poi;
        }
45
        /**
47       * This method removes the influence of an emotion from a poi's
         * attendance, EmotionAnxietyCalm and EmotionEuphoriaBored values.
49       *
         * @param poi - POI to be updated
51       * @param emo - Emotion to be removed
         *
53       * @return updated POI
         */
55      private static Pointofinterest removeEmotionValFromPOI(Pointofinterest
         poi, Emotion emo){
            Double newEmotionAnxietyCalm;
57          Double newEmotionEuphoriaBored;
            int newAttendance;
59
            //If attendance is greater than one, decrease it and perform
        calculations
61          if (poi.getAttendance()>1) {
                newAttendance = poi.getAttendance()-1;
63              newEmotionAnxietyCalm = (poi.getEmotionAnxietyCalm()*poi.
        getAttendance() - emo.getAnxietyCalm().doubleValue())/newAttendance;
                newEmotionEuphoriaBored = (poi.getEmotionEuphoriaBored()*poi.
        getAttendance() - emo.getEuphoriaBored().doubleValue())/newAttendance;
65          }
            //If not, then attendance and emotions values are necessarily 0
67          else {
                newAttendance = 0;
69              newEmotionAnxietyCalm = 0d;
                newEmotionEuphoriaBored = 0d;
71          }
            poi.setEmotionAnxietyCalm(newEmotionAnxietyCalm);
73          poi.setEmotionEuphoriaBored(newEmotionEuphoriaBored);
            poi.setAttendance(newAttendance);
75          return poi;
        }
77
    }
```

The proposed implementation of *addEmotionValToPOI* receives both a *Pointofinterest* and *Emotion* objects (line 17). As we can see by their respective import declarations (lines 3 and 4), both of these belong to the *hibernateMaps* package, which, as discussed in Section 5.2.2, provides classes that interface with the database. We encourage the reader to think of these objects as singular representations of POIs and emotions that can be retrieved, deleted, or saved to the database.

The method begins by declaring two doubles where the POI's new *emotionAnxietyCalm* and *emotionEuphoriaBored* attributes (refer to Figure 9.1) will be temporarily stored (lines 18 and 19). Since this method *adds* the influence of an emotion to a POI (we are not dealing with updates to existing influences yet), that means a new user is now in the surroundings of the POI. Therefore, a new attendance value is calculated by increasing the current attendance count by 1 (line 22).

To avoid unnecessary computation, the method checks if the POI's emotional averages are null (lines 25 and 32). If they are not, a new average is computed by multiplying the previous average by the previous attendance, adding the new emotion's value and dividing the result by the new attendance (lines 26 and 33). If they are, the new average is simply the new emotion's value (lines 29 and 36).

Finally, the POI's new values are set and the resulting *Pointofinterest* object is returned (lines 39–43). Notice that, as suggested by its name, the method does not perform an association between the POI and the emotion; it simply adds the emotion's values to the POI.

Turning our attention to *removeEmotionValFromPOI*, we see that it also receives a *Pointofinterest* and *Emotion* objects (line 55), and declares variables for temporarily storing the new values (lines 56–58). Since the influence from a previously existing emotion is being removed, the method does not verify if the POI's emotional averages are zero, since it is unlikely. However, as a small optimization and to avoid divisions by zero, the method does check if the POI's attendance count is greater than one (line 61). If it is not, the attendance and emotion values are necessarily zero and nothing needs to be computed (lines 67–71). In case it is greater than one, the attendance is decreased by one and the emotional averages are recomputed (lines 61–65). Again, the POI's new values are set and the resulting *Pointofinterest* object is returned (lines 72–75). Similarly to *addEmotionValToPOI*, the method does not remove the association between the POI and the emotion.

Let us build upon these two methods to implement *updatePOIsOfAddedEmotion* and *updatePOIsOfRemovedEmotion*:

```
   package com;
 2
   import java.util.Iterator;
 4 import java.util.List;

 6 import dao.DaoEmotion;
   import dao.DaoPointofinterest;
 8 import hibernateMaps.Emotion;
   import hibernateMaps.Pointofinterest;
10 import utilities.GlobalVariables;

12 public class ComEmotions {
       private static Pointofinterest addEmotionValToPOI(Pointofinterest poi,
        Emotion emo){
14         (...)
       }
16     private static Pointofinterest removeEmotionValFromPOI(Pointofinterest
        poi, Emotion emo){
           (...)
18     }

20     /**
        * Finds, associates and updates the
22      * appropriate POIs with a new emotion.
        *
24      * @param emo
        * @param latitude
26      * @param longitude
        */
```

```
28      public static void updatePOIsOfAddedEmotion(Emotion emo, Double
        latitude, Double longitude) {
            List<Pointofinterest> poisToUpdate = null;
30          DaoPointofinterest daopoi = new DaoPointofinterest();
            poisToUpdate = daopoi.getPoibyLocation(
32                  latitude, longitude,
                    GlobalVariables.EMOTION_AFFECT_POI_MILE_RADIUS);
34          Iterator<Pointofinterest> poisToUpdateit = poisToUpdate.iterator()
        ;

36          while(poisToUpdateit.hasNext())
            {
38              Pointofinterest poi = poisToUpdateit.next();
                daopoi.update(addEmotionValToPOI(poi, emo));
40              daopoi.addEmotion(poi, emo.getUuid());
            }
42      }

44      /**
         * Finds the POIs currently associated with an emotion and
46       * removes the corresponding influences.
         *
48       * @param emo
         */
50      public static void updatePOIsOfRemovedEmotion(Emotion emo) {
            List<Pointofinterest> poisToUpdate = null;
52          DaoEmotion demo = new DaoEmotion();
            DaoPointofinterest daopoi = new DaoPointofinterest();
54          poisToUpdate = demo.getAssociatedPOIs(emo.getUuid());
            Iterator<Pointofinterest> poisToUpdateit = poisToUpdate.iterator()
        ;
56
            while (poisToUpdateit.hasNext())
58          {
                Pointofinterest poi =poisToUpdateit.next();
60              daopoi.update(removeEmotionValFromPOI(poi, emo));
            }
62      }
    }
```

The *updatePOIsOfAddedEmotion* method receives an *Emotion* object, representing the emotion to add, and two doubles, for its latitude and longitude (line 28). It begins by declaring a *java.util.List* of *Pointofinterests* named *poisToUpdate*, which will hold all the POIs affected by this new emotion (line 29).

Next, it instantiates a *dao.DaoPointofinterest* object (line 30). As we explained back in Section 5.2.2, both the *hibernateMaps* and the *DAOs* interface with the database. The *DAOs'* function provides *actions* over specific *hibernateMaps* (e.g. find, save, delete, update). Therefore, the method uses a *DaoPointofinterest* to get all POIs around a certain location, within a range determined by *GlobalVariables.EMOTION_AFFECT_POI_MILE_RADIUS* (which, by default, is 200 meters or 0.12 miles) (lines 31–33).

Finally, *poisToUpdate* is iterated upon (using an iterator, defined in line 34). For each POI in range, we call *addEmotionValToPOI* and associate the new emotion through the *DaoPointofinterest's addEmotion* method (lines 36–41). Notice that *updatePOIsOfAddedEmotion()* does not create a new emotion in the database; it is merely responsible for finding and updating the related POIs.

The method *updatePOIsOfRemovedEmotion* works on a similar premise. It receives an *Emotion* object (line 50) and begins by declaring a *poisToUpdate* list (line 51). It then instantiates not only a *DaoPointofinterest* (line 52) but also a *DaoEmotion* object (line 53), which it uses to fetch all of the POIs currently associated with the emotion, from its UUID (line 54). The method finishes by calling *removeEmotionValFromPOI* for each associated POI (lines 57–61). The inquisitive reader may be wondering why the emotion associations are not being removed from within the *while* loop. That is because it is not necessary to do so; when the emotion is finally deleted from the database (supposedly after this method is called), Hibernate will also automatically remove any of its associations for us.

Finally, let us consider the last of *ComEmotion's* helper methods, *updatePOIsOfUpdatedEmotion*:

```
1  package com;

3  import (...);

5  public class ComEmotions {

7      (...)

9      /**
        * Updates the POIs of associated with an Emotion. The two Emotion
       objects should represent two states of the same emotion - outdated
       state and up-to-date state.
11      * Thus, they should have the same uuid.
        *
13      * @param oldEmo - object containing the values of the outdated
       emotion
        * @param newEmo - object containing new emotional values
15      * @param latitude - the emotion's latest latitude
        * @param longitude - the emotion's latest longitude
17      */
       public static void updatePOIsOfUpdatedEmotion(Emotion oldEmo, Emotion
       newEmo, Double latitude, Double longitude) {
19         DaoEmotion demo = new DaoEmotion();
           DaoPointofinterest daopoi = new DaoPointofinterest();
21         List<Pointofinterest> currentPOIs = demo.getAssociatedPOIs(oldEmo.
       getUuid());
           List<Pointofinterest> newPOIs = daopoi.getPoibyLocation(
23                 latitude, longitude,
                   GlobalVariables.EMOTION_AFFECT_POI_MILE_RADIUS);
25         Iterator<Pointofinterest> newPOISit = newPOIs.iterator();
           Iterator<Pointofinterest> currentPOIsit;
27
           while(newPOISit.hasNext())
29         {
               Pointofinterest newpoi = newPOISit.next();
31             /*
                * Using List.contains method does not work in this case
33              * Thus, we must individually compare the ids.
                */
35             boolean isACurrentPOI = false;
               //reset iterator
```

```
            currentPOIsit = currentPOIs.iterator();
            while(currentPOIsit.hasNext()) {
                Pointofinterest currentPOI = currentPOIsit.next();
                if (currentPOI.getIdPoi().equals(newpoi.getIdPoi())) {
                    /*
                     * Remove POI from the currentPOIs list.
                     * This is important, since we will be updating the remaining currentPOIs (those that are no longer associated with this emotion.)
                     */
                    currentPOIsit.remove();
                    isACurrentPOI = true;
                    break;
                }
            }

            if (isACurrentPOI) {
                System.out.println("CurrentPOIs contains " +newpoi.getIdPoi());
                //Ok, this POI was already in our list of associated POIs, update its means remove old value and recalculate mean with the new one
                Double newEmotionAnxietyCalm;
                Double newEmotionEuphoriaBored;

                if (newpoi.getEmotionAnxietyCalm() != null && newpoi.getEmotionAnxietyCalm().compareTo(0d)>0) {
                    newEmotionAnxietyCalm = ((newpoi.getEmotionAnxietyCalm()*newpoi.getAttendance() - oldEmo.getAnxietyCalm().doubleValue()) + newEmo.getAnxietyCalm().doubleValue())/newpoi.getAttendance();
                }
                else {
                    System.err.println(newpoi.getIdPoi() + " has null or 0 AnxietyCalm emotion values!");
                    newEmotionAnxietyCalm = newEmo.getAnxietyCalm().doubleValue();
                }
                if (newpoi.getEmotionEuphoriaBored() != null && newpoi.getEmotionEuphoriaBored().compareTo(0d)>0) {
                    newEmotionEuphoriaBored = ((newpoi.getEmotionEuphoriaBored()*newpoi.getAttendance() - oldEmo.getEuphoriaBored().doubleValue()) + newEmo.getEuphoriaBored().doubleValue())/newpoi.getAttendance();
                }
                else {
                    System.err.println(newpoi.getIdPoi() + " has null or 0 EuphoriaBored emotion values!");
                    newEmotionEuphoriaBored = newEmo.getEuphoriaBored().doubleValue();
                }

                newpoi.setEmotionAnxietyCalm(newEmotionAnxietyCalm);
                newpoi.setEmotionEuphoriaBored(newEmotionEuphoriaBored);
                daopoi.update(newpoi);
            }
            else {
                System.out.println("CurrentPOIs does NOT contain " +newpoi.getIdPoi());
```

```
              //well, this POI is new to this emotion, so let us
    consider recalculate its means accordingly
79            //the attendance increases by 1
              daopoi.update(addEmotionValToPOI(newpoi, newEmo));
81            //Ok, now we create the association between the newEmotion
     and the new POI
              daopoi.addEmotion(newpoi, newEmo.getUuid());
83          }
        }
85      /*
         * Now we should have iterated through all the newPOIs. Those
    remaining on the currentPOI list are old ones that should have their
    associations removed and means updated accordingly
87       */
        //reset the iterator
89      currentPOIsit = currentPOIs.iterator();

91      while(currentPOIsit.hasNext())
        {
93          Pointofinterest oldpoi =currentPOIsit.next();
            daopoi.update(removeEmotionValFromPOI(oldpoi, oldEmo));
95          daopoi.removeEmotion(oldpoi,newEmo.getUuid());
        }
97    }
}
```

This is a more complex method that implements most of the update logic from the pseudocode on page ???. The method requires two *Emotion* objects that represent the outdated state (from before the update, the *oldEmo* argument) and the up-to-date state (from the new *setEmotion* request, the *newEmo* argument) of the emotion. It also requires the emotion's latest location in terms of latitude and longitude (line 18). It begins by fetching the emotion's currently associated POIs (*currentPOIs,* line 21) and those that are in the range of the new location (*newPOIs,* line 22). For each new POI, the method checks if it was previously associated (line 40) and, if it was, removes it from the *currentPOIs* list (line 45).

The *isACurrentPOI* Boolean (defined in line 35) is used to mark these previously associated POIs; the corresponding *if* block updates the POI's means without changing its attendance (lines 51–75). The computations performed here follow a logic similar to the one used for *addEmotionValToPOI(),* back on page ???, but the emotional averages are updated by removing the old value before adding the new one. If the POI was not previously associated, *addEmotionValToPOI* is called and an association is created (lines 76–83). Finally, *removeEmotionValFromPOI* is called and associations are removed for each POI remaining within *currentPOIs* (lines 91–96).

We have finally implemented all of the necessary helper methods. All that is left is to actually use them to process incoming emotions. Let us implement the *setEmotionFromRequest* method, previously referenced by *SetEmotion*, on page ???:

```
  package com;
2
  import java.util.concurrent.TimeUnit;
4 import org.hibernate.HibernateException;
  import model.RequestSetEmotion;
6 import utilities.EmotionListener;
  import (...);
```

```
8
  public class ComEmotions {
10    /**
     * Processes a request to set new Emotion information. In order to avoid
       possible conflicts with Pruning tasks, this task is scheduled on a
      public scheduler
12   *
     * @param request
14   */
    public static void setEmotionFromRequest(final RequestSetEmotion request
      ) {
16    EmotionListener.scheduler.schedule(new Runnable() {
        @Override
18      public void run() {
          try {
20          DaoEmotion daoEmotion = new DaoEmotion();
            Emotion emotion = daoEmotion.find(request.getUuid());
22
            //If we do not have emotional information from that uuid...
24          if (emotion == null) {
              System.out.println("uuid: " + request.getUuid() + " is new,
    creating emotion");
26            emotion = new Emotion(request.getUuid(), request.
    getUserEuphoriaBored(), request.getUserAnxietyCalm());
              try {
28              daoEmotion.create(emotion);
                //update POIs within the range of the uuid location
30              updatePOIsOfAddedEmotion(emotion, request.getLatitude(),
    request.getLongitude());
              } catch (HibernateException e) {
32              System.out.println("Error: Internal error in DB");
              }
34          }
            //If this is updated information about an emotion...
36          else {
              System.out.println("uuid: " + request.getUuid() + ", updating
    emotion");
38            Emotion newEmotion = new Emotion(request.getUuid(), request.
    getUserEuphoriaBored(), request.getUserAnxietyCalm());
              //update associated POIs
40            updatePOIsOfUpdatedEmotion(emotion, newEmotion, request.
    getLatitude(), request.getLongitude());
              //update emotion
42            emotion.setAnxietyCalm(newEmotion.getAnxietyCalm());
              emotion.setEuphoriaBored(newEmotion.getEuphoriaBored());
44            emotion.setUpdatedAt(newEmotion.getUpdatedAt());
              daoEmotion.update(emotion);
46          }
          } catch (Exception e) {
48          e.printStackTrace();
          }
50      }
      }, 0, TimeUnit.MILLISECONDS);
52  }

54    (...)

56 }
```

Here, we receive a *RequestSetEmotion* (line 15) and create a new task for our *EmotionListener's* scheduler, to be run as soon as possible (line 16). First, we attempt to find an emotion in the database that corresponds to the request's UUID, through the *DaoEmotion's find* method (line 21). If no such emotion was found, this is a new emotional record that is consequently created and whose associated POI's are found and handled by *updatePOIsOfAddedEmotion* (lines 24–34). If an emotion with the request's UUID is already present, we use *updatePOIsOfUpdatedEmotion* to update the necessary POIs and then use *DaoEmotion's update* method to insert the new information (lines 36–46).

Our finished *ComEmotions* is presented below:

```
package com;
2
import java.util.Iterator;
4 import java.util.List;
import java.util.concurrent.TimeUnit;
6
import org.hibernate.HibernateException;
8
import utilities.EmotionListener;
10 import utilities.GlobalVariables;
import dao.DaoEmotion;
12 import dao.DaoPointofinterest;
import hibernateMaps.Emotion;
14 import hibernateMaps.Pointofinterest;
import model.RequestSetEmotion;
16

18 public class ComEmotions {
  /**
20   * Processes a request to set new Emotion information. In order to avoid
     possible conflicts with Pruning tasks, this task is scheduled on a
    public scheduler
   *
22   * @param request
   */
24  public static void setEmotionFromRequest(final RequestSetEmotion request
    ) {
    EmotionListener.scheduler.schedule(new Runnable() {
26      @Override
      public void run() {
28        try {
          DaoEmotion daoEmotion = new DaoEmotion();
30          Emotion emotion = daoEmotion.find(request.getUuid());

32          //If we do not have emotional information from that uuid...
          if (emotion == null) {
34            System.out.println("uuid: " + request.getUuid() + " is new,
    creating emotion");
            emotion = new Emotion(request.getUuid(), request.
    getUserEuphoriaBored(), request.getUserAnxietyCalm());
36            try {
              daoEmotion.create(emotion);
38              //update POIs within the range of the uuid location
              updatePOIsOfAddedEmotion(emotion, request.getLatitude(),
    request.getLongitude());
40            } catch (HibernateException e) {
              System.out.println("Error: Internal error in DB");
```

```
42              }
            }
44          //If this is updated information about an emotion...
            else {
46            System.out.println("uuid: " + request.getUuid() + ", updating
    emotion");
              Emotion newEmotion = new Emotion(request.getUuid(), request.
    getUserEuphoriaBored(), request.getUserAnxietyCalm());
48            //update associated POIs
              updatePOIsOfUpdatedEmotion(emotion, newEmotion, request.
    getLatitude(), request.getLongitude());
50            //update emotion
              emotion.setAnxietyCalm(newEmotion.getAnxietyCalm());
52            emotion.setEuphoriaBored(newEmotion.getEuphoriaBored());
              emotion.setUpdatedAt(newEmotion.getUpdatedAt());
54            daoEmotion.update(emotion);
            }
56        } catch (Exception e) {
            e.printStackTrace();
58        }
        }
60    }, 0, TimeUnit.MILLISECONDS);
    }
62
    /**
64   * Finds the POIs currently associated with an emotion and
     * removes the corresponding influences.
66   *
     * @param emo
68   */
    public static void updatePOIsOfRemovedEmotion(Emotion emo) {
70    List<Pointofinterest> poisToUpdate = null;
      DaoEmotion demo = new DaoEmotion();
72    DaoPointofinterest daopoi = new DaoPointofinterest();
      poisToUpdate = demo.getAssociatedPOIs(emo.getUuid());
74    Iterator<Pointofinterest> poisToUpdateit = poisToUpdate.iterator();

76    while (poisToUpdateit.hasNext())
      {
78      Pointofinterest poi =poisToUpdateit.next();
        daopoi.update(removeEmotionValFromPOI(poi, emo));
80    }
    }
82
    /**
84   * Finds, associates and updates the appropriate POIs with a new emotion
     .
     *
86   * @param emo
     * @param latitude
88   * @param longitude
     */
90   public static void updatePOIsOfAddedEmotion(Emotion emo, Double
     latitude, Double longitude) {
     List<Pointofinterest> poisToUpdate = null;
92   DaoPointofinterest daopoi = new DaoPointofinterest();
     poisToUpdate = daopoi.getPoibyLocation(
94       latitude, longitude,
         GlobalVariables.EMOTION_AFFECT_POI_MILE_RADIUS);
```

```
96      Iterator<Pointofinterest> poisToUpdateit = poisToUpdate.iterator();

98      while(poisToUpdateit.hasNext())
        {
100       Pointofinterest poi = poisToUpdateit.next();
          daopoi.update(addEmotionValToPOI(poi, emo));
102       daopoi.addEmotion(poi, emo.getUuid());
        }
104   }

106   /**
       * Updates the POIs of associated with an Emotion. The two Emotion
      objects should represent two states of the same emotion - outdated
      state and up-to-date state.
108    * Thus, they should have the same uuid.
       *
110    * @param oldEmo - object containing the values of the outdated
      emotion
       * @param newEmo - object containing new emotional values
112    * @param latitude - the emotion's latest latitude
       * @param longitude - the emotion's latest longitude
114    */
      public static void updatePOIsOfUpdatedEmotion(Emotion oldEmo, Emotion
      newEmo, Double latitude, Double longitude) {
116     DaoEmotion demo = new DaoEmotion();
        DaoPointofinterest daopoi = new DaoPointofinterest();
118   List<Pointofinterest> currentPOIs = demo.getAssociatedPOIs(oldEmo.
      getUuid());
      List<Pointofinterest> newPOIs = daopoi.getPoibyLocation(
120       latitude, longitude,
          GlobalVariables.EMOTION_AFFECT_POI_MILE_RADIUS);
122   Iterator<Pointofinterest> newPOISit = newPOIs.iterator();
      Iterator<Pointofinterest> currentPOIsit;
124
      while(newPOISit.hasNext())
126   {
        Pointofinterest newpoi = newPOISit.next();
128     /*
         * Using List.contains method does not work in this case
130      * Thus, we must individually compare the ids.
         */
132     boolean isACurrentPOI = false;
        //reset iterator
134     currentPOIsit = currentPOIs.iterator();
        while(currentPOIsit.hasNext()) {
136       Pointofinterest currentPOI = currentPOIsit.next();
          if (currentPOI.getIdPoi().equals(newpoi.getIdPoi())) {
138         /*
             * Remove POI from the currentPOIs list.
140          * This is important, since we will be updating the remaining
       currentPOIs (those that are no longer associated with this emotion.)
             */
142         currentPOIsit.remove();
            isACurrentPOI = true;
144         break;
          }
146     }

148     if (isACurrentPOI) {
```

```
          System.out.println("CurrentPOIs contains " +newpoi.getIdPoi());
150       //Ok, this POI was already in our list of associated POIs, update
        its means remove old value and recalculate mean with the new one
          Double newEmotionAnxietyCalm;
152       Double newEmotionEuphoriaBored;

154       if (newpoi.getEmotionAnxietyCalm() != null && newpoi.
        getEmotionAnxietyCalm().compareTo(0d)>0) {
            newEmotionAnxietyCalm = ((newpoi.getEmotionAnxietyCalm()*newpoi.
        getAttendance() - oldEmo.getAnxietyCalm().doubleValue()) + newEmo.
        getAnxietyCalm().doubleValue())/newpoi.getAttendance();
156       }
          else {
158         System.err.println(newpoi.getIdPoi() + " has null or 0
        AnxietyCalm emotion values!");
            newEmotionAnxietyCalm = newEmo.getAnxietyCalm().doubleValue();
160       }
          if (newpoi.getEmotionEuphoriaBored() != null && newpoi.
        getEmotionEuphoriaBored().compareTo(0d)>0) {
162         newEmotionEuphoriaBored = ((newpoi.getEmotionEuphoriaBored()*
        newpoi.getAttendance() - oldEmo.getEuphoriaBored().doubleValue()) +
        newEmo.getEuphoriaBored().doubleValue())/newpoi.getAttendance();
          }
164       else {
            System.err.println(newpoi.getIdPoi() + " has null or 0
        EuphoriaBored emotion values!");
166         newEmotionEuphoriaBored = newEmo.getEuphoriaBored().doubleValue
        ();
          }
168
          newpoi.setEmotionAnxietyCalm(newEmotionAnxietyCalm);
170       newpoi.setEmotionEuphoriaBored(newEmotionEuphoriaBored);
          daopoi.update(newpoi);
172     }
        else {
174       System.out.println("CurrentPOIs does NOT contain " +newpoi.
        getIdPoi());
          //well, this POI is new to this emotion, so let us consider
        recalculate its means accordingly
176       //the attendance increases by 1
          daopoi.update(addEmotionValToPOI(newpoi, newEmo));
178       //Ok, now we create the association between the newEmotion and the
         new POI
          daopoi.addEmotion(newpoi, newEmo.getUuid());
180     }
      }
182   /*
       * Now we should have iterated through all the newPOIs. Those
      remaining on the currentPOI list are old ones that should have their
      associations removed and means updated accordingly
184    */
      //reset the iterator
186   currentPOIsit = currentPOIs.iterator();

188   while(currentPOIsit.hasNext())
      {
190     Pointofinterest oldpoi =currentPOIsit.next();
        daopoi.update(removeEmotionValFromPOI(oldpoi, oldEmo));
192     daopoi.removeEmotion(oldpoi,newEmo.getUuid());
```

```
194     }
        }

196     /**
         * This method adds the influence of an emotion to a poi's
198      * attendance, EmotionAnxietyCalm and EmotionEuphoriaBored values.
         *
200      * @param poi - POI to be updated
         * @param emo - Emotion to be added
202      *
         * @return updated POI
204      */
        private static Pointofinterest addEmotionValToPOI(Pointofinterest poi,
         Emotion emo){
206     Double newEmotionAnxietyCalm;
        Double newEmotionEuphoriaBored;
208
        //increase attendance by 1
210     int newAttendance = poi.getAttendance()+1;

212     //avoid nulls and check if emotion values are greater than 0 to avoid
        unnecessary computation
        if (poi.getEmotionAnxietyCalm() != null && poi.getEmotionAnxietyCalm()
        .compareTo(0d)>0) {
214       newEmotionAnxietyCalm = (poi.getEmotionAnxietyCalm()*poi.
        getAttendance() + emo.getAnxietyCalm().doubleValue())/newAttendance;
        }
216     else {
          newEmotionAnxietyCalm = emo.getAnxietyCalm().doubleValue();
218     }

220     if (poi.getEmotionEuphoriaBored() != null && poi.
        getEmotionEuphoriaBored().compareTo(0d)>0) {
          newEmotionEuphoriaBored = (poi.getEmotionEuphoriaBored()*poi.
        getAttendance() + emo.getEuphoriaBored().doubleValue())/newAttendance;
222     }
        else{
224       newEmotionEuphoriaBored = emo.getEuphoriaBored().doubleValue();
        }
226
        poi.setEmotionAnxietyCalm(newEmotionAnxietyCalm);
228     poi.setEmotionEuphoriaBored(newEmotionEuphoriaBored);
        poi.setAttendance(newAttendance);
230
        return poi;
232     }

234     /**
         * This method removes the influence of an emotion from a poi's
236      * attendance, EmotionAnxietyCalm and EmotionEuphoriaBored values.
         *
238      * @param poi - POI to be updated
         * @param emo - Emotion to be removed
240      *
         * @return updated POI
242      */
        private static Pointofinterest removeEmotionValFromPOI(Pointofinterest
         poi, Emotion emo){
244       Double newEmotionAnxietyCalm;
```

```
        Double newEmotionEuphoriaBored;
246     int newAttendance;

248     //If attendance is greater than one, decrease it and perform
        calculations
        if (poi.getAttendance()>1) {
250       newAttendance = poi.getAttendance()-1;
          newEmotionAnxietyCalm = (poi.getEmotionAnxietyCalm()*poi.
        getAttendance() - emo.getAnxietyCalm().doubleValue())/newAttendance;
252       newEmotionEuphoriaBored = (poi.getEmotionEuphoriaBored()*poi.
        getAttendance() - emo.getEuphoriaBored().doubleValue())/newAttendance;
        }
254     //If not, then attendance and emotions values are necessarily 0
        else {
256       newAttendance = 0;
          newEmotionAnxietyCalm = 0d;
258       newEmotionEuphoriaBored = 0d;
        }
260     poi.setEmotionAnxietyCalm(newEmotionAnxietyCalm);
        poi.setEmotionEuphoriaBored(newEmotionEuphoriaBored);
262     poi.setAttendance(newAttendance);

264     return poi;
        }
266 }
```

With this, we have finished our processing of incoming emotions. Let us now focus on pruning outdated emotions from the database.

9.1.5 Pruning Outdated Emotions

As we previously discussed, pruning outdated emotions is a periodic task. How often should it be performed? For this tutorial, we are aiming at providing near real-time emotional information to HappyWalk's users. As such, we choose the arbitrary lifespan of **5 minutes** as the time to live of emotional data, which can be regarded as a reasonable compromise between accuracy and performance. Lifespan is translated into the *GlobalVariables.OUTDATED_EMOTION_MILLIS* variable, which the reader is welcome to change and experiment with.

Let us create a new class named *PruneTask* within the *utilities* package. By using a combination of the *DaoEmotion's* and *ComEmotions'* helper methods, the implementation of this pruning class becomes trivial:

```
    package utilities;
2
    import hibernateMaps.Emotion;
4
    import java.util.Date;
6   import java.util.ArrayList;
    import java.util.List;
8
    import com.ComEmotions;
10
    import utilities.GlobalVariables;
12  import dao.DaoEmotion;

14
```

```
public class PruneTask implements Runnable {
  @Override
  public void run() {
    System.out.println("Refreshing and Purging Data...");
    long current = new Date().getTime();

    List<Emotion> listemotions = new ArrayList<Emotion>();
    try {
      DaoEmotion emodao = new DaoEmotion();
      listemotions = emodao.findAll();

      current-= GlobalVariables.OUTDATED_EMOTION_MILLIS;
      for(int i=0;i<listemotions.size();i++)
      {
        Emotion emo=listemotions.get(i);
        if(emo.getUpdatedAt().getTime()<current){
          //Emotion is outdated, update POI values
          ComEmotions.updatePOIsOfRemovedEmotion(emo);
          //remove emotion
          emodao.delete(emo);
          System.out.println("Emotion uuid: "+emo.getUuid()+" is too old
    so it has been deleted!");
        }
      }

    } catch (Exception e) {
      e.printStackTrace();
    }
  }
}
```

First, we get the current time (line 19), save it into a variable (*long current*) and subtract the value of *GlobalVariables.OUTDATED_EMOTION_MILLIS* (line 26). Then, we iterate over all available emotions and verify if they have expired. For those whose lifespan has reached its end, we call *ComEmotions.updatePOIsOfRemovedEmotion* and subsequently remove them (lines 27–37).

To finalize, we need to actually schedule our *PruneTask*. This can be done within *EmotionListener*, right after our scheduler is instantiated:

```
package utilities;

import java.util.concurrent.Executors;
import java.util.concurrent.ScheduledExecutorService;
import java.util.concurrent.TimeUnit;

import javax.servlet.ServletContextEvent;
import javax.servlet.ServletContextListener;

public class EmotionListener implements ServletContextListener {
  /*
   * public scheduler that will execute all tasks related
   * with Emotions.
   */
  public static ScheduledExecutorService scheduler;

  @Override
  public void contextInitialized(ServletContextEvent event) {
    scheduler = Executors.newSingleThreadScheduledExecutor();
```

```
        scheduler.scheduleAtFixedRate(
21              new PruneTask(), 0, GlobalVariables.OUTDATED_EMOTION_MILLIS,
        TimeUnit.MILLISECONDS);
    }
23
    @Override
25  public void contextDestroyed(ServletContextEvent event) {
        scheduler.shutdownNow();
27  }
}
```

With this, our server should now be ready to process emotional information. Test it out by running it as it was previously explained on page ???. Check the console log for any glaring exceptions (there should be none, but some warnings and information messages are to be expected). The console should also display the message *Refreshing and Purging Data...* once during the server's startup, and every *GlobalVariables.OUTDATED_EMOTION_MILLIS* milliseconds thereafter.

In the next section we will resume our work on the Android application and finish its *EmotionTasker*.

9.2 Finishing up *EmotionTasker*

In this section we will finish up the *EmotionTasker*, by handling the output of the ANN (Section 9.2.1) and posting new emotion inferences (Section 9.2.2). Before we proceed, as a first step, let us consider when to use the results of our neural network instead of asking the user for feedback. One possibility is to use our previous *wMeanEuclideanDistance* value; as the reader may recall from page ???, *wMeanEuclideanDistance* represents an averaged sum of the distance between previous emotional classifications and the user's feedback. This directly relates to how accurate our neural network has been behaving. Thus, we can define a simple rule that states that our neural network results will only be considered if *wMeanEuclideanDistance* is greater than a certain percentage of the maximum Euclidean distance:

```
package hitlexamples.happywalk.service;
2
import (...)
4
public class EmotionTasker {
6
    (...)
8
    class EmotionRecognitionTask implements Runnable{
10          private double[] outputs;
            private double[] inputs;
12
            @Override
14          public void run() {
                //Only run if we have location, since we need it for the
        neural net!
16              if (hWServ.getHwLocationListener().getActualposition() != null
        ) {
                    try {
18                      /* Check if it is time to request user feedback */
```

```
                if((lastEmoFeedbackReq + baseTimeToNextEmoFdbckReq) <
    System.currentTimeMillis()) {
20                  fetchInputsAndCompute();
                    //only fire notification if the service is still
    running!
22                  if (hWServ.isRunning()) {
                        showEmotionFeedbackNotification(inputs,
    outputs);
24                      postNotificationRemovalTask();
                    }
26              } else {
                    /* first, check if our neural network has been
    behaving well enough for us to consider its output */
28                  float maxEuclideanDistance = (float) Math.sqrt(
    GlobalVariables.NN_OUTPUTS);
                    if (wMeanEuclideanDistance < GlobalVariables.
    MARGIN_PERCNT_MAX_EUCLD_DIST_ACCPT * maxEuclideanDistance) {
30                      //TODO: here, we will use our NN output
                    }
32              }
            } catch (Exception e) {
34              e.printStackTrace();
            }
36      }
        //TODO: post a new emotion inference
38  }

40  private void postNotificationRemovalTask() {
        (...)
42  }

44  private void fetchInputsAndCompute() throws NoCurrentPosition {
        (...)
46  }
}
```

First, we verify if the *HwLocationListener* has location information (line 16); if not, there is no point in performing emotion recognition, since we cannot feed location information to the neural network. We perform this verification here even when *collectInputs()* throws an exception with this exact purpose (as seen on page ???). This is because it is bad coding practice to rely on Java exceptions to control the execution flow. Thus, *collectInputs()* only throws a *NoCurrentPosition* exception in the event that, owing to some unexpected circumstance, we lose location information just before collecting sensor data.

We then check if it is time to request user feedback, by checking if the current time is greater than the sum of the time when we last sent a feedback request plus *baseTimeToNextEmoFdbckReq* (line 19). In case it is, we fetch sensory inputs and compute a result using the ANN (line 20). We also display a notification asking for user feedback and its respective notification removal task (see page ???). However, we only fire the notification in case our service is still running (lines 22–25). This is important, since the user may have closed the application while we were still acquiring sensory data. If this is the case, *EmotionRecognitionTask* would be completely oblivious to this fact. Firing notifications while our background service is not running is asking for trouble. If it is not yet time to ask for user feedback, we determine whether or not to use our neural network results by resorting to the rule discussed above (lines 26–32).

The *TODO* comments above (lines 30 and 37) denote the two tasks that we still need to tackle. First, we need to handle our ANN output and find a way to make it useful to the user. Second, we need to figure out how to post new emotion inference tasks. We will consider each of these challenges in their own subsections.

9.2.1 Handling ANN Output

In case our neural network has been behaving well, we need to act upon its result. As the reader may remember from Section 5.1, the objective here is to show a notification that suggests walking exercise whenever the data from the smartphone's sensors indicates a negative state of mind. To do so, we will first need to determine if the ANN's output represents a negative emotion or not. We will then need to implement a second type of notification, different from the one used for emotional feedback. Whenever a negative emotion is detected this new notification shall, instead of requesting feedback, suggest walking exercise, and show our map.

Let us first determine the type emotion our ANN output represents:

```
1  package hitlexamples.happywalk.service;

3  import (...)

5  public class EmotionTasker {

7    (...)

9    public static int getTypeOfEmotion(double[] emotionArray) {
           /*
11         for code readability, let us use an (x,y) representation of the
           emotion color map:
13
           (0.0)____________
15          y |      Eph      |
              |               |
17            |Anx        Clm |
              |               |
19            |____Brd_____|
                          x   (1,1)
21          */

23         double y = emotionArray[GlobalVariables.
     NN_OUTPUT_ARRAY_INDEX_EUPHORIC_BORED];
           double x = emotionArray[GlobalVariables.
     NN_OUTPUT_ARRAY_INDEX_ANXIOUS_CALM];
25         int typeOfEmotion;

27         if (y < 0.5 && x < 0.5) {
               if (y>x){
29                 //anxiety
                   typeOfEmotion = GlobalVariables.EMOTION_ANXIETY;
31             }
               else {
33                 //euphoria
                   typeOfEmotion = GlobalVariables.EMOTION_EUPHORIA;
35             }
           }
37         else if (y < 0.5 && x >= 0.5)
```

```
39          {
                if (y>x){
41                  //calmness
                    typeOfEmotion = GlobalVariables.EMOTION_CALMNESS;
43              }
                else {
45                  //euphoria
                    typeOfEmotion = GlobalVariables.EMOTION_EUPHORIA;
47              }
            }
49          else if (y >= 0.5 && x < 0.5)
            {
51              if (y>x){
                    //anxiety
53                  typeOfEmotion = GlobalVariables.EMOTION_ANXIETY;
                }
55              else {
                    //boredom
57                  typeOfEmotion = GlobalVariables.EMOTION_BOREDOM;
                }
59          }
            else
61          {
                if (y>x){
63                  //boredom
                    typeOfEmotion = GlobalVariables.EMOTION_BOREDOM;
65              }
                else {
67                  //calmness
                    typeOfEmotion = GlobalVariables.EMOTION_CALMNESS;
69              }
            }
71          return typeOfEmotion;
        }

73      /**
         * checks if an emotion is "negative"
75       * @param typeOfEmotion an int representing the type of emotion
         * @return true if emotion is negative / false if it isnt
77       */
        private boolean emotionIsNegative(int typeOfEmotion) {
79          boolean emotionIsNegative = false;
            for (int i = 0; i<GlobalVariables.NEGATIVE_EMOTIONS.length; i++) {
81              if (typeOfEmotion == GlobalVariables.NEGATIVE_EMOTIONS[i]) {
                    emotionIsNegative = true;
83                  break;
                }
85          }
            return emotionIsNegative;
87      }

89      (...)
    }
```

This can be done through the above *EmotionTasker's* helper methods *getTypeOfEmotion()* (lines 9–71) and *emotionIsNegative()* (lines 78–87).

As we can see by the comment in lines 10–21, the method *getTypeOfEmotion()* determines the position of the neural network's output within the coordinates of our *EmotionSpace* view (refer to Figure 8.1) and returns a globally defined integer value representative of the corresponding emotion (lines 27–69). As for *emotionIsNegative()*, it makes use of a list of negative emotions kept within the *GlobalVariables.NEGATIVE_EMOTIONS* array (line 81).

Note that, while *emotionIsNegative()* is a private method, *getTypeOfEmotion()* is a public and static method. This is because we will reuse this method later on.

The notification to be shown in this case is slightly different from the emotion feedback notification we previously implemented. Let us define a new string resource to contain this "normal emotion notification" message in the proper XML resource file, *src/main/res/values/strings.xml*:

```
<resources>
2     <string name="app_name">HappyWalk</string>

4     (...)

6     <!-- Notifications -->
      <string name="serviceNotifContent">is walking with you</string>
8     <string name="emotionFeedbackNotifContent">Can you tell me how you
      feel?</string>
      <string name="emotionNormalNotifContent">How are you feeling? What
      about a walk?</string>
10
</resources>
```

The newly defined *emotionNormalNotifContent* resource now contains the sentence *How are you feeling? What about a walk?* We can use this resource to create a "normal" notification through a new method, *showNormalEmotionNotification()*:

```
1  package hitlexamples.happywalk.service;

3  import hitlexamples.happywalk.activities.MapsActivity;
   import (...)
5
   public class EmotionTasker {
7
     (...)
9
       private void showNormalEmotionNotification(int typeOfEmotion) {
11         //First, cancel previous notification removal tasks
           hWServiceHandler.removeCallbacks(currentNotifRemovTask);
13         //Now, prepare a Bundle with the necessary information
           Bundle bnd = new Bundle();
15         /* put a timestamp on this bundle to avoid the user clicking
       notifications that have been fired a long time ago. */
           lastEmotionNotifMillis = System.currentTimeMillis();
17         bnd.putLong(GlobalVariables.BND_EXTRA_EMOTION_TIMESTAMP_KEY,
                   lastEmotionNotifMillis);
19         bnd.putInt(GlobalVariables.BND_EXTRA_EMOTION_TYPE_NOTIF_KEY,
       typeOfEmotion);
           bnd.putInt(GlobalVariables.BND_EXTRA_REQ_CODE_KEY,
21                 GlobalVariables.AREQ_EMOTION_NORMAL_NOTIF);

23         // We will show our map, to promote walking when emotions are
       negative
           Intent intent = new Intent(hWServ, MapsActivity.class);
```

```
25          intent.putExtras(bnd);

27          PendingIntent resultPendingIntent =
                    PendingIntent.getActivity(
29                          hWServ,
                            0,
31                          intent,
                            PendingIntent.FLAG_UPDATE_CURRENT
33                  );

35          NotificationCompat.Builder mNotifyBuilder = new NotificationCompat
        .Builder(hWServ)
                    .setTicker(hWServ.getResources().getString(R.string.
        app_name) + " " + hWServ.getResources().getString(R.string.
        emotionNormalNotifContent))
37                  .setContentTitle(hWServ.getResources().getString(R.string.
        app_name))
                    .setContentText(hWServ.getResources().getString(R.string.
        emotionNormalNotifContent))
39                  .setSmallIcon(R.drawable.emot_notif_icon)
                    .setContentIntent(resultPendingIntent)
41                  .setOngoing(true)
                    .setSound(RingtoneManager.getDefaultUri(RingtoneManager.
        TYPE_NOTIFICATION));
43
            hWServ.getNotificationManager().notify(
45                  hWServ.hHNotificNum,
                    mNotifyBuilder.build());
47      }

49    (...)
  }
```

This method is very similar to *showEmotionFeedbackNotification()*, previously discussed on page ???; however, it displays the new message *How are you feeling? What about a walk?*, as defined by the *R.string.emotionNormalNotifContent* resource. Notice that this notification also calls the *MapsActivity* class instead of *EmotionFeedback*. This is because our intention is to show a map to the user, with interesting POIs that serve as possible destinations for his walking exercise. We will further customize this motivational action through a dialog box in Section 9.3.

Let us now use this new method to complete the associated logic within *EmotionRecognitionTask*:

```
  package hitlexamples.happywalk.service;
2
  import (...)
4
  public class EmotionTasker {
6
    (...)
8
    class EmotionRecognitionTask implements Runnable{
10          private double[] outputs;
            private double[] inputs;
12
            @Override
14          public void run() {
```

```
            //Only run if we have location, since we need it for the
neural net!
16          if (hWServ.getHwLocationListener().getActualposition() != null
) {
                try {
18                  /* Check if it is time to request user feedback */
                    if((lastEmoFeedbackReq + baseTimeToNextEmoFdbckReq) <
System.currentTimeMillis()) {
20                      fetchInputsAndCompute();
                        //only fire notification if the service is still
running!
22                      if (hWServ.isRunning()) {
                            showEmotionFeedbackNotification(inputs,
outputs);
24                          postNotificationRemovalTask();
                        }
26                  } else {
                        /* first, check if our neural network has been
behaving well enough for us to consider its output */
28                      float maxEuclideanDistance = (float) Math.sqrt(
GlobalVariables.NN_OUTPUTS);
                        if (wMeanEuclideanDistance < GlobalVariables.
MARGIN_PERCNT_MAX_EUCLD_DIST_ACCPT * maxEuclideanDistance) {
30                          fetchInputsAndCompute();
                            //fire a regular emotion notification, in case
 emotion is negative
32                          int typeOfEmotion = getTypeOfEmotion(outputs);

                            //only fire notification if the service is
still running!
34                          if (hWServ.isRunning()) {
                                if (emotionIsNegative(typeOfEmotion)) {
36                                  showNormalEmotionNotification(
typeOfEmotion);
                                    postNotificationRemovalTask();
38                              }
                                sendEmotionToServer(outputs);
40                          }
                        }
42                  }
                } catch (Exception e) {
44                  e.printStackTrace();
                }
46          }
            //TODO: post a new emotion inference
48      }

50      private void postNotificationRemovalTask() {
            (...)
52      }

54      private void fetchInputsAndCompute() throws NoCurrentPosition {
            (...)
56      }
}
```

The code between lines 30–40 completes the first *TODO* comment we had left in the code from page ???.

We begin by fetching sensory inputs and computing a result (line 30). If we are going to use our ANN's output, we first determine which type of emotion it represents (line 32). In case our service is still running and the emotion is negative, a notification motivating the user to go for a walk is shown, together with a removal task which will revert it after its expiration time has passed (lines 34–38). Regardless of the type of emotion that was detected, the result is sent to the server in order to update POI information (line 39).

Now that we can handle the ANN's output, let us consider the issue of posting New Emotion Inferences.

9.2.2 Posting New Emotion Inferences

On this section we will be looking at handling the remaining *TODO* task from line 37 back on page ???. One may be tempted to think that posting new emotion inferences should be simple enough. After all, we have previously posted runnable tasks using the HappyWalkService's handler. That was the case when we posted new *NeuralNetworkTrainingTasks* and *UpdateEmotionAccuracyTasks* back in Section 8.3. However, using the same approach for *EmotionRecognitionTasks* may cause problems.

EmotionRecognitionTasks differ from other tasks in the sense that they are to be executed periodically, with large time intervals (as we defined back on page ??, once or twice an hour). In theory, we could use our handler's *postDelayed* method to run this task whenever we saw fit. On the other hand, to avoid draining the battery, most Android devices that are left idle for few minutes go into a sleep state where the CPU is inactive. Unfortunately for us, Android handlers cannot guarantee their operation in this condition. Instead of running after the desired time period, handlers will resume their operation whenever the device awakes again. This defeats the purpose of background emotion inference and of our dynamic feedback strategy, since we would only be able to acquire information and fire notifications whenever the user handled the device.

Fortunately, Android offers a way of submitting cron jobs to be executed at specific times. This may be achieved through a combination of Android's *AlarmManager*, a *WakefulBroadcastReceiver*, an *IntentService*, and *Wakelocks*. In this section, we will use each of these elements to schedule our *EmotionRecognitionTasks*.

The *AlarmManager* is a class that provides access to the system alarm services. Alarms are the basis for what is used for Android's alarm clocks: they must be run at specific times, independently of the sleep state of the device. Alarms are also associated with *intents* that are processed whenever they go off. Therefore, our objective is to define a class that receives these intents and will, thereafter, initiate the process of starting a new *EmotionRecognitionTask*.

In this book we will be using a *WakefulBroadcastReceiver*, a special type of *BroadcastReceiver*. *BroadcastReceivers* are classes specifically dedicated to receiving and processing intents. A *WakefulBroadcastReceiver* is special in the sense that it prevents a device from going back to sleep during this processing. It does so through the use of *Wakelocks*.

Wakelocks are objects that indicate that the application needs to have the device's CPU to stay on. We could, theoretically, use a wakelock to *always* have the device on while HappyWalk was running. However, as the reader may imagine, this would be a highly effective way of killing the device's battery very quickly. Therefore, wakelocks should be used sparingly.

BroadcastReceivers do not usually perform work by themselves. Their job is to pass the work on to a service. We will be using an *IntentService,* whose purpose is to handle single asynchronous intent requests and stop themselves after their work is done. Thus, our *IntentService* will interface with our *EmotionTasker* to post a new *EmotionRecognitionTask*.

We will approach this challenge by first acquiring the necessary permissions to use wakelocks. Then, we will create our own extension of a *WakefulBroadcastReceiver,* which will start our *IntentService.* This *IntentService* will acquire a connection to our main *HappyWalkService,* request a new emotion inference, and release the *WakefulBroadcastReceiver's* wakelock. We will then turn our attention back to the *EmotionTasker* class, where we will acquire a new wakelock and implement a method that handles the posting of a new *EmotionRecognitionTask*. Then, we will finalize *EmotionRecognitionTask* by allowing it to determine when it should be run again, making it schedule the next emotion inference, and release the wakelock. Finally, we conclude our *EmotionTasker* by kickstarting the emotion recognition process during initialization and stopping it when the app is closed.

Before we continue, let us first give our app the ability to use wakelocks. Edit the file *happywalk/app/src/main/AndroidManifest.xml* and add the following permission:

```
1  <?xml version="1.0" encoding="utf-8"?>
   <manifest xmlns:android="http://schemas.android.com/apk/res/android"
3      xmlns:tools="http://schemas.android.com/tools"
       package="hitlexamples.happywalk" >
5
       <uses-permission android:name="android.permission.INTERNET" />
7      <uses-permission android:name="android.permission.ACCESS_NETWORK_STATE
       " />
       <uses-permission android:name="android.permission.
       WRITE_EXTERNAL_STORAGE" />
9      <uses-permission android:name="com.google.android.providers.gsf.
       permission.READ_GSERVICES" />
       <!-- Permission to use the wakelock -->
11     <uses-permission android:name="android.permission.WAKE_LOCK" />

13     (...)

15 </manifest>
```

Now, let us create a new class under the package *hitlexamples.happywalk.service* and name it *EmotionWakefulReceiver*. It shall extend the *WakefulBroadcastReceiver* class to process the appropriate incoming intents:

```
1  package hitlexamples.happywalk.service;

3  import android.content.Context;
   import android.content.Intent;
5  import android.support.v4.content.WakefulBroadcastReceiver;

7  public class EmotionWakefulReceiver extends WakefulBroadcastReceiver {
       public static final int REQUEST_CODE = 33245;
9
       @Override
11     public void onReceive(Context receiveContext, Intent intent) {
           // Call upon the EmotionIntentService.
```

```
13          Intent service = new Intent(receiveContext, EmotionIntentService.
        class);

15          // Start the service, holding a Wakelock
            startWakefulService(receiveContext, service);
17      }
    }
```

We first define a public static *REQUEST_CODE,* which is a way of identifying our intents (line 8). Then, we override the *onReceive* method (lines 10 and 11), where we start a wakeful *EmotionIntentService* which we still need to implement (lines 12–16).

Create yet another class under the package *hitlexamples.happywalk.service* and name it *EmotionIntentService.* This class shall extend the *IntentService* class:

```
   package hitlexamples.happywalk.service;
 2
   import android.app.IntentService;
 4 import android.content.ComponentName;
   import android.content.Context;
 6 import android.content.Intent;
   import android.content.ServiceConnection;
 8 import android.os.IBinder;

10 public class EmotionIntentService extends IntentService{

12     private ServiceConnection hwConnection;
       private HappyWalkService hWService;
14     private Intent incomingIntent;

16     public EmotionIntentService() {
           super("EmotionIntentService");
18         hwConnection = new ServiceConnection() {
               public void onServiceConnected(ComponentName className,
       IBinder service) {
20                 hWService = ((HappyWalkService.HappyWalkBinder) service).
       getService();
                   //The service shall begin a new Emotion Recognition task
22                 hWService.getEmotionTasker().postEmotionRecognitionTask();

                   /* Since the EmotionTasker acquires its own wakelock, we
       no longer need EmotionWakefulReceiver to hold one for us. */
24                 EmotionWakefulReceiver.completeWakefulIntent(
       incomingIntent);
                   // Our work is done.
26                 unbindService(hwConnection);
                   hWService = null;
28             }

30             @Override
               public void onServiceDisconnected(ComponentName componentName)
        {
32                 unbindService(hwConnection);
                   hWService = null;
34                 EmotionWakefulReceiver.completeWakefulIntent(
       incomingIntent);
               }
36         };
```

```
38      }

40      @Override
        protected void onHandleIntent(Intent intent) {
42          /* This is the intent that we will need to pass
            to EmotionWakefulReceiver, in order to release its wakelock */
44          incomingIntent = intent;

46          //The main HappyWalk service should already be running, we simply
        need to bind
            bindService(new Intent(this, HappyWalkService.class), hwConnection
        , Context.BIND_AUTO_CREATE);
48      }
    }
```

In its constructor, we instantiate a *ServiceConnection hwConnection* object (line 18), similar to what we previously did for *EmotionFeedback* on page ???. This object connects us to our *HappyWalkService* (line 20), from which we can get a reference to *EmotionTasker* to post a new *EmotionRecognitionTask* (line 22). We will do so through a *postEmotionRecognitionTask()* method, which we will implement next.

The objective of *postEmotionRecognitionTask()* will be to pass on the work to our *HappyWalkService* handler, which runs it on HappyWalk's background thread. The problem here is that, from our current thread's perspective, we don't really know when the work will be complete. Therefore, we should make sure that *EmotionTasker* acquires its own wakelock. As such, for now, we can tell our *EmotionWakefulReceiver* to release its wakelock through the *completeWakefulIntent()* method (line 24). Afterwards, *EmotionIntentService's* work is complete.

Now, we just need to actually attempt to bind to the *HappyWalkService*. We do this by overriding the *onHandleIntent()* method, which is called whenever a new intent is received (lines 40–48).

The only thing left to do is to let the application know that our new receiver and service exist. To do so, we need to edit the file *happywalk/app/src/main/AndroidManifest.xml* again:

```
    <?xml version="1.0" encoding="utf-8"?>
2   <manifest xmlns:android="http://schemas.android.com/apk/res/android"
        xmlns:tools="http://schemas.android.com/tools"
4       package="hitlexamples.happywalk" >

6           (...)

8           <service
                android:name=".service.HappyWalkService"
10              android:enabled="true"
                android:exported="false" >
12          </service>

14          <receiver
                android:name=".service.EmotionWakefulReceiver">
16          </receiver>

18          <service
                android:name=".service.EmotionIntentService"
20              android:exported="false" >
```

```
        </service>
22
        (...)
24
    </application>
26
</manifest>
```

Now that our *EmotionWakefulReceiver* and *EmotionIntentService* are implemented, we can focus back our attention on *EmotionTasker*. Let us begin by implementing the *postEmotionRecognitionTask()* method:

```
1   package hitlexamples.happywalk.service;

3   import android.os.PowerManager;
    import (...)
5
    public class EmotionTasker {
7       (...)
        private EmotionRecognitionTask emotionRecog;
9       PowerManager.WakeLock wakeLock;

11      (...)

13      public EmotionTasker(HappyWalkService hWServ) {
            this.hWServ = hWServ;
15          this.hWServiceHandler = hWServ.getHappyWalkServiceHandler();
            //preparing sensor manager to fetch data
17          try {
                esSensorManager = ESSensorManager.getSensorManager(hWServ);
19              esSensorManager.setGlobalConfig(GlobalConfig.
        PRINT_LOG_D_MESSAGES, false);
            } catch (ESException e) {
21              e.printStackTrace();
            }
23          //fetch the base emotion feedback time interval
            restoreEmoFeedbackValsFromPreferences();
25          //initialize Neural Network
            initNetwork();
27          //restore neural network weights from our preferences, if we have
        them
            restoreNNWeightsFromPreferences();
29
            emotionRecog = new EmotionRecognitionTask();
31          PowerManager powerManager = (PowerManager) hWServ.getSystemService
        (hWServ.POWER_SERVICE);
            wakeLock = powerManager.newWakeLock(PowerManager.PARTIAL_WAKE_LOCK
        , "EmotionTaskerWakeLock");
33      }

35      (...)

37      /**
         * Posts a new emotion recognition task
39       */
        public void postEmotionRecognitionTask() {
41          /* We need the device's CPU to remain
            awake while we perform our emotion recognition */
43          wakeLock.acquire();
```

```
            hWServiceHandler.post(emotionRecog);
45      }

47      (...)

49  }
```

To begin, we first declare a *PowerManager.WakeLock wakeLock* object (line 9), which we initialize, together with *emotionRecog*, on the constructor (lines 30–32). The *postEmotionRecognitionTask()* method itself is very simple; it acquires a wakeLock and posts an *EmotionRecognitionTask* (lines 40–45).

This is not enough, however, since we also need to keep our emotion recognitions working while HappyWalk is running. Therefore, let us go back to the *EmotionRecognitionTask* class and make it post itself again in the future:

```
   package hitlexamples.happywalk.service;
2
   import android.app.AlarmManager;
4  import (...)

6  public class EmotionTasker {

8      (...)

10     /**
        * Returns a pseudo-random value (in Milliseconds) that represents
12      * the amount of time until the next emotion recognition task
        */
14     private long nextEmotionExecutionMillis() {
           Random rand = new Random();
16         int randomNum = rand.nextInt(
                   (GlobalVariables.RECOG_EMOTION_MAX_MINUTES -
       GlobalVariables.RECOG_EMOTION_MIN_MINUTES) + 1) + GlobalVariables.
       RECOG_EMOTION_MIN_MINUTES;
18         return (long) 1000*60*randomNum;
       }
20
     class EmotionRecognitionTask implements Runnable{
22         private double[] outputs;
           private double[] inputs;
24
           @Override
26         public void run() {
               //Only run if we have location, since we need it for the
       neural net!
28             if (hWServ.getHwLocationListener().getActualposition() != null
       ) {
                   try {
30
                       (...)
32
                   } catch (Exception e) {
34                     e.printStackTrace();
                   }
36             }
               //post a new emotion inference
38             scheduleEmotionRecog(System.currentTimeMillis() +
       nextEmotionExecutionMillis());
```

```
            /* Since we have posted our nex emotion inference, we
40          no longer need to keep the device's CPU awake. */
            wakeLock.release();
42      }

44      /**
         * Schedules the next emotion recognition through the
46       * AlarmManager class
         * @param triggerAtMillis - Time when the alarm should trigger
48       */
        protected void scheduleEmotionRecog(long triggerAtMillis) {
50          // Construct an intent that will execute the AlarmReceiver
            Intent intent = new Intent(hWServ.getApplicationContext(),
    EmotionWakefulReceiver.class);
52          // Create a PendingIntent to be triggered when the alarm goes
    off
            final PendingIntent pIntent = PendingIntent.getBroadcast(
    hWServ, EmotionWakefulReceiver.REQUEST_CODE, intent, PendingIntent.
    FLAG_UPDATE_CURRENT);
54          AlarmManager alarm = (AlarmManager) hWServ.getSystemService(
    hWServ.ALARM_SERVICE);
            // First parameter uses the wall clock time in UTC
56          // Interval is calculated based on nextEmotionExecutionMillis
            alarm.set(AlarmManager.RTC_WAKEUP, triggerAtMillis, pIntent);
58      }

60      protected void cancelEmotionRecog() {
            Intent intent = new Intent(hWServ.getApplicationContext(),
    EmotionWakefulReceiver.class);
62          final PendingIntent pIntent = PendingIntent.getBroadcast(
    hWServ, EmotionWakefulReceiver.REQUEST_CODE, intent, PendingIntent.
    FLAG_UPDATE_CURRENT);
            AlarmManager alarm = (AlarmManager) hWServ.getSystemService(
    hWServ.ALARM_SERVICE);
64          alarm.cancel(pIntent);
        }
66
        private void postNotificationRemovalTask() {
68          (...)
        }
70
        private void fetchInputsAndCompute() throws NoCurrentPosition {
72          (...)
        }
74  }
```

We begin by writing a new method or *EmotionTasker* named *nextEmotionExecutionMillis()* (lines 14–19). It is responsible for determining when the next *EmotionRecognitionTask* will run. Notice that this is different from checking when the next emotion feedback request is to be sent (back on page ???, we did this by checking the value of *baseTimeToNextEmoFdbckReq*). Here, we want to run automatic emotion recognitions **once or twice an hour, randomly determined within these constraints in order to avoid user habituation** (see page ??). Therefore, *nextEmotionExecutionMillis()* returns a pseudo-random value (in milliseconds) that represents the amount of time until the next emotion recognition task.

We use this method at the end of our *EmotionRecognitionTask* by summing it with *currentTimeMillis()* and passing the result to *scheduleEmotionRecog()* (line 38).

The *scheduleEmotionRecog()* method (lines 49–58) constructs a *PendingIntent* that contains a reference to our *EmotionWakefulReceiver* and its *REQUEST_CODE* (line 53). It then schedules it through Android's *AlarmManager* class (lines 54–57). We also implemented a *cancelEmotionRecog()* method, which cancels any pending intents with the corresponding *REQUEST_CODE* that have yet to be delivered (lines 60–65).

Since our emotional recognition work is complete, we release the *EmotionTasker's* wakelock so that the device can return to sleep (line 41). Notice that the code between lines 37–41 completes the second *TODO* comment we had left in the code from page ???.

To conclude our *EmotionTasker*, we still need to implement a way to actually trigger the initial *EmotionRecognitionTask*. Let us write methods to start and stop the emotion recognition process:

```
package hitlexamples.happywalk.service;
2
import (...)
4
public class EmotionTasker {
6
    (...)
8
    /**
10   * Begins the process of scheduling emotion recognition tasks
     */
12  public void startEmotionRecognitionTasks() {
        /*First, check if we should perform emotion recog right now. */
14      long timeToNextEmoRecog = nextEmotionExecutionMillis();

16      /*Let us compare with the time when the last emotion recog was
    performed. */
        if (lastEmotionNotifMillis + timeToNextEmoRecog < System.
    currentTimeMillis()) {
18          //enough time has passed
            postEmotionRecognitionTask();
20      }
        else {
22          /* not enough time has passed.
            Let us schedule a emotion recog using the remaining time */
24          emotionRecog.scheduleEmotionRecog(lastEmotionNotifMillis +
    timeToNextEmoRecog);
        }
26  }

28  public void stopEmotionRecognitionTasks() {
        if (emotionRecog != null) {
30          hWServiceHandler.removeCallbacks(emotionRecog);
            emotionRecog.cancelEmotionRecog();
32      }
    }
34
    (...)
36
}
```

The *startEmotionRecognitionTasks()* method first checks if emotion recognition tasks should initiate as soon as it is called (line 17). If so, the first emotion recognition task is triggered (line 19); if not, it is scheduled appropriately (line 24). On the other hand, *stopEmotionRecognitionTasks()* cancels any pending emotion recognitions (lines 28–33).

Let us now use these methods within *HappyWalkService*. As the reader may remember, we previously instantiated an *emotionTasker* object back on page ???. We can now use it to start and stop the emotion recognition process within the service's *onCreate()* and *onDestroy()* methods:

```
package hitlexamples.happywalk.service;

import (...)

public class HappyWalkService extends Service {

    (...)

  @Override
    public void onCreate() {

        (...)

        //Prepare our worker thread
        hWServiceThread = new Thread(new Runnable() {
            public void run() {

        (...)

                // Instantiate Emotion Tasker
                emotionTasker = new EmotionTasker(HappyWalkService.this);
                //start the emotion recognition tasks
                emotionTasker.startEmotionRecognitionTasks();

                (...)

                Looper.loop();
            }
        });
        hWServiceThread.start();
    isRunning = true;
    }

    @Override
    public void onDestroy() {
        isRunning = false;
        if (hWServiceHandler != null) {
            //stop emotion recognition tasks
            emotionTasker.stopEmotionRecognitionTasks();
       /*
        * This uses the Handler to send a "quit" message to the Looper,
        * thus terminating our background Thread.
        */
            hWServiceHandler.getLooper().quit();
        }
        (...)
    }

    (...)
}
```

As shown above, we start the emotion recognition process right after *emotionTasker's* instantiation (line 23) and stop it as the service is destroyed (line 39). With this, we have finally finished our *EmotionTasker* and, with it, HappyWalk's ability to perform emotion recognition. As a reference, we provide a final overview of the class's full code in Appendix A.

In the next section, we will finalize this tutorial by using what we have built so far in order to provide positive reinforcement to the user.

9.3 Providing Positive Reinforcement

When describing our HiTL approach in Section 5.1, it was mentioned that the app should show some sort of "positive notification" that motivates the user to go for walks, as well as heatmaps representing the near real-time context of nearby POIs.

How can we translate these requirements into actual functionality? The approach we are going to describe in this section will use the *MapsActivity* class to display a dialog box containing a motivational message, similar to the one presented in Figure 9.10. For completeness, sake, we will implement a dialog box that handles all the four types of emotions, not just the negative ones. As such, the reader is welcome to expand the usage of this motivational messages to positive cases in the future. As for the heatmaps, we will resort to existing functionality already implemented within HappyWalk. The next two subsections will focus on each of these tasks.

9.3.1 Creating a Motivational Dialog Box

As we described back on page ???, our "normal emotion notification" is displayed whenever a negative emotion is detected, redirecting the user towards the *MapsActivity* whenever pressed. How should the activity respond to these notifications? As the reader might have noticed, the method that creates this notification (*EmotionTasker's showNormalEmotionNotification()* defines an *extras Bundle* that accompanies the *Intent* to call *MapsActivity*:

```
1   private void showNormalEmotionNotification(int typeOfEmotion) {
        //First, cancel previous notification removal tasks
3       hWServiceHandler.removeCallbacks(currentNotifRemovTask);
        //Now, prepare a Bundle with the necessary information
5       Bundle bnd = new Bundle();
        /* put a timestamp on this bundle to avoid the user clicking
    notifications that have been fired a long time ago. */
7       lastEmotionNotifMillis = System.currentTimeMillis();
        bnd.putLong(GlobalVariables.BND_EXTRA_EMOTION_TIMESTAMP_KEY,
9               lastEmotionNotifMillis);
        bnd.putInt(GlobalVariables.BND_EXTRA_EMOTION_TYPE_NOTIF_KEY,
    typeOfEmotion);
11      bnd.putInt(GlobalVariables.BND_EXTRA_REQ_CODE_KEY,
                GlobalVariables.AREQ_EMOTION_NORMAL_NOTIF);
13
        // We will show our map, to promote walking when emotions are
    negative
15      Intent intent = new Intent(hWServ, MapsActivity.class);
        intent.putExtras(bnd);
```

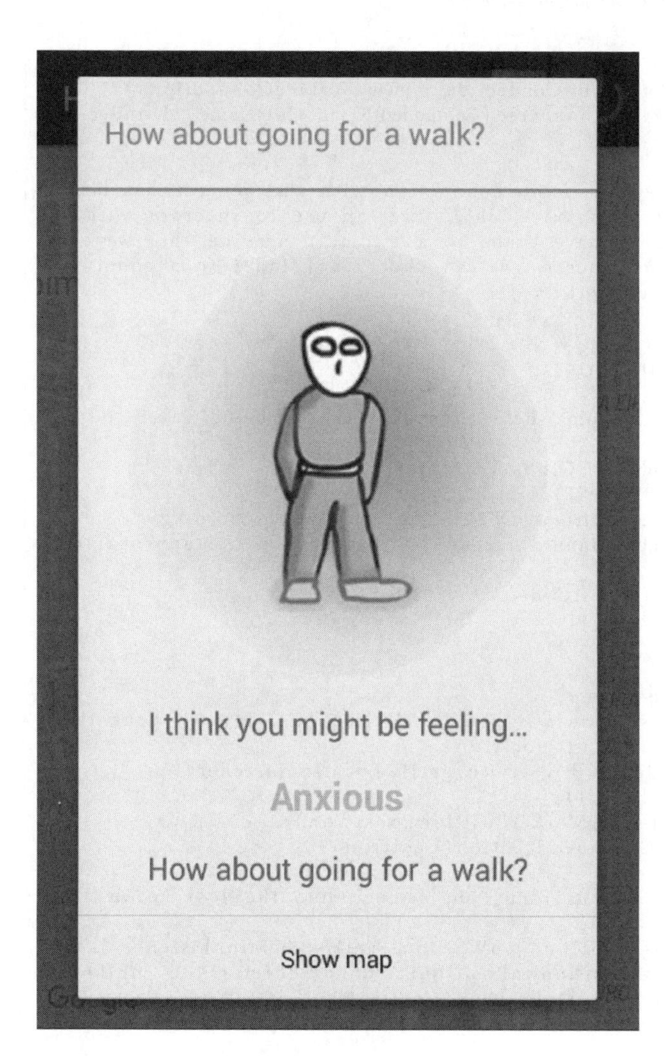

Figure 9.10 The emotion alert dialog.

This bundle uses several *GlobalVariables* keys to identify and include a timestamp, the type of emotion that was detected and a "request code". The value of the extra's key *GlobalVariables.BND_EXTRA_REQ_CODE_KEY* can be used by *MapsActivity* to know what part of the application requested it. Therefore, we can take advantage of this value to determine how to act when the user presses the normal notification. The handling of request codes is performed right after *MapsActivity* has connected to our background service:

```
1  package hitlexamples.happywalk.activities;

3  import (...)

5  public class MapsActivity extends ActionBarActivity {

7      (...)
```

```
    private ServiceConnection hwConnection = new ServiceConnection() {
        public void onServiceConnected(ComponentName className, IBinder
    service) {
            /*
             This is called when the connection with the service has been
    established, giving us a service object we can use to interact with
    the service. Because we have bound to a explicit service that we know
     is running in our own process, we can cast its IBinder to a concrete
    class and directly access it.
            */
            hWService = ((HappyWalkService.HappyWalkBinder)service).
    getService(MapsActivity.this);

            /* Now we check from where this activity is being initiated
    from */
            if (requestCode == GlobalVariables.
    AREQ_POI_DESCRIPTION_REQUEST) {
                /* if we come from POI description, we do nothing.
                * The camera should already be focused on the appropriate
    POI*/
                //revert requestCode
                requestCode = 99;
            }
            else {
                LatLng position;
                //If our service was previously running, let us focus the
    camera on the user's current position.
                if((position = hWService.getHwLocationListener().
    getActualposition()) !=null) {
                    updateUserMarkerPosition(position);
                    focusCameraOnPosition(position);
                }
                //if not, let us focus our camera onto the last known
    position
                else if ((position = hWService.getHwLocationListener().
                        getLastKnownPositionFromSharedPrefs()) != null){
                    focusCameraOnPosition(position);
                }
            }
        }
  }

  (...)
}
```

Notice that *MapsActivity* currently only checks for a request code of type *GlobalVariables.AREQ_POI_DESCRIPTION_REQUEST*, which corresponds to returning from a POI description activity. Let us add an additional check for the normal notification's request code:

```
package hitlexamples.happywalk.activities;

import (...)

public class MapsActivity extends ActionBarActivity {

    (...)
```

```
8
    private ServiceConnection hwConnection = new ServiceConnection() {
10      public void onServiceConnected(ComponentName className, IBinder
    service) {

12          (...)

14              /* Now we check from where this activity is being initiated
    from */
                if (requestCode == GlobalVariables.
    AREQ_POI_DESCRIPTION_REQUEST) {
16                  /* if we come from POI description, we do nothing.
                    * The camera should already be focused on the appropriate
    POI*/
18                  //revert requestCode
                    requestCode = 99;
20              }
                else {
22                  if (requestCode == GlobalVariables.
    AREQ_EMOTION_NORMAL_NOTIF) {
                        if (getIntent().getExtras().containsKey(
    GlobalVariables.BND_EXTRA_EMOTION_TYPE_NOTIF_KEY)) {
24                          // TODO: show motivation dialog here
                        }
26                      else {
                            throw new AssertionError("No typeOfEmotion in
    Intent extras");
28                      }
                        //revert notification
30                      hWService.showNotification(false);
                        //revert requestCode
32                      requestCode = 99;
                    }
34                  LatLng position;
                    //If our service was previously running, let us focus the
    camera on the user's current position.
36                  (...)
                }
38          }

40          public void onServiceDisconnected(ComponentName className) {
                (...)
42          }
    };
44
    (...)
46 }
```

In case *MapsActivity* has been called from a normal notification, we want to display a motivational dialog box, as suggested by the *TODO* comment in line 24. How should this dialog work?

With the objective of implementing our motivational dialog, we will create a new *EmotionAlertDialog* class which will receive the type of emotion as an argument. Let us begin by creating and editing a new layout resource file for our dialog as *app/res/layout/negative_emotion_alert*. As usual, right-click on the *app/res/layout/* folder → *New* → *Layout resource file*. Use a *RelativeLayout* and name the file as *negative_emotion_alert* (refer to page ??? for how to edit the newly created file):

```
<?xml version="1.0" encoding="utf-8"?>
<RelativeLayout xmlns:android="http://schemas.android.com/apk/res/android"
    android:layout_width="wrap_content" android:layout_height="
    wrap_content">


    <ImageView
        android:layout_width="wrap_content"
        android:layout_height="wrap_content"
        android:id="@+id/alert_negative_emotion_iview"
        android:layout_marginTop="30dp"
        android:layout_centerHorizontal="true"
        android:src="@drawable/alert_negative_emotion"
        android:layout_alignParentLeft="true"
        android:layout_alignParentBottom="false"
        android:layout_alignParentRight="true"
        android:layout_alignParentTop="false" />

    <TextView
        android:layout_width="wrap_content"
        android:layout_height="wrap_content"
        android:textAppearance="?android:attr/textAppearanceMedium"
        android:text="@string/emotAlertDiag1"
        android:id="@+id/alert_negative_emotion_textview1"
        android:layout_alignParentBottom="false"
        android:layout_alignParentLeft="true"
        android:layout_alignParentRight="true"
        android:layout_alignParentEnd="false"
        android:layout_below="@+id/alert_negative_emotion_iview"
        android:layout_marginTop="15dp"
        android:gravity="center_vertical|center_horizontal" />

    <TextView
        android:layout_width="wrap_content"
        android:layout_height="wrap_content"
        android:textAppearance="?android:attr/textAppearanceMedium"
        android:text="@string/emotAlertDiag2"
        android:id="@+id/alert_negative_emotion_textview2"
        android:gravity="center_vertical|center_horizontal"
        android:layout_alignParentTop="false"
        android:layout_centerHorizontal="true"
        android:layout_below="@+id/alert_negative_emotion_emotion_textView
    "
        android:layout_alignParentLeft="true"
        android:layout_alignParentRight="true"
        android:layout_marginTop="15dp"
        android:layout_marginBottom="15dp" />

    <TextView
        android:layout_width="wrap_content"
        android:layout_height="wrap_content"
        android:textAppearance="?android:attr/textAppearanceLarge"
        android:text="Emotion"
        android:id="@+id/alert_negative_emotion_emotion_textView"
        android:layout_alignParentTop="false"
        android:layout_centerHorizontal="true"
        android:layout_below="@+id/alert_negative_emotion_textview1"
        android:layout_alignParentLeft="true"
```

```
        android:layout_alignParentRight="true"
58      android:gravity="center_vertical|center_horizontal"
        android:layout_marginTop="15dp"
60      android:textStyle="bold" />

62 </RelativeLayout>
```

The above layout should represent the format of the dialog previously shown in Figure 9.10. We begin by defining an *ImageView* to contain the image referenced in the *android:src* property, *@drawable/alert_negative_emotion*, which should already exist within the project (lines 6–16). We then define several *TextViews* to contain our text (lines 18–60). We will later modify the contents of *alert_negative_emotion_emotion_textView* (line 52) programmatically, to present the neural network's output.

A few associated strings also need to be defined within *app/res/values/strings.xml*:

```
<resources>
2
  (...)
4
    <!--Emotion Feedback -->
6   <string name="emotFeedButton">This is how I feel</string>
    <string name="emotAlertDiag1">I think you might be feeling...</string>
8   <string name="emotAlertDiag2">How about going for a walk?</string>
    <string name="emotAlertDiagButton">Show map</string>
10  <string name="emotAlertEuphoric">Euphoric</string>
    <string name="emotAlertAnxious">Anxious</string>
12  <string name="emotAlertBored">Bored</string>
    <string name="emotAlertCalm">Calm</string>
14 </resources>
```

Let us also define some colors to be associated with each emotion within *app/res/values/colors.xml*. Notice that, for extendability's sake, we already define a color for each type of emotion despite the fact that our notification only appears for negative ones:

```
<?xml version="1.0" encoding="utf-8"?>
2 <resources>
    <color name="darkText">#ff000000</color>
4   <color name="scVpoiDescBck">#d2f7cf</color>
    <color name="whiteText">#ffffff</color>
6   <color name="happyWalkGreen1">#3e8c39</color>
  <!-- Emotion Colors -->
8   <color name="emotionColor_euphoria">#fe0101</color>
    <color name="emotionColor_anxiety">#d9d619</color>
10  <color name="emotionColor_boredom">#0cd900</color>
    <color name="emotionColor_calmness">#00a2ff</color>
12 </resources>
```

We can now create a new *EmotionAlertDialog* class within the package *hitlexamples.happywalk.activities*:

```
package hitlexamples.happywalk.activities;
2
import android.app.AlertDialog;
4 import android.app.Dialog;
import android.content.DialogInterface;
```

```
6   import android.os.Bundle;
    import android.support.annotation.NonNull;
8   import android.support.v4.app.DialogFragment;
    import android.view.LayoutInflater;
10  import android.view.View;
    import android.widget.TextView;
12
    import hitlexamples.happywalk.R;
14  import hitlexamples.happywalk.utilities.GlobalVariables;

16  public class EmotionAlertDialog extends DialogFragment {
        @NonNull
18      @Override
        public Dialog onCreateDialog(Bundle savedInstanceState) {
20          AlertDialog.Builder builder = new AlertDialog.Builder(getActivity
        ());
            LayoutInflater inflater = getActivity().getLayoutInflater();
22          View emotionAlertView = inflater.inflate(R.layout.
        negative_emotion_alert, null);

24         TextView emotionText =
                   (TextView) emotionAlertView.findViewById(
26                         R.id.alert_negative_emotion_emotion_textView);

28          int typeOfEmotion = getArguments().
                    getInt(GlobalVariables.BND_EXTRA_EMOTION_TYPE_NOTIF_KEY);
30
            setUpTextViewFromEmotion(emotionText, typeOfEmotion);
32
            builder.setView(emotionAlertView)
34                  .setTitle(R.string.emotAlertDiag2)
                    .setPositiveButton(R.string.emotAlertDiagButton,
36                          new DialogInterface.OnClickListener() {
                                @Override
38                              public void onClick(DialogInterface dialog,
                                                    int which) {
40                                  EmotionAlertDialog.this.dismiss();
                                }
42                          });
            return builder.create();
44      }

46      private void setUpTextViewFromEmotion(TextView emotionText, int
        typeOfEmotion) {
            switch (typeOfEmotion) {
48              case GlobalVariables.EMOTION_EUPHORIA:
                    emotionText.setText(getResources().getText(R.string.
        emotAlertEuphoric));
50                  emotionText.setTextColor(getResources().getColor(R.color.
        emotionColor_euphoria));
                    break;
52              case GlobalVariables.EMOTION_ANXIETY:
                    emotionText.setText(getResources().getText(R.string.
        emotAlertAnxious));
54                  emotionText.setTextColor(getResources().getColor(R.color.
        emotionColor_anxiety));
                    break;
56              case GlobalVariables.EMOTION_BOREDOM:
```

```
                emotionText.setText(getResources().getText(R.string.
        emotAlertBored));
58              emotionText.setTextColor(getResources().getColor(R.color.
        emotionColor_boredom));
                break;
60          case GlobalVariables.EMOTION_CALMNESS:
                emotionText.setText(getResources().getText(R.string.
        emotAlertCalm));
62              emotionText.setTextColor(getResources().getColor(R.color.
        emotionColor_calmness));
                break;
64      }
    }
66 }
```

This class is an extension of the Android support library's *DialogFragment* (*android.support.v4.app.DialogFragment*, lines 8 and 16). As evidenced by its code, the *EmotionAlertDialog* inflates the layout *R.layout.negative_emotion_alert* (line 22). It then acquires a reference to a *TextView* object with id *R.id.alert_negative_emotion_emotion_textView* (lines 24–26) and sets its text and color according to the received type of emotion, through the method *setUpTextViewFromEmotion()* (line 31). This method consists of a *switch* statement that evaluates the type of emotion and calls the *TextView's setText()* and *setTextColor()* methods, providing them with the string and color resources we defined above (lines 46–65).

We build the view using an *AlertDialog.Builder* object, defined in line 20. We set the dialog's view to the *emotionAlertView* object we had previously inflated (line 33), and reuse the *emotAlertDiag2* string resource as a title (line 34). Notice that the *AlertDialog's* builder provides us with a *setPositiveButton()* method that we can use to define an "acknowledgment-type" button. We use this to our advantage and set the *emotAlertDiagButton* string resource as the button's text, followed by an implementation of the *DialogInterface.OnClickListener* interface where we simply dismiss the dialog (lines 35–42). Finally, we return the result of the builder's *create()* method (line 43).

The resulting *AlertDialog* is identical to the one shown in Figure 9.10. As soon as the user presses the button "Show map", the dialog closes and the app displays the map, as normal.

Finally, let us use this new class to complete the *TODO* left in *MapsActivity*:

```
   package hitlexamples.happywalk.activities;
2
   import (...)
4
   public class MapsActivity extends ActionBarActivity {
6
       (...)
8
       private ServiceConnection hwConnection = new ServiceConnection() {
10         public void onServiceConnected(ComponentName className, IBinder
       service) {

12         (...)

14             else {
```

```
                    if (requestCode == GlobalVariables.
    AREQ_EMOTION_NORMAL_NOTIF) {
16                      if (getIntent().getExtras().containsKey(
    GlobalVariables.BND_EXTRA_EMOTION_TYPE_NOTIF_KEY)) {
                            showEmotionAlertDialog(getIntent().getExtras().
    getInt(GlobalVariables.BND_EXTRA_EMOTION_TYPE_NOTIF_KEY));
18                      }
                        else {
20                          throw new AssertionError("No typeOfEmotion in
    Intent extras");
                        }
22                      //revert notification
                        hWService.showNotification(false);
24                      //revert requestCode
                        requestCode = 99;
26                  }

28                  (...)
                }
30          }

32          public void onServiceDisconnected(ComponentName className) {
                (...)
34          }
        };
36
        private void showEmotionAlertDialog(int typeOfEmotion) {
38          //pass type of emotion to the new emotion alert dialog
            EmotionAlertDialog emotionAlertDialog = new EmotionAlertDialog();
40          Bundle alertBnd = new Bundle();
            alertBnd.putInt(GlobalVariables.BND_EXTRA_EMOTION_TYPE_NOTIF_KEY,
    typeOfEmotion);
42          emotionAlertDialog.setArguments(alertBnd);
            //show the alert dialog
44          emotionAlertDialog.show(getSupportFragmentManager(),"
    EmotionAlertDialog");
        }
46
        (...)
48  }
```

In the code presented above, we have replaced the *TODO* comment from page ??? with a call to a method named *showEmotionAlertDialog()* (line 17). This method initializes, passes the emotion on to, and shows a new instance of our *EmotionAlertDialog* (lines 37–45).

9.3.2 Enabling the Emotion Heatmaps

One final piece is missing in our HiTL application: the displaying of emotional information on the map. This will be achieved by displaying heatmaps whose color is associated with the average emotion being felt at a certain POI. Heatmaps are currently managed by the classes *GeoCluster* and *GeoClusterer*. Since the clustering of POIs can be a rather complex problem that remains outside of the HiTL focus of this tutorial, most tasks described in this section should already be implemented.

However, we still need to tell our *GeoCluster* class how to calculate the type of emotion. Fortunately, we already implemented this in our *EmotionTasker.getTypeOfEmotion()*

method, which we previously exposed as *public static* back on page ???. Therefore, let us edit the code of the *hitlexamples.happywalk/cluster/GeoCluster* class, particularly within its already existing *getHeatMapImageFromEmotion()* method:

```
package hitlexamples.happywalk.cluster;

import hitlexamples.happywalk.service.EmotionTasker;
import (...);

/**
 * GeoCluster class.
 * contains single Marker object that is also
 * stored in the Geoclusterer's "MarkerGeo" HashMap".
 */
public class GeoCluster {

    (...)

    private int getHeatMapImageFromEmotion(double[] meanEmotion)
    {
        /*
        Emotions are mapped to colors in accordance to the
    emotion_color_map, which follows (in a somewhat limited way) the
    Wright Theory on the Colour Affects System:

        red -> euphoria
        anxiety -> yellow
        boredom -> green
        calmness -> blue
         */
        int imageDescriptor;
        int typeOfEmotion = EmotionTasker.getTypeOfEmotion(meanEmotion);

        switch(typeOfEmotion) {
            case GlobalVariables.EMOTION_EUPHORIA:
                imageDescriptor = R.drawable.heatmap_red;
                break;
            case GlobalVariables.EMOTION_ANXIETY:
                imageDescriptor = R.drawable.heatmap_yellow;
                break;
            case GlobalVariables.EMOTION_BOREDOM:
                imageDescriptor = R.drawable.heatmap_green;
                break;
            case GlobalVariables.EMOTION_CALMNESS:
                imageDescriptor = R.drawable.heatmap_blue;
                break;
            default :
                imageDescriptor = R.drawable.heatmap_red;
                break;
        }
        return imageDescriptor;
    }

  (...)

}
```

For the sake of simplicity, all the reader should need to do is uncomment a line that properly sets the following variable (see line 26):

int typeOfEmotion = EmotionTasker.getTypeOfEmotion(meanEmotion);

Nevertheless, the curious reader is more than welcome to further explore the provided source code.

Our next objective is to allow for heatmaps to be toggled each time the user presses a button on the *MapsActivity's* action bar. To do so, we will use the method *toggleHeatMaps()* of the *MapActivity's GeoClusterer* object.

Let us define this new button in the resource file *res/menu/map_menu.xml*:

```
<?xml version="1.0" encoding="utf-8"?>
<menu xmlns:android="http://schemas.android.com/apk/res/android"
    xmlns:happywalk="http://schemas.android.com/apk/res-auto">

    <item android:id="@+id/menuHeatmapToggle"
        android:icon="@drawable/emotion_color_map"
        android:title="@string/menuHeatmap"
        happywalk:showAsAction="ifRoom"/>

    <item android:id="@+id/menuExit"
        android:icon="@drawable/off"
        android:title="@string/menuExit"
        happywalk:showAsAction="ifRoom"/>
</menu>
```

Notice that we reuse *@drawable/emotion_color_map* (the background of *EmotionSpace*) as the button's icon, in line 6. Let us also define a title string for our new menu button in *res/values/strings.xml*:

```
<resources>

  (...)

    <!-- Menu -->
    <string name="menuExit">Exit</string>
    <string name="menuHeatmap">Emotion Heatmap</string>

    (...)

</resources>
```

Now, we simply have to change the *onOptionsItemSelected()* method of *MapsActivity*:

```
package hitlexamples.happywalk.activities;

import (...)

public class MapsActivity extends ActionBarActivity {

    (...)

    /**
     * This method handles menu items
     */
    @Override
    public boolean onOptionsItemSelected(MenuItem item) {
```

```
            // Handle item selection
15          switch (item.getItemId()) {
                case R.id.menuHeatmapToggle:
17                  clusterer.toggleHeatMaps();
                    return true;
19              case R.id.menuExit:
                    performExit();
21                  return true;
                default:
23                  return super.onOptionsItemSelected(item);
            }
25      }
    }
```

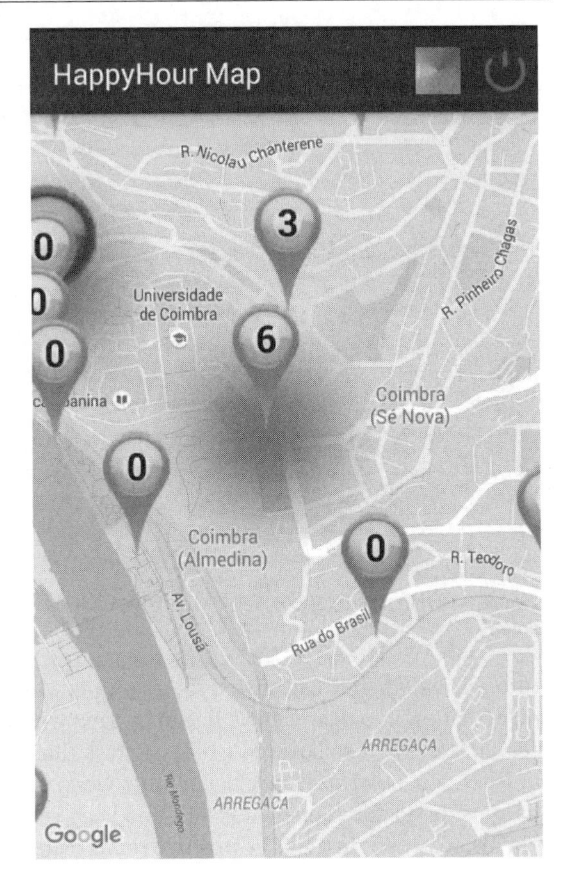

Figure 9.11 The emotion heatmaps

We added a new *case* label within the method's *switch* block (lines 16–18). Now, each time the user presses this new button, the app should show or hide the associated heatmaps, as shown in Figure 9.11.

9.4 In Summary...

We have finally completed our smartphone-based HiTLCPS. Figure 9.12 shows the final state of our HiTLCPS. It illustrates all of the tasks we have performed through the

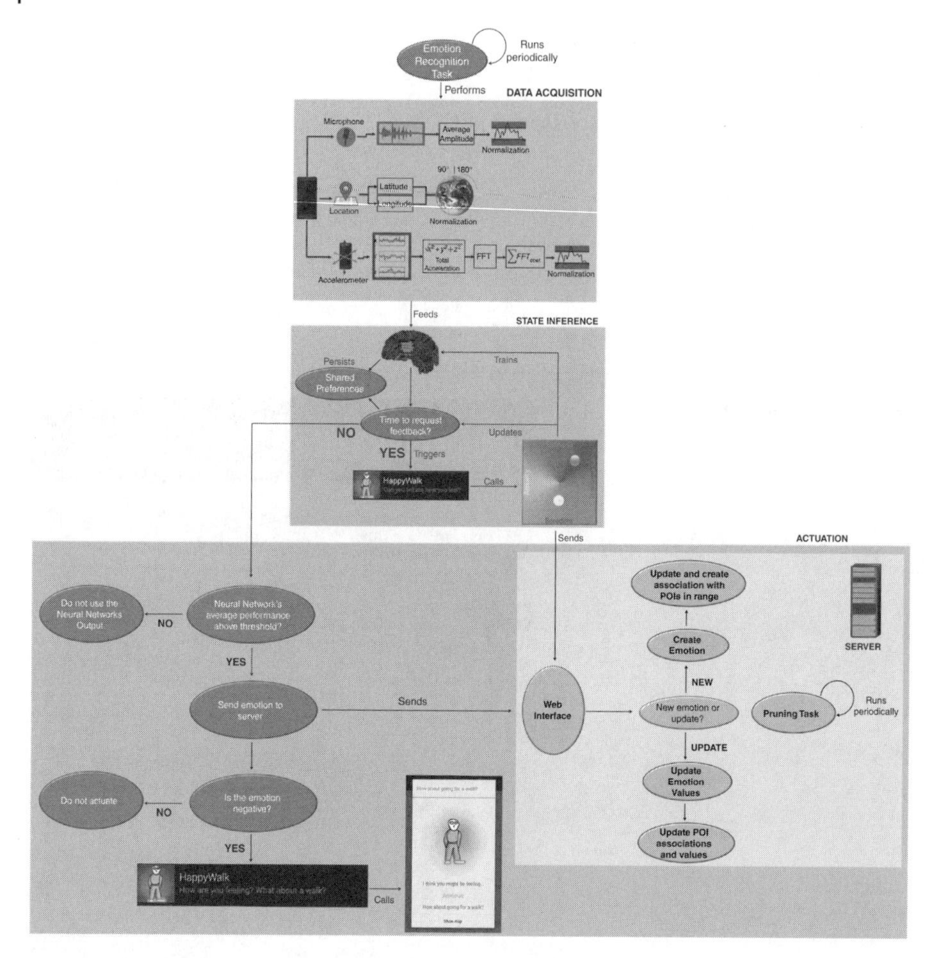

Figure 9.12 Final state of our HiTLCPS at the end of Chapter 9. (*See insert for color representation of the figure.*)

tutorial, including those related to *Actuation*, which we covered in this chapter. Focusing on actuation in particular, we began by saving and updating incoming emotions on the server side. We also covered a pruning task that removed outdated emotions. Then, we finished our *EmotionTasker* by defining when the results of our neural network should be considered and how to kickstart and keep running the emotion recognition process. Finally, we implemented the positive reinforcement feature in the form of a notification and an alert dialog.

Nevertheless, there are still a number of improvements that could be made: support more emotional states, display a dialog requesting the enabling of location for proper emotion inference, supporting smartwatch heart-rate data... the possibilities are endless. We leave these as exercises for the curious reader who might be willing to experiment further with HappyWalk.

Despite its simplicity, we hope this tutorial has given the reader a more complete notion of how the theoretical concepts of *Data Acquisition*, *State Inference*, and *Actuation* may be applied in actual systems.

Part III

Future of Human-In-the-Loop Cyber-Physical Systems

After looking at their trajectory, theory, and practice in Parts I and II, in this part of the book we will conclude our journey through human-in-the-loop cyber-physical systems by discussing their likely evolution, given the state of the art. In Chapter 10 we will delve into the requirements and challenges faced by emerging and future HiTL applications. Both developers and users should be aware of them, as they are crucial for any human-centric application. Afterwards, in Chapter 10, we will look at the constraints that affect HiTL systems and review the lessons learned throughout the book, in order to understand how the remaining challenges may shape the next steps of HiTL research and development.

A Practical Introduction to Human-in-the-Loop Cyber-Physical Systems, First Edition.
David Nunes, Jorge Sá Silva and Fernando Boavida.

Companion Website URL: www.wiley.com/go/nunesloop

10

Requirements and Challenges for HiTL Applications

Many roadblocks stand in the path of future HiTLCPSs. In this chapter, we will attempt to identify some of the major requirements and challenges that still need to be addressed before we start to see a truly human-aware IoA. These include resilience, security, and privacy, standard communications, localization, state inference, and safety.

10.1 Resilience

It is important to extend current research by providing resilient and performance-controlled solutions for IoA environmental interactions. Instead of targeting previously planned and static deployments, new performance-controlled systems will need to be designed in an adaptable way in order to operate in dynamic environments and to enable coordinated HiTL control. This also has to be achieved while keeping the system performance under acceptable levels, even in the presence of mobility and a diversity of faults. These requirements raise a number of new challenges that must be addressed to enable the successful implementation of the IoA and HiTL paradigms in real-world situations, particularly in critical ones (e.g. industrial management or healthcare).

Key innovation still needs to be achieved in terms of performance-aware models and mechanisms that enhance the overall performance and management of HiTL systems. An inherent ability for handling faults in these naturally distributed environments also needs to be considered. To allow performance-controlled HiTLCPSs to meet dependability targets it is, therefore, necessary to incorporate performance, fault-tolerant, and self-healing mechanisms into their design, deployment, and execution tasks. These mechanisms will ensure end-to-end performance in environments where HiTL control is an important feature.

When sensitive data retrieved by sensors is transmitted through critical environments, security challenges are also raised. If not protected, data may be unduly accessed, corrupted, or even destroyed, reducing the safety required in critical environments. Consequently, security mechanisms should also be investigated and added to the design.

We consider the term "resilience" to encompass a combination of a number of features such as the ones discussed above. As shown in Figure 10.1, these include dependability, security, and privacy, as well as the required means to provide overall robustness and performance to the system.

A Practical Introduction to Human-in-the-Loop Cyber-Physical Systems, First Edition.
David Nunes, Jorge Sá Silva and Fernando Boavida.

Companion Website URL: www.wiley.com/go/nunesloop

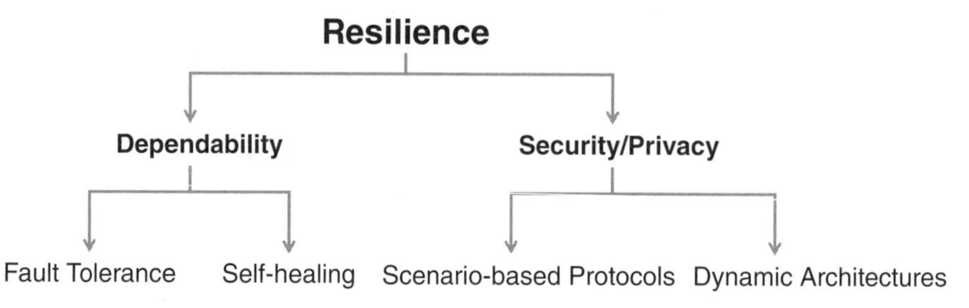

Figure 10.1 The HiTL resilience paradigm.

Dependability is achieved by integrating, at development time, static approaches based on fault-tolerance mechanisms, and by providing, at runtime, dynamic approaches based on self-healing. These approaches will probably be, however, dependent on the deployment technology; IoT may still be too young for researchers to have a proper grasp of how to plan and design proper dependability. Nevertheless, some projects, such as the RELYonIT [143], are already laying the groundwork for how to bring dependability into IoT.

10.2 Security and Privacy

Security and privacy, on the other hand, have long been addressed in academic research, with several secure protocols having been proposed over the years. However, most of these protocols have only been evaluated in an isolated fashion and not in the context of overall HiTLCPS security, often because of the lack of concrete applications or deployment scenarios. In fact, most of the current state-of-the art IoT security protocols are purely academic, as they are based on theoretical or simulation results. Thus, security in IoA can be achieved by developing new protocols targeted to the specific needs of HiTL scenarios and applications.

While secure systems exist for serving humans and carrying out human-oriented tasks, and are designed and built by humans, most of them often compromise usability for the sake of security, requiring the use of passwords and authentication methods that are cumbersome and/or too difficult to remember but should never be written down. Thus, security often becomes a hurdle that has to be worked around, which translates into difficulties in integrating humans into HiTLCPSs. These difficulties often stem from bad system design, which does not consider the interaction between humans and systems, restricting analysis and designs exclusively to the computation side. To solve this problem, it is necessary to rethink the underlying technology and keep users central in threat model security design, aligning and inferring security state changes from humans' actions and states, and communicating these changes back to the human [144]. This idea was reinforced in [145], where security in mobile and pervasive systems was rethought in an HiTL perspective. The authors argued that pervasive systems introduce human-driven security vulnerability that traditional usability design cannot address, which raises the need for better understanding the roles and relationships of humans in the context of pervasive systems security. In particular, they highlighted mobility and sociability as two new sources of security vulnerability that need to be addressed. There are also some dissident views on the human role in security, as presented in [146]. In this work, a framework for reasoning about HiTL was presented, arguing that humans often fail in their security roles, and thus secure system designers should find ways of keeping humans out of the loop. Automated security measures tend to be more accurate, predictable, and, unlike humans, do not get tired or bored. Thus, secure systems should "just work", without any human intervention. However, for tasks where feasible or cost-effective alternatives to humans are not available, secure system designers should engineer their systems to support humans in the loop while maximizing their chances of performing their security-critical functions successfully.

It is our opinion that future HiTLCPSs should adopt a position midway between these oppossing views. Humans should always be a part of the security loop, but security mechanisms should not critically depend on invasive and direct human intervention, such as intrusive notifications requesting human judgment. Instead, HiTLCPSs should follow the same principles of ubiquity and pervasiveness that are typical of the IoT and CPSs paradigms, and "just work" while still considering the human. This means that HiTLCPSs should automatically evaluate the human context, including psychological states, emotions, position, movements, and actions to modulate security and even privacy mechanisms.

As for privacy, there is a need to define models and architectures for HiTLCPSs, for supporting dynamic policies that are adapted and tailored to each individual. As different applications of IoA have different security and privacy requirements, an architecture should be able to guarantee a variety of privacy levels. As the desired level might change even during the deployment of an IoA system, it should be possible to dynamically configure and control both security and privacy levels. Thus, we believe the major challenge concerning security and privacy is to define an adaptable and manageable architecture for use in real-time scenarios that combine humans and IoT. Instead of an isolated analysis of different protocols or communication layers, this architecture should consider the security and privacy of real application deployments.

Let us consider a real-world HiTL scenario where privacy and security settings may dynamically adapt to the human situation, and a hypothetical HiTL mobile app that allows users to share their location with their friends in real time, continuously monitoring their states, including location. Users can easily see which of their friends are nearby and figure out who is up for grabbing a drink. However, users might not want to share this information continuously, and only share it when they are out for a drink and only with a specific group of people. In this case, the user's privacy settings should dynamically adapt to their position, avoiding the sharing of location when location is sensitive. On the other hand, if the user is in a crowded and public place, the messages exchanged with the server might not require particularly high levels of protection since this information is not particularly sensitive. Thus, security mechanisms can be relaxed, in order to save the mobile's devices processing and battery power. Other locations, such as the user's home or workplace, might invoke the need for additional levels of privacy and encryption, in order to avoid abuse by malicious entities. For example, thieves could attempt to exploit a user's location information to more easily discover when homes are empty, thus facilitating their illegal activity.

Future work should, therefore, define, implement, and evaluate a set of resilient protocols, techniques, and tools for performance-controlled supervision of cooperating humans and technology, with HiTL control. The devised resilient supervision should be designed for providing safe and mobile HiTL interaction and cooperation in various scenarios, including safety-critical environments. This work should rely on the complementary use of design-time and run-time approaches for obtaining compliant solutions that enable the provision of performance-controlled services even in the presence of changes that may occur to the system, its environment, or its requirements.

10.3 Standard Communications

The current Internet already displays high heterogeneity in terms of devices and communication protocols. This heterogeneity will become even more pronounced if we consider all the human elements (human as a set of sensors, human nature, human as actuator, human as a communications node, and human as a processing node) described in Chapter 3. Thus, it is important to find processes and protocols that can support communication between all these elements, human and otherwise. These heterogeneous processes and protocols must be able to allow communication between devices that are highly heterogeneous in terms of processing capabilities, size, and function, such as robotic elements, wireless sensor nodes, body-coupled sensors, smartphones, etc. Additionally, communication must remain reliable even in the face of the highly crowded wireless spectrum, where different kinds of communication technologies co-exist. Supporting persistent, reliable, and interoperable connections, in addition to mobility and different kinds of wireless mediums, is a very demanding challenge.

The dynamic management of network connections is an important task for HiTLCPSs, since different human contexts set dissimilar requirements for network communications. For example, most mobile devices use wireless interfaces to communicate, but amongst these we have a multitude of possibilities: GPRS, WiFi, *Bluetooth*, *ZigBee*, etc. This heterogeneity in terms of types of devices and communication interfaces has its advantages, since each wireless technology adapts better to specific conditions: *Bluetooth* is great for replacing cables in short-distance communications, while RFID and near-field communication are better when devices come into near-physical contact; WiFi and *ZigBee* are excellent technologies for mid-range wireless communication, used to cover areas from single buildings to large campuses, while cellular communications allow for long-range packet-oriented mobile communication. Thus, HiTLCPSs should be able to determine the necessary levels of quality of service and quality of experience depending on the human context, and manage the device's wireless technologies that better adapt to the current situation.

Since the availability of wireless connections and their signal strength changes considerably as users move, the multiple network interfaces of their devices can allow for more flexibility and greater Internet coverage. In fact, the importance of network switching, also known as multi-homing, in face of heterogeneous mobile environments has long been noted by academic research [147]. One reason for this is the ever-increasing mobile traffic. A recent study by Cisco [148] forecasts staggering increases in mobile networking: traffic from wireless devices will exceed traffic from wired devices in 2019 and WiFi and mobile devices will account for 66% of all IP traffic. Globally, mobile data traffic will increase ten-fold between 2014 and 2019, and will grow three times faster than fixed IP traffic from 2014 to 2019.

To counter this issue, some researchers have suggested that the usage of multiple wireless mediums and network interfaces will contribute not only to a better distribution of network traffic but also to the increase of Internet coverage and connectivity for mobile environments [149], and even to the energy efficiency of devices [150, 151]. However, mobile handoff is a complex problem, with a large amount of research and effort targeting it. Most of this complexity comes from the fact that there are multiple layers to be considered, since proper handoff requires several different types of tasks that need to be performed.

To deal with the low-level "Link and Network Technology Information" layer, working at the data-link and physical levels of the OSI model, the IEEE 802.21 working group finalized the first standard for handovers in heterogeneous networks, called Media-Independent Handovers (IEEE 802.21-2008) [152]. The latest draft version of the standard was accepted as a new standard by the IEEE-SA Standards Board in November 2008 and published in January 2009 [153]. It provided a framework for efficiently discovering the networks in range, as well as their respective capabilities and current link conditions. HiTL handover mechanisms, conceptually existing in upper layers, have to rely on limited and rudimentary information, such as signal quality, for understanding the state of surrounding networks. If mobile devices could collect timely and consistent information about the state of all available networks in range, much more accurate and efficient network selection mechanisms would be possible. In this context, IEEE 802.21-2008 aimed at being a standard framework for enhancing the efficiency of handover decision makers. The standard contributes to seamless handovers by specifying mechanisms to gather and distribute notifications about changes in link conditions and available access networks. Nevertheless, the scope of IEEE 802.21-2008 is restricted to this gathering of technology-independent information. Actual technical solutions for performing intratechnology handovers, handover policies, or security mechanisms are not considered. Thus, while the efforts made by the IEEE 802.21 group are certainly important for achieving efficient handover mechanisms, they do not directly embrace the development of these mechanisms themselves. The difficulty in switching between networks is to change the network flow without disrupting the corresponding application. Simple "brute force" switching, where one network is simply disabled and another enabled, causes periods of interruption of connectivity, which lead to losses of data and latency. While UDP (User Datagram Protocol) is a protocol traditionally used for applications that can handle this intermittent connectivity (e.g. VoIP, video streaming), its lack of ordering and error correction schemes make it unfeasible for people-centric sensing scenarios, since their robustness depends on the quality of sensed data. Unfortunately, TCP does not play well with "brute force" handoffs, since acknowledgments from the mobile host may not be delivered. TCP misinterprets the loss of data as a congestion problem and tries to re-send the missing packets, exponentially increasing the time between each unsuccessful retransmission (exponential backoff), while also reducing its window size when the network is congested. This results in long delays before retransmitting a data packet, even after connection has been reestablished [154]. Since 99.7% of all mobile traffic is TCP [155], this is a serious problem. In order to achieve seamless handoff, which are necessary for many HiTLCPSs, it is necessary to perform make-before-break handoff, that is activating a new network interface and deactivating the current one—dynamically and without interrupting existing connections. Smooth interface switching brings a multitude of advantages to HiTLCPSs since it increases the connectivity range and decreases intermittency.

Initial work on this area attempted to solve the problem of intermittent switching by proposing changes to the IP protocol, as with Mobile IP [156, 157]. Despite being a mature standard, Mobile IP has yet to achieve widespread adoption, and there are several drawbacks to the protocol that might explain its limited deployment. One of these drawbacks has to do with handoff operations, which are usually riddled with large delays [158] and high data loss rates [159, 160], making it an unfit protocol for an HiTLCPS's real-time traffic. While there are some extensions for improving handoff performance (namely FMIPv6 [158] and HMIPv6 [161]) and promoting flow mobility through PMIPv6 [162], these features come at the cost of modifying the standard to a great extent. This results in additional hurdles for achieving widespread deployment, since it is necessary for each of these approaches to be adopted on top of the standard by all parties. Additionally, both standard Mobile IPv6 and its extensions are highly dependent on the existence of proxies or gateways, which, considering the huge size of the current Internet, would be an extremely expensive endeavor in terms of additional hardware or implementation of software agents and other changes to currently deployed systems [155]. Finally, a large part of the Internet is made of "middleboxes" (routers, firewalls, and NATs) that modify TCP/IP headers to distribute traffic and that are not prepared to handle new protocols that they have not been designed to. It is also necessary to consider that changing the standard network protocols (IP/TCP) may not be compatible with existing applications and operating systems, also resulting in backward compatibility issues. This leads us to believe that network-level solutions such as Mobile IP are not sufficient to effectively solve the problem of seamless handoffs in a practical way.

There are, however, approaches that do not depend on additional infrastructure or on changes to current network protocols. The authors of [155] performed a characterization of IP traffic on smartphones for three months. They found that 99.7% of all mobile traffic is TCP, and that most TCP connections in current applications are short-lived (two seconds or less). Propelled by these results, they devised a client-side handoff technique based on the manipulation of the device's routing tables. This same approach was also employed in MultiNets [163] to create a system which is capable of seamlessly switching between wireless network interfaces on mobile devices, while considering different switching policies: energy saving, data offloading, and performance. Nevertheless, this handoff approach does not work for long-lived TCP flows: older TCP flows that are still connected to the previous interface are broken whenever the connection lives past the handoff time and the original network's signal is lost before the transmission finishes. This is quite a common occurrence, particularly in cases where links are transient (e.g. when the user is on a bus). While the authors make the assumption that this is not a problem for most applications, we believe that the need for reliable connections in HiTLCPSs makes such an assumption fall short.

A recent protocol, devised by the IETF, is capable of performing smooth handoffs while being compatible with regular TCP/IP. MultiPath TCP (MPTCP) is a modified version of TCP that implements multipath transport by pooling multiple TCP paths (or "subflows") within a transport connection in such a way that they appear as a single logical resource to the application [164]. Multiple disjoint network interfaces might be used simultaneously, thus avoiding intermittent connectivity by providing multiple connection paths that protect end hosts from failure of any single one. The protocol follows the same model as regular TCP and maintains backward compatibility with existing TCP APIs; changes are made exclusively at the OS level. MPTCP was designed to remain compatible with the Internet as it exists today, that is to be able to flow freely through existing middleboxes, supporting both IPv4 and IPv6 connections interchangeabl.

Previous extensions to known protocols, such as the Stream Control Transmission Protocol (SCTP) [165], also supported multihoming and multipath functions [166, 167, 168]. However, most middleboxes, such as firewalls and NAT devices, that are pervasive in our home and enterprise networks do not support any transport protocol that they haven't been designed for in advance. Since these entities require considerable knowledge of the network and transport layers to be able to handle ports and addresses, and keep track of the connection state, their unawareness of SCTP leads to the protocol being consequently blocked. SCTP is also fundamentally different from TCP in the sense that it is a message-oriented protocol, transporting data in the form of a sequence of messages, rather than a continuous stream of bytes. This means that applications would need to be modified to support MultiPath SCTP, while MultiPath TCP can work with current applications as they are. Additionally, SCTP uses a more complex packet format than TCP, and a cyclic redundancy checksum that is expensive to compute in software, making SCTP less appropriate for mobile scenarios where energy is constrained [169].

An interesting take on this issue was the Socialnets initiative [170], which aimed at enabling wireless devices, such as mobile phones or sensors, to socially network with each other in order to disseminate information. Instead of depending on end-to-end connectivity to transmit information, the idea is to take advantage of human social interactions, which allows for the existence of temporary short-range connections between devices worn by different people in close physical proximity with each other. Using these short-range connections between devices of people within the same social group, it is possible to share information relevant to diverse aspects of everyday human life. Thus, the project was dedicated to the creation of social-inspired opportunistic network protocols for HiTLCPSs, which could be used as a basis for sharing information pertaining to real-world social activities.

More recently, the attention of researchers and industry players has began to focus on the so-called 5G mobile technologies. From its conceptualization, the core tenets of IoT are being integrated within the next generation of mobile communication technologies. 5G will be ready for the massive amount of data advent from M2M communication. According to the book *Fundamentals of 5G Mobile Networks* [171], it is envisaged that this ultra-broadband and ultra-low latency wireless infrastructure will be able to provide fiber-like experience for mobile users as early as 2020. Advances expected in small-cell technologies will be aggregated with advanced antennas (mmWave, massive multiple-input and multiple-output) and additional spectrum to meet the 5G challenge.

One could argue that the advent of mobile 5G technologies will attenuate or even solve this problem. We believe that such a notion is merely an illusion; much like the increased 3G bandwidth allowed for much more complex network applications when compared to 2G, 5G will act as a mere "band-aid" compared to 4G. This is particularly apparent in the face of the tremendous evolution of smartphones and their increasing demand for advanced multimedia applications, such as ultra-high definition and 3D video or augmented reality.

Since the wireless spectrum is finite, and as international bandwidth demand continues to grow, the problem of allocating enough bandwidth for everyone becomes critical. Unless we witness the surge of a new disruptive technology in the next few years, once the densification limits of current cells are reached it will be increasingly difficult to further increase spectral efficiency levels. Thus, wider spectrum and a more efficient utilization of available resources remain the way forward [171].

In conclusion, future research work will need to consider the existing communication processes and protocols at the different OSI layers to evaluate their feasibility into real HiTLCPSs. If existing solutions are unable to support seamless interfacing between IoT devices, robots, WSNs, humans, and smartphones, while supporting reliability and mobility, new kinds of protocols and communication paradigms might have to be considered.

10.4 Localization

Determining the location of CPS elements, especially mobile nodes or humans, is critical for some types of IoA applications, since many times the data is meaningless if the location where it was generated is unknown. Location is also often critically important for supporting HiTL control-loop decisions in a timely manner. Thus, localization helps in the determination of the location of collected data, coming from people, animals, robots, or vehicles. It plays an important role in many types of HiTL-CPS scenarios, ranging from monitoring healthcare patients, people-centric sensing, mobile applications, monitoring of workers within hazardous environments, robotic drone positioning, smart homes, etc. The localization problem has been considered since the 1960s, resulting in the GPS location system that is widely in use today [172]. While GPS is an excellent solution for outdoor localization, it is not adequate for many types of devices. For example, cost and energy consumption constraints in wireless sensor nodes makes localization using GPS inefficient in most WSNs. In addition, there are some cases in which GPS is not feasible, such as indoor locations, underground tunnels, or places with a lot of obstacles. The accuracy of civil GPS units may also not satisfy the requirements of some HiTL applications. Despite some previous attempts at embedding GPS receivers into constrained devices [173] by offloading processing to the cloud, the resulting accuracy is still low (35m). Numerous other approaches for achieving localization have previously been proposed, notably those based on the "closest beacon principle" [174], WiFi-based positioning systems [175], Kalman filters [176], multilateration [177], and even machine learning techniques [178, 179].

A critical problem in localization is the accuracy and stability of the measurement methods, which is even more exacerbated in HiTLCPS, since these values may influence the result of the entire control-loop decision. Consequently, it is necessary to have scalable, low-cost, and near real-time localization systems which can lead to acceptable accuracy using the commonly available measurements for controlled HiTLCPSs. Thus, future research will need to study current localization methods in order to find appropriate solutions for determining location in different situations (outdoor, indoor) and for different elements (humans, robots, smartphones, sensor nodes).

10.5 State Inference

The accuracy and reliability of the inference of human states, is critical for HiTLCPSs [45]. This is a very broad requirement that includes the detection of many states related to the human, be it activities or actions, commands, intents, attention level, physiological parameters, psychological states, or emotions. Some of these aspects of human nature are more challenging than others.

As discussed in Section 4.1.2, the detection of physiological parameters is a topic that has long been debated in research, and we have plenty of devices capable of detecting a wide range of parameters, ranging from one's heart rate to the electromagnetic waves generated by neural activity. In terms of activity detection, current approaches can achieve high levels of accuracy for narrow ranges of activities in specific scenarios, such as medical environments or daily activities. However, despite many activity detection solutions reaching high levels of accuracy, these results are only valid for a limited number of activities and for a limited audience. The standard practice in most sensing systems is the use of unchanging classification models trained prior to deployment. When dealing with large-scale HiTLCPSs, this poses a big problem, since the target audience is highly heterogeneous (e.g. an elderly person walks in a way that is very different from how a young person does). Some recent research has attempted to address these issues through the personalization of existing classification models through manually provided training data [180], and by incorporating inter-person similarity into the process of classifier training and allowing crowd-sourced sensor data to personalize classifiers [93]. Another gap in current activity recognition research is the problem of flexibility of activities. The way a certain activity is performed may change over time: a person may develop quirks or get more efficient without even realizing it. The personalization of existing classification models shouldn't depend on manually provided training examples or labeling, it should be a transparent process that happens during daily life. Additionally, most research focuses on achieving high accuracy rates for a limited number of activities. In HiTLCPSs the number of activities of interest may be very high and change over time: it is limiting to develop a system that only handles a few of activities. A more interesting solution would allow the collaborative identification of new activities by users. This requires HiTL control to detect new types of activities that are not envisioned at the time of deployment. This requirement brings a number of challenges yet to be addressed: how to scale the introduction of new activities? How to avoid redundant labeling? How to perform lightweight classifier training in a fashion not too taxing on mobile hardware? These are important challenges that need to be addressed in order to achieve good contextual analysis in future HiTLCPSs.

On the other hand, the quantitative detection of the more abstract aspects of human nature, such as feelings and emotions, is less established. The emotions and psychological states are crucial aspects for improving relations, learning, health, and the quality of life of human beings. These emotional processes have a crucial value for determining human behavior in HiTLCPSs because they are the primary source of human motivation. The literature on emotion is very extensive, and there have been controversies even in its definition. The word "emotion" has its foundations in Latin *emovere*, a word that derives from *movi*, which means to "put in motion". Thus, first of all, emotion means movement, and without emotions nothing progresses. A more scientific definition of emotion can be that it is a psychological construction where cognitive, physiological, and subjective components interact. Several psychology researchers have focused on the problem of definition emotion. Early researchers proposed various models that grouped emotions into several categories. For example, Ortony *et al.* [181] established an architecture of conditions and variables which influenced emotions. In another attempt of emotion classification, Ekman studied human facial expressions and associated them with a set of six basic emotions, through the Facial Action Coding System [182], which is now widely used in the field of psychology, animation, and robotics. A circumplex model of emotion was first proposed by Russell [183], where emotions were distributed in a two-dimensional circular space, ranging from "miserable" to "pleased" and from "sleepy" to "aroused". The work of psychologist Magda Arnold, then followed by Richard Lazarus [184], resulted in the "appraisal theory", which states that emotions derive from our own evaluation of physical events, which then cause different reactions in different people. For example, if a certain event is perceived as positive, it will trigger a response that will evoke positive emotions; on the other hand, negative perceptions of reality will result in negative emotions.

The area of HiTLCPSs will have to build upon the fundamental works of psychology as a basis for accurate and reliable emotional classification, relating each emotion to associated physiological signals such as skin conductivity, blood pressure, heart rate, and breathing rate. Such areas are currently very active in computer science and engineering. Body and wireless sensors that measure these vital signals, video cameras for facial recognition, and several other devices are frequently used to capture emotional states. Nowadays, even the use of EEG sensors in an obtrusive way for measuring emotions is a realistic possibility, thanks to portable EEG devices such as the Emotiv [185]. Identifying physical activities and measuring quantitative emotion states is, however, still more challenging than measuring the associated physical parameters, being one of the major challenges for future HiTLCPS research.

10.6 Safety

Safety in actuation can become a primary concern in real-world HITLCPS deployments. This is particularly true when dealing with robotic actuation. As discussed in Section 4.1.3, robots are becoming progressively integrated in HiTLCPSs and involved in increasingly more complex and less structured environments and activities, including interaction with people for task execution. This means that there is a critical need for novel safety mechanisms that can ensure a safe and effective cooperation between humans and robots, that is robots need to start considering the "human-in-the-loop" aspect of working tasks.

10.7 In Summary...

In Table 10.1 we summarize the identified requirements and challenges faced by emerging and future HiTLCPSs and IoA.

Now that we have identified their requirements and challenges, let us consider the constraints that affect HiTL systems and review the lessons we have learned throughout the book.

Table 10.1 Summary of the identified HiTL requirements and challenges.

Requirement/Challenge	*Description*
	Resilience
Adaptability	It is necessary for operation in dynamic environments and to enable coordinated HiTL control
Performance	It must be kept under acceptable levels, even in the presence of mobility and a diversity of faults.
Dependability	HiTLCPSs should incorporate fault tolerance and self-healing mechanisms
	Security and Privacy
Lack of real-world research	Many protocols have only been evaluated in an isolated fashion, or are academic and based on theoretical or simulation results
Usability	It is important to avoid cumbersome authentication methods and keep users central in threat model security design
Human-driven vulnerabilities	There is a need to better understand the roles and relationships of humans in the context of pervasive systems, particularly in terms of mobility and sociability
Independence from human intervention	Humans should be a part of the security loop, but without a need for direct human intervention
Adaptable and manageable architecture	Security and privacy policies should be adapted and tailored to each individual in real-time
	Standard Communications
Heterogeneity	Communication must occur between humans and devices that are highly heterogeneous in terms of processing capabilities, size, and function
Reliability	Communication must remain reliable even in face of the highly crowded wireless spectrum, where different kinds of communication technologies co-exist
Dynamic management	Different contexts set dissimilar requirements for networking and each wireless technology adapts better to specific conditions
Mobile handoff	Multiple wireless mediums and network interfaces allow for more flexibility, but switching between them can cause periods of interruption of connectivity, which lead to losses of data and latency
Opportunistic networking	Temporary short-range connections between devices in close physical proximity can share delay-tolerant information relevant to diverse aspects of human context
Finite wireless bandwidth	It will be increasingly difficult to further increase spectral efficiency levels. Thus, a wider spectrum and more efficient utilization of available resources will become essential
	Localization
Cost and energy consumption	Limited and battery-powered devices struggle with the use of power-hungry location technologies (e.g. GPS)
Indoor location	GPS is not feasible indoors and, despite extensive research, indoor localization techniques remain challenging

(Continued)

Table 10.1 (Continued)

Requirement/Challenge	*Description*
	State Inference
Accuracy and reliability	There is a need for precise and consistent detection of human activities, commands, intents, physiological parameters, and psychological states or emotions
Heterogeneity of people	HiTLCPSs target people of different sexes, ages, cultural backgrounds, and health status
Flexibility and scope	The number of human activities of interest can be very high and dynamic, and the way each individual activity is performed can change over time
Emotions and psychological states	The quantitative detection of more abstract aspects of human nature, such as feelings and emotions, is highly difficult and yet to be fully understood
	Safety
Robotic actuation	It is necessary to ensure a safe and effective cooperation between humans and robots

11

Human-in-the-Loop Constraints

As suggested in Chapter 2, although many of the developments we have discussed so far happened in parallel and overlapped with each other, it is quite possible to identify a certain convergence. We believe that technological progress will always revert back to its origins: the adaptation of the environment to human beings, be it an ancient terrain that became a cultivated field or a world filled with intelligent devices that work together to accommodate human needs.

Throughout this book, we have observed many limitations in the current state of the art. Several limitations are of a technical nature, requiring additional research effort in order to be overcome. However, there are also limitations of a more ethical nature that relate to the public's acceptance of these new types of technological paradigms. We dedicate this chapter to the identification of both types of limitations–technical and non-technical–which, in fact, can be looked at as lessons learned from the by now long journey we initiated with the writing of this book.

11.1 Technical Limitations

Despite all of the development in terms of base technologies, only now are we beginning to devise how sensing, state inference, and actuation can be combined together in HiTLCPSs, as evidenced by the research projects described in Sections 4.1 and 4.2. In general, most of these projects still assume architectures within environments that are well known and static. We believe that future IoT environments will be mobile, dynamic, and reactive, where humans and technology will have to react in real time to stimuli from the environment, in order to guide their actions [186]. To this end, WSNs allow for the monitoring of environmental conditions, helping IoT devices, robots, and humans to react much more effectively to changes.

Additionally, most current scenarios do not fully consider humans, their behaviour, and their psychological state as integral parts of the system. Humans are still mostly seen as an external final user, and rarely directly interfere in the control loop of working tasks. As far as we know, there is no significant work that fully utilizes the potential of the human element to support the control system itself. In all previous projects there is a very well-defined border between humans and the system, instead of a tightly coupled integration. As we saw in Chapter 3, humans can play various roles within HiTLCPSs, ranging from actuators co-helped by robotic elements and acting on information

A Practical Introduction to Human-in-the-Loop Cyber-Physical Systems, First Edition.
David Nunes, Jorge Sá Silva and Fernando Boavida.

Companion Website URL: www.wiley.com/go/nunesloop

collected by the sensor networks to intermediate nodes in multi-hop communication processes. They can also become an element of environmental monitoring (through the sensors carried by them, e.g. on smartphones or smartshirts).

The pieces of work presented in [74], [124], and [16] are quite good demonstrations of the potential of HiTLCPSs. Still, we have some reservations about the feasibility of the presented approaches for widespread deployment. The use of vision-based systems is very prone to noise and limitations in image processing, only working for very controlled environments (e.g. recognizable objects limited to those programmed into the system). Brain/computer interfaces based on EEG signals are not practical, since electrodes are usually very cumbersome to wear and thus not suitable for day-to-day HiTL applications. This leads us to believe that near-future HiTLCPSs will most likely be based on more pervasive and mobile technology. In particular, the smartphone is a ubiquitous sensing platform that is already used by millions of people around the globe, every day. Theis device gives us the sensing power and computational capabilities that might be key for the first generation of massive HiTL deployments coming in the next few years. Still, there are few actual applications of smartphones and HiTLCPSs. While [118] did use HiTL concepts to limit the current mobile data demand, the actuation aspect was limited to suggestions and incentives on a smartphone's graphical user interface, and aspects such as robotics and direct actuation were not considered.

Let us attempt to condense all of these technical limitations and challenges in to a single model, shown in Figure 11.1. This model presents the various processes associated with HiTL control. A human is integrated into a CPS through **"human-in-the-loop intelligence"**, responsible for receiving input from the human sensors and also for influencing the system's control loop depending on the inferred context. This intelligence's specific implementation should follow the general principles and requirements introduced in Chapter 10, to guarantee reliable and secure human-context monitoring. In particular, we consider the issues of privacy and reliability as two of the most important requirements, largely responsible for the current lack of HiTLCPSs in real scenarios.

In a first step, determining a human's state requires the **acquisition of data**, through sensors (carried by the human or present in the environment) and/or from information present in social networks. This information can relate to several aspects of physical

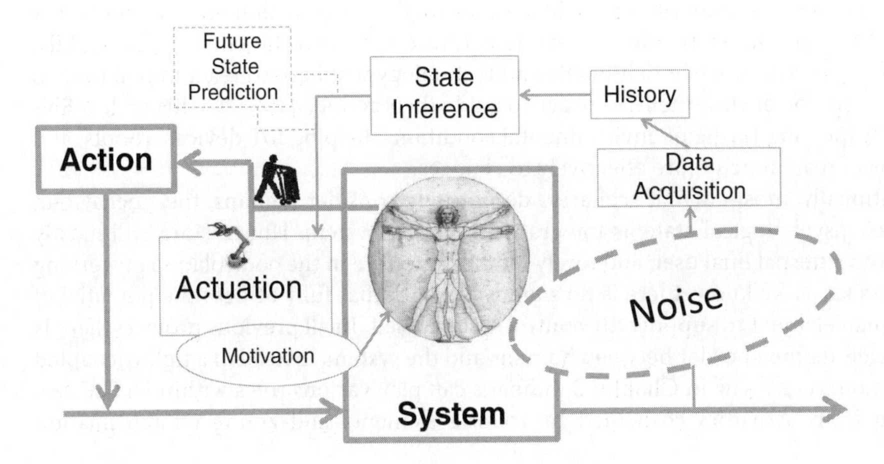

Figure 11.1 Lessons learned towards human-in-the-loop control.

reality, such as the human's thought patterns through EEG, who their friends are, their heart rate, movement through accelerometers, positioning through GPS, and facial expressions through video-cameras. Assessing physical reality through sensory data is the cornerstone of HiTL control, since every other aspect of the system is related to the raw data acquired from the sensors.

History, or memory, is another important aspect that is closely relate to the acquisition of data. In fact, research has shown how previous human states may provide important insights for inference mechanisms [187]. This historical data can also be used by delay-tolerant mechanisms in non-critical applications, setting a meaningful state whenever the real-time connection to sensory data is interrupted.

Perhaps one of the most critical aspects of HiTLCPSs is the reliable **inference of human state**. State inference mechanisms need to adapt to the current context as well as the human's preferences and historical behavior, integrating this information into the control-loop as feedback to determine the actions of the HiTLCPS. This is extremely difficult and implies a need for reliable and secure mechanisms for modeling, detecting, and possibly predicting human nature, as discussed in Section 4.1.2.

There are two types of **actuation** in HiTL controls. A system actuation is based on the system's current status and the inference of human state. For example, an HiTL-enabled HVAC system should only adapt the room temperature in the presence of humans. Human actuation relates to the actions of humans within the HiTL system, since they can themselves actuate whenever necessary. Motivation is a crucial aspect of this type of actuation and one of the most important research challenges. Future HiTLCPSs need to provide the necessary motivation and benefits for humans to act in a way that benefits the overall system and refrain from adopting greedy or prejudicial attitudes.

Finally, **noise** reminds us how real-world environments are far from idealized academic-controlled testbeds. For example, HiTLCPSs based on speech and video-captured gestures have to deal with challenges such as ambient noise, moving background clutter, and object segmentation. The acquisition of human vital signs is also prone to problems in terms of signal-to-noise ratios, where many signal frequencies result from internal physiological functions that have nothing to do with what needs to be acquired.

Another source of noise relates to human variability. The human species has a high genetic variance and thrives in many different environments with highly disparate cultural backgrounds, which results in many possible phenotypes. Age, physical disability, and interperson variability also need to be taken into account. While current research in HiTL state inference can reach high levels of accuracy, as discussed in Section 4.1.2, these results are mostly limited in terms of number of human activities, psychological states, and audience. On the other hand, future HiTLCPSs will most likely address a highly heterogeneous target audience. This personalization of existing state-inference models should follow a ubiquitous approach, and not overly depend on manually providing training examples or on the collaborative labeling by the system user. To promote usability, it should be a transparent process that happens naturally, as the user lives his/her daily life.

The identification of new human states that were not predicted at the time of deployment may also be important. However, this brings yet another realm of unresolved challenges. It is necessary to scale the learning of new states, avoid redundant labeling, perform training in a lightweight fashion, ensure security and privacy, and take advantage of collaboration between users while avoiding overlapping efforts.

All of these are important challenges for HiTLCPSs that have yet to be properly addressed by research in the field.

11.2 Ethical limitations

As discussed in Section 4.1, much of the necessary technology for supporting HiTLCPSs is already in place. But then, why are current IoT and CPSs still unable to integrate the human element into the control loop? As previously discussed, we believe that reliability is one of the major factors that influence the current lack of real-world deployments. Reliable and consistent inference of a human's state are essential for the adoption of HiTLCPS in real industrial, medical, or social scenarios. The inability to do so can have severe consequences on the effectiveness of the entire system. Reliable networking is also crucial for HiTLCPS, since these systems are often distributed and need to share information between many devices.

There is, however, another important factor that needs to be taken into consideration: the introduction of radically new technologies is usually accompanied by a considerable dose of skepticism. Thus, reliability is only relevant if the market accepts the underlying technology. This is crucial, since this new paradigm of human-centric technologies has already been previously met with some reservation. As evidenced by Section 2.2, current attempts at creating social-networking HiTLCPSs show that users place a high importance on their privacy and on the security of their personal information. In fact, these privacy concerns have been present since the beginning of social networking. Facebook, for example, has been the target of criticism since its beginnings because of its reliance on the user's willingness to share information as the key point of its business. In fact, according to an AP-CNBC pool [188] with a sample of 1004 people, 59% of Facebook users have little to no trust in Facebook to keep their information private. This apparent lack of trust reflects just how closely people follow intrusive practices, further exemplifying how privacy concerns are one of the biggest obstacles to the growth of social networking and, by extension, to HiTLCPSs.

Concrete examples of how such reservations may affect the introduction smartphone-based HiTLCPSs can be seen in the cases of Highlight and SceneTap, previously presented in Section 4.2.3. Concerning Highlight, Baig [189] argued that such encounters are sometimes best "left to fate" and that the application "may tell others too much about you", but still praised its functionality and novel features.

As for SceneTap, skeptics advocate privacy concerns and have raised questions around the facial detection technologies used to collect information, since they are employed without people's consent. The application met a troubling launch in May 2012, where it was supposed to be supported by 25 San Francisco bars of which ten dropped out after angry calls and an editorial that called the service "creepy". The app has also been criticized for its gender filtering options, letting people find bars with a larger proportion of men or women in a certain age range [190]. In addition to the ethical concerns, SceneTap was also plagued with technical limitations, as pointed out by Anderson [191]: the app apparently had accuracy problems with its facial recognition software, with several bars showing high attendance levels when in fact they were "as dead as can be". She also claimed that the software failed to register her presence when she attempted to enter a bar very slowly.

Putting skepticism aside, it is difficult to deny that the idea of someone else monitoring our every step and activity is very disturbing. However, it is also true that this problem does not reside entirely with the existence of HiTLCPS frameworks. For example, Sauvik *et al.* [192] discussed the possibility of current smartphones posing a security threat to the user, claiming that accelerometers and other sensors within the device could be used without the user's consent. They have also shown how activity recognition algorithms can be used to obtain sensitive information about users without their knowledge, by having them identify pre-defined general activities or even make the user's phone learn to identify new ones. Hence, the existence of smartphone-based HiTLCPSs does not impede this type of privacy invasion, although it might make it easier to accomplish.

Still, it would have been, perhaps, unthinkable in a pre-social-networking era that people would enjoy publishing their personal information in a public database for their peers to see and comment on. Yet, little by little, we have reached the stage where huge social networks and photo sharing are the norm. Despite all the past and ongoing privacy concerns and surrounding criticism, both the number of users and their engagement in mobile social networks continue to increase [73].

Nevertheless, security and privacy are two other critical requirements, in addition to reliability, for HiTLCPSs. Industrial processes, medical data, and sensitive personal information need to be protected from unauthorized exploitation. As already discussed, protecting confidential information is often not only a business requirement but, in many cases, also an ethical and legal requirement.

Another important ethical consideration relates to the use of robotics in HiTLCPSs. As mentioned in Section 4.1.3, robotics is growing at a progressively faster pace and there are some who believe its role in future HiTLCPSs may not be completely optimistic. For example, while robotics enables automation, this may in turn result in human unemployment. In fact, futuristic journalist Kevin Kelly predicts that a wave of automation centered on artificial cognition, cheap sensors, machine learning, and distributed intelligence will likely result in 70% of today's occupations being replaced by automation before the end of this century. Starting with assembly line and warehouse work, agriculture picking, cleaning, "it doesn't matter if you are a doctor, lawyer, architect, reporter, or even programmer: The robot takeover will be epic" [107].

Brynjolfsson and McAffee provide an interesting insight into this matter, arguing that despite the improvement of technology in areas that used to be typically human-oriented, such as pattern recognition, people will still have vital roles to play [67]. As an example, they refer to Garry Kasparov's experience in "freestyle" chess tournaments, where teams combining average-skilled humans and machines dominated both strong computers and human grandmasters [193]. As pointed out in Diego Rasskin-Gutman's book, *Chess Metaphors,* what computers are good at is where humans are weak, and vice versa [194]. This is evidence of the importance of human–machine collaboration in the years to come, the cornerstone of HiTLCPSs. Brynjolfsson and McAffee continue their discussion on these "uniquely human" abilities that will remain essential, even in the face of the continued automation of routine tasks by technological advancement. Despite their impressive calculation capabilities, there has yet to exist a machine that is capable of human creativity and intuition. The ability to create and innovate through new and meaningful ideas preoccupies the pinnacle of AI (AI) research, and is the one task that humans still excel in comparison to machines. Additionally, evolution has shaped humans into highly responsive beings that can quickly adapt to new situations, while current machines simply cannot react outside of the frame of their programming. As evidenced by Brynjolfsson and McAffee, "[The supercomputer] Watson is an amazing *Jeopardy!* player, but would be defeated by a child at *Wheel of Fortune, The Price is Right*, or any other TV game show unless it was substantially reprogrammed by its human creators" [67]. Thus, human–machine collaboration will most likely become increasingly critical in the next few decades, at least until machines evolve to a point where they reach (or surpass) "human-like" intelligence. As memorization skills become increasingly redundant owing to the assistance of modern search engines, it is this human ability to quickly combine information from different sources and to react to new situations that will remain essential in future HiTLCPSs.

Precursors of this human–machine interaction are already among us. Baxter, a workbot from Rethink Robotics, is an early example of a new class of industrial robots created to work alongside humans [195]. Baxter has several characteristics that make it more "human-aware" than most of its ancestors. It is capable of showing where it is looking by shifting drawn eyes on its "head". It is also capable of perceiving humans and avoid injuring them, using force-feedback mechanisms that tell it it is colliding with a person or another bot. This "human-like" body language is an innovation that allows humans to understand and predict the robot's intentions, which may in turn reduce the previous mistrust placed in robotic companions [108, 109]. Equally important is Baxter's capability of learning through imitation: to train it, one simply grabs its arms and guides them through the correct motions and sequence. This mode of operation is remarkably different from traditional industrial robotics, which requires highly educated personnel to program even the simplest tasks. Considering all of these tendencies, it is very likely that, in the future, people will be paid "based on how well they work with robots" [107].

Another good example of how robotics and HiTL are becoming intimately related is Pepper, a humanoid robot designed by Aldebaran Robotics and SoftBank Mobile who is capable of reading human emotions [196]. Unlike the Baxter workbot, Pepper is an emotional robot, not a functional robot. It was designed with the purpose of making people happy, making them grow, enhancing their life, and facilitating their relationships. To do so, it is capable of communicating through "voice, touch and emotions", maintaining conversations, and "having fun". For example, Pepper is capable of understanding laughter and associate it with "good mood". It does this through knowledge of "universal emotions" (joy, surprise, anger, doubt, and sadness) and its ability to analyze facial expression, body language, and language. Its also adapts its behavior according to the situation, and employs machine learning to better understand humans, being capable of learning its user's tastes.

It is very likely that Baxter and Pepper are just the first signs of a new technological revolution that is quickly approaching. On one hand, expecting AI to evolve until it becomes "humanlike" is "the same flawed logic as demanding that artificial flying be birdlike, with flapping wings" [67]. It has already been proven that tremendously complex programs, despite being based on simple instructions, are already able to outperform human thinking. On the other hand, innovative, creative, and imaginative machines are yet to be seen and it is unlikely that humans will be replaced in this department in the decades to come… or is it?

Developments in AI research are continuously casting shadows of doubt on the prowess of the human mind. One of the last bastions of the human mind against AI evolution was the ancient game of Go: originated in ancient China, more than 2500 years ago, Go remains the oldest board game still played today. Unlike chess, which has long been beaten by machines, Go had, until recently, been considered a tremendous challenge for AI. As put by mathematician I. J. Good in 1965 [197]:

> In order to programme a computer to play a reasonable game of Go [...] it is necessary to formalise the principles of good strategy, or to design a learning programme. The principles are more qualitative and mysterious than in chess, and depend more on judgment. So I think it will be even more difficult to programme a computer to play a reasonable game of Go than of chess.

In fact, for a very long time, most computer Go programs were considered worse than an average player with just a few years, experience; Go is a game that may take an entire lifetime to master. However, that all changed very recently.

In January 2016, Google published a paper on how they managed to build an AI, named AlphaGo, that won a match against a professional Go player: the European champion Fan Hui [198]. Hui, ranked *2-dan*, lost a five-game match at the Google DeepMind office in London in October, without handicaps. In March 2016, AlphaGo and South Korea's Lee Sedol, considered one of the highest-ranking Go players of the last decade, played an historic five-game match. AlphaGo won the match (4–1), making it the first time a computer Go program defeated a world-class human player on even terms [199]. Mankind may have just lost its last bastion against computer intelligence, at least as far as board games are concerned.

However, what about scenarios where machine intelligence can have a more dire impact on human lives? This is the case for autonomous vehicles (AVs), which are about to become a tremendously important type of CPS. Projects such as Google's Self-Driving Car [200] are racing ahead, trying to create fully automatic cars that simply require no human driving whatsoever. It is rather interesting to consider the possibility of such vehicles adapting their behavior or their interior environment to their occupants' desires. Should a self-driving car attempt to go faster if its owner is in a hurry? What about slow leisure drives across the country? Could a driverless car be capable of adapting its driving behavior and route in order for its occupants to enjoy the scenery?

One of the most prominent reasons in favor of automated vehicles is the possible reduction of deaths from traffic accidents. However, in these HiTL scenarios, the consequence of failure certainly has a greater impact than losing a freestyle chess match or mis-detecting a human emotion in a smartphone app. What if, in its attempt to please its owner by changing its driving condition, the vehicle ends up being forced to choose between two evils? In a situation where an accident is inevitable, should the vehicle run over pedestrians or sacrifice its passengers? This dilemma is interestingly discussed by Bonnefon *et al.* in their report *The Social Dilemma of Autonomous Vehicles* [201]. The authors pointed out how the potential consumers may be more willing to ride in AVs that protect their passengers at all costs.

How can this selfishness factor be input into the design of decision-making algorithms for AVs? Will car manufacturers favor AI that values the desires and safety of their passengers over other individuals? According to Bonnefon *et al.*, manufacturers and regulators will need to accomplish three potentially incompatible objectives: being consistent, not causing public outrage, and not discouraging buyers [201]. These are difficult ethical decisions that will have a profound impact on the adoption of this and other types of HiTLCPS technology that can influence human integrity.

From all of this, we can safely say that while intelligent HiTLCPSs will most likely "think" very differently from us, at the same time they will further integrate humans and their intuition into their own control-loop tasks. Without a doubt, for better or for worse, HiTLCPSs are here to stay and will become increasingly more prominent and ubiquitous in our daily lives. In the face of this, it is now high time that we, humans, take decisions and act, and do not limit ourselves to observing the long-term consequences of such systems and how they will transform our world and the way we live.

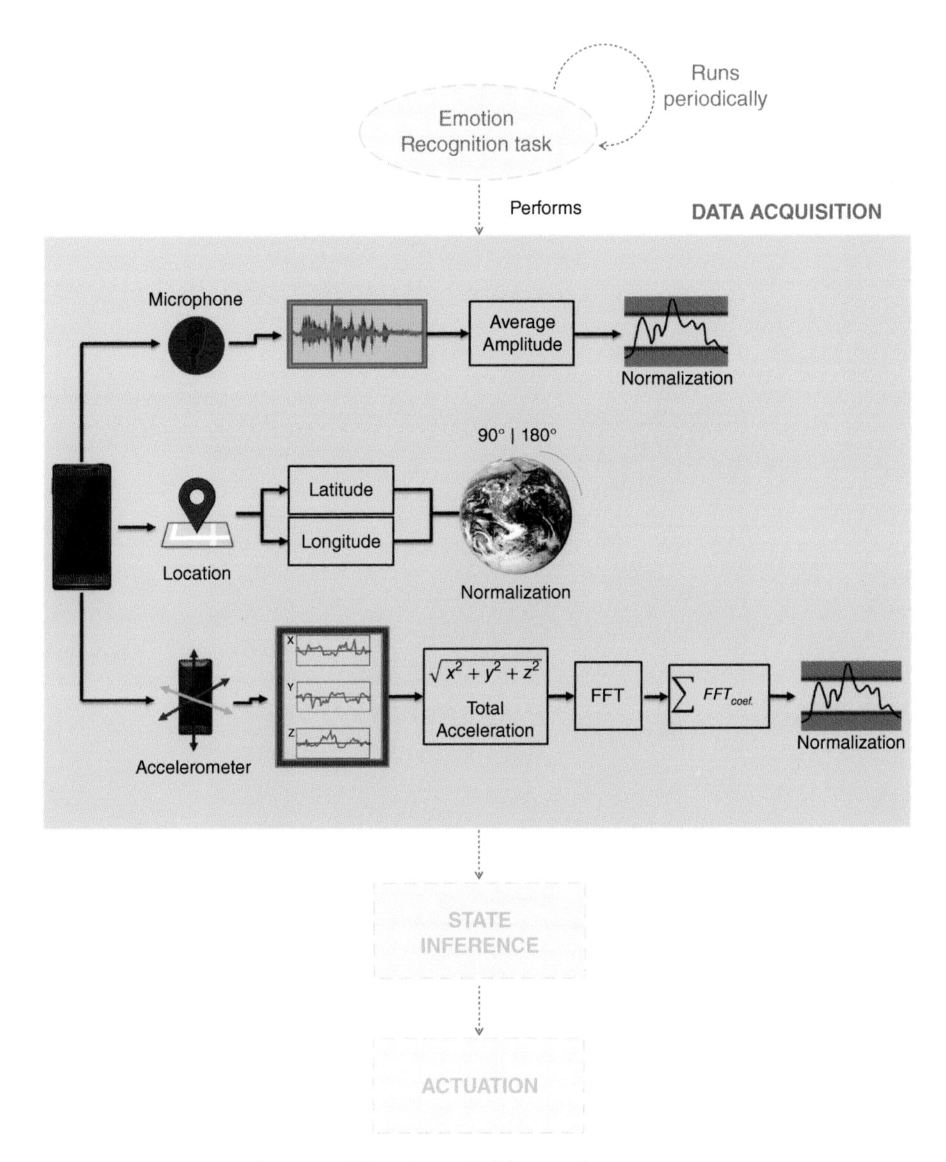

Figure 7.7 Current state of our HiTLCPS at the end of Chapter 7.

A Practical Introduction to Human-in-the-Loop Cyber-Physical Systems, First Edition.
David Nunes, Jorge Sá Silva and Fernando Boavida.

Companion Website URL: www.wiley.com/go/nunesloop

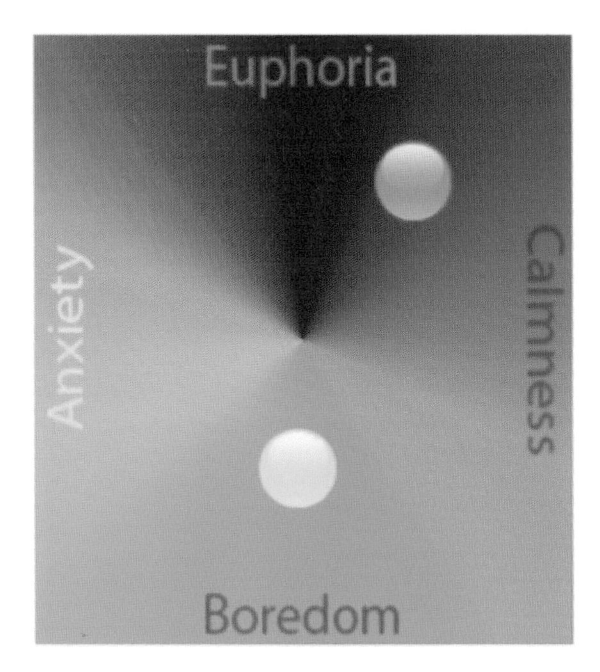

Figure 8.5 Our goal for the *EmotionSpace* view.

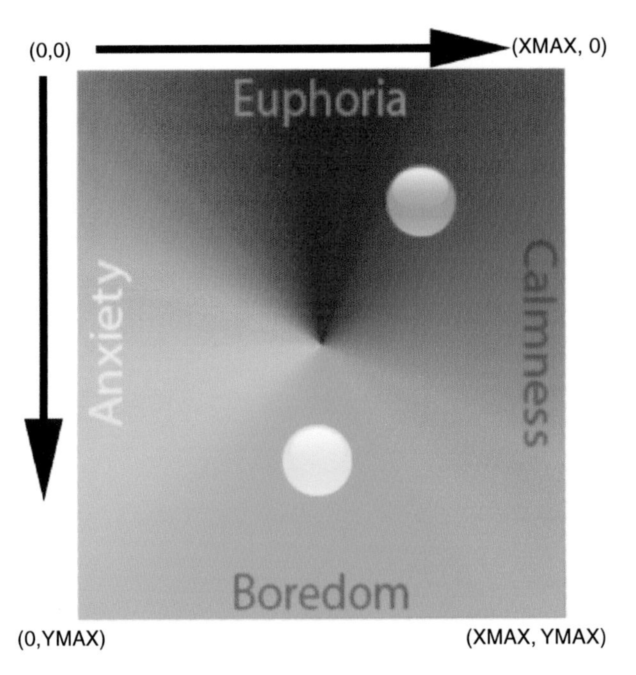

Figure 8.12 The coordinates of the *EmotionSpace* view.

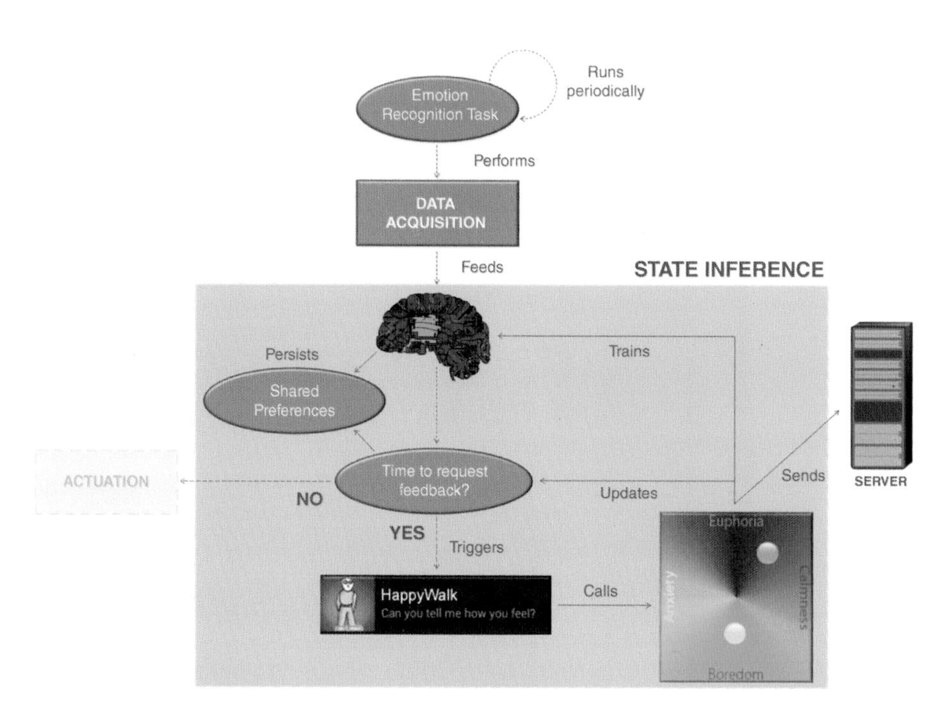

Figure 8.15 Current state of our HiTLCPS at the end of Chapter 8.

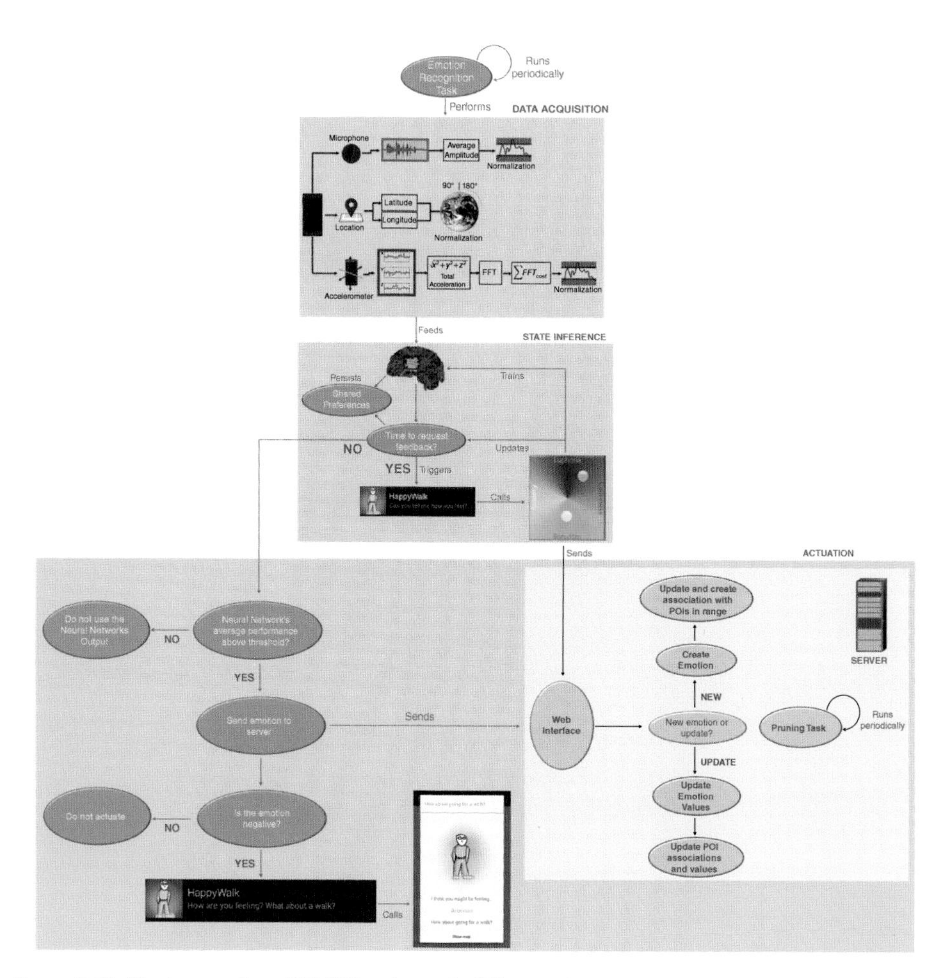

Figure 9.12 Final state of our HiTLCPS at the end of Chapter 9.

A

EmotionTasker's full code

This appendix provides a final overview of HappyWalk's *EmotionTasker* class so that it may serve as a reference to the reader.

```
package hitlexamples.happywalk.service;

import android.app.AlarmManager;
import android.app.PendingIntent;
import android.content.Intent;
import android.content.SharedPreferences;
import android.media.RingtoneManager;
import android.os.Bundle;
import android.os.Handler;
import android.os.PowerManager;
import android.preference.PreferenceManager;
import android.support.v4.app.NotificationCompat;
import android.util.Log;

import com.google.android.gms.maps.model.LatLng;
import com.ubhave.sensormanager.ESException;
import com.ubhave.sensormanager.ESSensorManager;
import com.ubhave.sensormanager.config.GlobalConfig;
import com.ubhave.sensormanager.data.pull.AccelerometerData;
import com.ubhave.sensormanager.data.pull.MicrophoneData;
import com.ubhave.sensormanager.sensors.SensorUtils;

import org.encog.engine.network.activation.ActivationSigmoid;
import org.encog.ml.data.basic.BasicMLDataSet;
import org.encog.neural.networks.BasicNetwork;
import org.encog.neural.networks.layers.BasicLayer;
import org.encog.neural.networks.training.propagation.resilient.
    ResilientPropagation;

import java.util.Random;

import hitlexamples.happywalk.R;
import hitlexamples.happywalk.activities.EmotionFeedback;
import hitlexamples.happywalk.activities.MapsActivity;
import hitlexamples.happywalk.emotion.processors.HwAccelerometerProcessor;
import hitlexamples.happywalk.emotion.processors.HwLocationProcessor;
import hitlexamples.happywalk.emotion.processors.HwMicrophoneProcessor;
import hitlexamples.happywalk.exceptions.NoCurrentPosition;
import hitlexamples.happywalk.tasks.TaskSendEmotion;
```

A Practical Introduction to Human-in-the-Loop Cyber-Physical Systems, First Edition.
David Nunes, Jorge Sá Silva and Fernando Boavida.

Companion Website URL: www.wiley.com/go/nunesloop

```
    import hitlexamples.happywalk.utilities.GlobalVariables;
40
    public class EmotionTasker {
42      private HappyWalkService hWServ;
        private Handler hWServiceHandler;
44      private BasicNetwork network;
        private ESSensorManager esSensorManager;
46      private EmotionRecognitionTask emotionRecog;
        PowerManager.WakeLock wakeLock;
48
        /*this variable keeps track of the time when we fired our last
50      emotion notification */
        private long lastEmotionNotifMillis = 0;
52      private NotificationRemovalTask currentNotifRemovTask;

54      /*
        we keep the current base time, the last time and the
56      previous euclidean distance for emotion feedback requests
        in memory, to avoid constantly accessing the disk
58       */
        private long baseTimeToNextEmoFdbckReq = 0;
60      private long lastEmoFeedbackReq = 0;
        private float wMeanEuclideanDistance = 0;
62
        public EmotionTasker(HappyWalkService hWServ) {
64          this.hWServ = hWServ;
            this.hWServiceHandler = hWServ.getHappyWalkServiceHandler();
66          //preparing sensor manager to fetch data
            try {
68              esSensorManager = ESSensorManager.getSensorManager(hWServ);
                esSensorManager.setGlobalConfig(GlobalConfig.
        PRINT_LOG_D_MESSAGES, false);
70          } catch (ESException e) {
                e.printStackTrace();
72          }
            //fetch the base emotion feedback time interval
74          restoreEmoFeedbackValsFromPreferences();
            //initialize Neural Network
76          initNetwork();
            //restore neural network weights from our preferences, if we have
        them
78          restoreNNWeightsFromPreferences();

80          emotionRecog = new EmotionRecognitionTask();
            PowerManager powerManager = (PowerManager) hWServ.getSystemService
        (hWServ.POWER_SERVICE);
82          wakeLock = powerManager.newWakeLock(PowerManager.PARTIAL_WAKE_LOCK
        , "EmotionTaskerWakeLock");
        }
84
        /**
86       *This method initializes a Neural Network with
         * two hidden layers, three neurons in the first and two
88       * neurons in the second
         */
90      private void initNetwork(){
            network = new BasicNetwork();
92          network.addLayer(new BasicLayer(null, true, GlobalVariables.
        NN_INPUTS));
```

```
        network.addLayer(new BasicLayer(new ActivationSigmoid(), true,
    GlobalVariables.NN_HL1_NEURONS));
94      network.addLayer(new BasicLayer(new ActivationSigmoid(), true,
    GlobalVariables.NN_HL2_NEURONS));
        network.addLayer(new BasicLayer(new ActivationSigmoid(), false,
    GlobalVariables.NN_OUTPUTS));
96      network.getStructure().finalizeStructure();
        network.reset();
98  }

100 /**
     * Begins the process of scheduling emotion recognition tasks
102  */
    public void startEmotionRecognitionTasks() {
104     /*First, check if we should perform emotion recog right now. */
        long timeToNextEmoRecog = nextEmotionExecutionMillis();
106
        /*Let us compare with the time when the last emotion recog was
    performed. */
108     if (lastEmotionNotifMillis + timeToNextEmoRecog < System.
    currentTimeMillis()) {
            //enough time has passed
110         postEmotionRecognitionTask();
        }
112     else {
            /* not enough time has passed.
114         Let us schedule a emotion recog using the remaining time */
            emotionRecog.scheduleEmotionRecog(lastEmotionNotifMillis +
    timeToNextEmoRecog);
116     }
    }
118
    /**
120  * Posts a new emotion recognition task
     */
122 public void postEmotionRecognitionTask() {
        /* We need the device's CPU to remain
124     awake while we perform our emotion recognition */
        wakeLock.acquire();
126     hWServiceHandler.post(emotionRecog);
    }
128
    public void stopEmotionRecognitionTasks() {
130     if (emotionRecog != null) {
            hWServiceHandler.removeCallbacks(emotionRecog);
132         emotionRecog.cancelEmotionRecog();
        }
134 }

136 /**
     * Handles user emotion feedback. Training of the neural network only
    takes place if the emotion notification has not expired.
138  * This avoids the possible issue of users providing feedback long
    after our notification has been shown.
     */
140 public void processUserFeedback(double[] inputs, double[] outputs,
    double[] idealOutput, long emotionTimestamp) {
        /*
```

```
142         We request a timestamp instead of relying on our internal
        lastEmotionNotifMillis because there is a possibility that a new
        emotion was inferred while the user was still using the feedback
        screen. Thus, while the lastEmotionNotifMillis works for notifications
        , there are no guarantees that it is associated with the emotion we
        are about to process right now. Hence, timestamps are necessary.

144         If no new notifications were created in the meantime, we can
        cancel the notification removal tasks and revert the notification to
        its normal state.
         */
146         if (emotionTimestamp == lastEmotionNotifMillis) {
                //cancel notification removal task
148             hWServiceHandler.removeCallbacks(currentNotifRemovTask);
                //revert notification
150             hWServ.showNotification(false);
            }
152         //perform training and send to server, if emotion has not expired
            if (System.currentTimeMillis() - emotionTimestamp <
        GlobalVariables.EXPIRE_EMOTION_MILLIS) {
154             hWServiceHandler.post(new NeuralNetworkTrainingTask(inputs,
        idealOutput));
                sendEmotionToServer(idealOutput);
156             //update emotion feedback frequency
                hWServiceHandler.post(new UpdateEmotionAccuracyTask(outputs,
        idealOutput));
158         }
            else {
160             Log.d("NEURALNETWORK TRAINING", "Expired emotion, discarding
        feedback...");
            }
162     }

164     private void restoreNNWeightsFromPreferences() {
            //Try to get network from preferences first
166         SharedPreferences pref = PreferenceManager.
        getDefaultSharedPreferences(hWServ);
            if (pref.contains(GlobalVariables.PREF_NEURALNETWORK_WEIGHT_KEY))
        {
168             String weights = pref.getString(GlobalVariables.
        PREF_NEURALNETWORK_WEIGHT_KEY, null);
                if(weights!=null) {
170                 String[] weights_string_array = weights.split(",");
                    double[] weights_array = new double[weights_string_array.
        length];
172                 int i = 0;
                    for (String value : weights_string_array) {
174                     weights_array[i++] = Double.parseDouble(value);
                    }
176                 network.decodeFromArray(weights_array);
                }
178         } else {
                Log.d("NEURAL NETWORK", "No weights were found; neural network
        has been reset");
180         }
        }
182
        private void saveNNWeightsToPreferences(String weights) {
```

```
184        SharedPreferences pref = PreferenceManager.
    getDefaultSharedPreferences(hWServ);
           SharedPreferences.Editor editor = pref.edit();
186        editor.putString(GlobalVariables.PREF_NEURALNETWORK_WEIGHT_KEY,
    weights);
           editor.commit();
188    }

190    private void restoreEmoFeedbackValsFromPreferences() {
           SharedPreferences pref = PreferenceManager.
    getDefaultSharedPreferences(hWServ);
192
           /* if, for some reason, we cannot find this value, we use
    getDefaultBaseEmoFeedbackTime() */
194        baseTimeToNextEmoFdbckReq = pref.getLong(GlobalVariables.
    PREF_EMOTION_BASE_FEEDBACK_MILLIS_KEY,
                   getDefaultBaseEmoFeedbackTime());
196
           /* if, for some reason, we cannot find this value, set it to zero
    to ensure a feedback request in the next emotion detection task */
198        lastEmoFeedbackReq = pref.getLong(GlobalVariables.
    PREF_LAST_EMOTION_FEEDBACK_REQ_KEY, 0);

200        // same thing for lastEmotionNotifMillis
           lastEmotionNotifMillis = pref.getLong(GlobalVariables.
    PREF_LAST_EMOTION_NOTIF_KEY, 0);
202
           /* by default, we define half of largest distance possible to
    ensure that new neural nets will require some training before having
    their results taken into account
204         */
           float maxEuclideanDistance = (float) Math.sqrt(GlobalVariables.
    NN_OUTPUTS);
206        wMeanEuclideanDistance = pref.getFloat(GlobalVariables.
    PREF_NEURALNETWORK_EUCLDIST_WGT_AVG_KEY,
                   maxEuclideanDistance/2);
208
       }
210
       /**
212     * Stores the current value for the base and last emotion feedback
       times into the shared preferences
        */
214    public void saveEmoFeedbackValsToSharedPreferences() {
           SharedPreferences pref = PreferenceManager.
    getDefaultSharedPreferences(hWServ);
216        SharedPreferences.Editor editor = pref.edit();
           editor.putLong(GlobalVariables.
    PREF_EMOTION_BASE_FEEDBACK_MILLIS_KEY, baseTimeToNextEmoFdbckReq);
218        editor.putLong(GlobalVariables.PREF_LAST_EMOTION_FEEDBACK_REQ_KEY,
      lastEmoFeedbackReq);
           editor.putLong(GlobalVariables.PREF_LAST_EMOTION_NOTIF_KEY,
    lastEmotionNotifMillis);
220        editor.putFloat(GlobalVariables.
    PREF_NEURALNETWORK_EUCLDIST_WGT_AVG_KEY, wMeanEuclideanDistance);
           editor.commit();
222    }

224    /**
```

```
     * Returns the default baseTimeToNextEmoFdbckReq value
226  * by averaging between the maximum and minimum values.
     */
228 private long getDefaultBaseEmoFeedbackTime() {
        //number of milliseconds in an hour = 3600000
230     return (long) ((GlobalVariables.RECOG_MAX_HOURS_WITHOUT_FEEDBACK +
     GlobalVariables.RECOG_MIN_HOURS_WITHOUT_FEEDBACK) / 2
                * 3600000);
232 }

234 /**
     * The inputs are collected through the ubhave module
236  * We use ESSensorManager's default sense window time
     * @return - an array of doubles containing the normalized (0-1)
238  * collected inputs. The indexes are defined in GlobalVariables
     */
240 private double[] collectInputs() throws NoCurrentPosition{
        double[] inputs = null;
242     LatLng actualPosition;
        /*
244     first, check if we have location information. This is required for
     performing emotion recognition
         */
246     if((actualPosition = hWServ.getHwLocationListener().
    getActualposition()) != null) {
            try {
248             //normalize location data
                double[] normalizedLocation = HwLocationProcessor.
    normalizeLatLng(actualPosition);
250             //collect and process microphone and accelerometer data
                MicrophoneData micData = (MicrophoneData) esSensorManager.
    getDataFromSensor(SensorUtils.SENSOR_TYPE_MICROPHONE);
252             double averageMicValue = HwMicrophoneProcessor.
    getAverageAmplitude(micData);
                averageMicValue = HwMicrophoneProcessor.
    normalizeAvgAmplitude(averageMicValue);
254
                AccelerometerData accData = (AccelerometerData)
    esSensorManager.getDataFromSensor(SensorUtils.
    SENSOR_TYPE_ACCELEROMETER);
256             double normFCTCoeffSum = HwAccelerometerProcessor.
    getNormalizedFCTCoeffSum(HwAccelerometerProcessor.getTotalAcceleration
    (accData));

258             //insert inputs into array
                inputs = new double[GlobalVariables.NN_INPUTS];
260             inputs[GlobalVariables.NN_INPUT_ARRAY_INDEX_LATITUDE] =
    normalizedLocation[HwLocationProcessor.LATITUDE_INDEX];
                inputs[GlobalVariables.NN_INPUT_ARRAY_INDEX_LONGITUDE] =
    normalizedLocation[HwLocationProcessor.LONGITUDE_INDEX];
262             inputs[GlobalVariables.NN_INPUT_ARRAY_INDEX_NOISE] =
    averageMicValue;
                inputs[GlobalVariables.NN_INPUT_ARRAY_INDEX_MOVEMENT] =
    normFCTCoeffSum;
264         } catch (ESException e) {
                e.printStackTrace();
266         }
        }
268     else{
```

```
        //There is no location information.
270     throw new NoCurrentPosition("No current position available, cannot perform emotion classification.");
    }
272     return inputs;
}
274
/**
276  * Returns a pseudo-random value (in Milliseconds) that represents
 * the amount of time until the next emotion recognition task
278  */
private long nextEmotionExecutionMillis() {
280     Random rand = new Random();
    int randomNum = rand.nextInt(
282             (GlobalVariables.RECOG_EMOTION_MAX_MINUTES - GlobalVariables.RECOG_EMOTION_MIN_MINUTES) + 1) + GlobalVariables.RECOG_EMOTION_MIN_MINUTES;
    return (long) 1000*60*randomNum;
284 }

286 private void showEmotionFeedbackNotification(double[] inputs, double[] outputs) {
    //First, cancel previous notification removal tasks
288     hWServiceHandler.removeCallbacks(currentNotifRemovTask);
    //Now, prepare a Bundle with the information to be passed to EmotionFeedback
290     Bundle bnd = new Bundle();
    bnd.putDoubleArray(GlobalVariables.BND_EXTRA_EMOTION_INPUT_ARRAY_KEY,
292             inputs);
    bnd.putDoubleArray(GlobalVariables.BND_EXTRA_EMOTION_OUTPUT_ARRAY_KEY,
294             outputs);
    /* put a timestamp on this bundle to avoid expired feedback. We store this same value within lastEmotionNotifMillis long, to keep track of when the last emotion feedback notification was sent.*/
296     lastEmotionNotifMillis = System.currentTimeMillis();
    bnd.putLong(GlobalVariables.BND_EXTRA_EMOTION_TIMESTAMP_KEY, lastEmotionNotifMillis);
298     bnd.putInt(GlobalVariables.BND_EXTRA_REQ_CODE_KEY,
            GlobalVariables.AREQ_EMOTION_FEEDBACK_NOTIF);
300
    Intent intent = new Intent(hWServ, EmotionFeedback.class);
302     intent.putExtras(bnd);

304     PendingIntent resultPendingIntent =
            PendingIntent.getActivity(
306                     hWServ,
                    0,
308                     intent,
                    PendingIntent.FLAG_UPDATE_CURRENT
310             );

312     NotificationCompat.Builder mNotifyBuilder = new NotificationCompat.Builder(hWServ)
            .setTicker(hWServ.getResources().getString(R.string.app_name) + " " + hWServ.getResources().getString(R.string.emotionFeedbackNotifContent))
```

```
314                 .setContentTitle(hWServ.getResources().getString(R.string.
    app_name))
                    .setContentText(hWServ.getResources().getString(R.string.
    emotionFeedbackNotifContent))
316                 .setSmallIcon(R.drawable.emot_notif_icon)
                    .setContentIntent(resultPendingIntent)
318                 .setOngoing(true)
                    .setSound(RingtoneManager.getDefaultUri(RingtoneManager.
    TYPE_NOTIFICATION));
320
        hWServ.getNotificationManager().notify(
322             hWServ.hHNotificNum,
                mNotifyBuilder.build());
324 }

326 private void showNormalEmotionNotification(int typeOfEmotion) {
        //First, cancel previous notification removal tasks
328     hWServiceHandler.removeCallbacks(currentNotifRemovTask);
        //Now, prepare a Bundle with the necessary information
330     Bundle bnd = new Bundle();
        /* put a timestamp on this bundle to avoid the user clicking
    notifications that have been fired a long time ago. */
332     lastEmotionNotifMillis = System.currentTimeMillis();
        bnd.putLong(GlobalVariables.BND_EXTRA_EMOTION_TIMESTAMP_KEY,
334             lastEmotionNotifMillis);
        bnd.putInt(GlobalVariables.BND_EXTRA_EMOTION_TYPE_NOTIF_KEY,
    typeOfEmotion);
336     bnd.putInt(GlobalVariables.BND_EXTRA_REQ_CODE_KEY,
                GlobalVariables.AREQ_EMOTION_NORMAL_NOTIF);
338
        // We will show our map, to promote walking when emotions are
    negative
340     Intent intent = new Intent(hWServ, MapsActivity.class);
        intent.putExtras(bnd);
342
        PendingIntent resultPendingIntent =
344             PendingIntent.getActivity(
                        hWServ,
346                     0,
                        intent,
348                     PendingIntent.FLAG_UPDATE_CURRENT
                );
350
        NotificationCompat.Builder mNotifyBuilder = new NotificationCompat
    .Builder(hWServ)
352             .setTicker(hWServ.getResources().getString(R.string.
    app_name) + " " + hWServ.getResources().getString(R.string.
    emotionNormalNotifContent))
                .setContentTitle(hWServ.getResources().getString(R.string.
    app_name))
354             .setContentText(hWServ.getResources().getString(R.string.
    emotionNormalNotifContent))
                .setSmallIcon(R.drawable.emot_notif_icon)
356             .setContentIntent(resultPendingIntent)
                .setOngoing(true)
358             .setSound(RingtoneManager.getDefaultUri(RingtoneManager.
    TYPE_NOTIFICATION));

360     hWServ.getNotificationManager().notify(
```

```
362             hWServ.hHNotificNum,
                mNotifyBuilder.build());
    }
364
    public static int getTypeOfEmotion(double[] emotionArray) {
366     /*
        for code readability, let us use an (x,y) representation of the
368     emotion color map:

370     (0.0)___________
         y |    Eph     |
372        |            |
           |Anx     Clm |
374        |            |
           |____Brd_____|
376                   x  (1,1)
         */
378
        double y = emotionArray[GlobalVariables.
NN_OUTPUT_ARRAY_INDEX_EUPHORIC_BORED];
380     double x = emotionArray[GlobalVariables.
NN_OUTPUT_ARRAY_INDEX_ANXIOUS_CALM];
        int typeOfEmotion;
382
        if (y < 0.5 && x < 0.5) {
384         if (y>x){
                //anxiety
386             typeOfEmotion = GlobalVariables.EMOTION_ANXIETY;
            }
388         else {
                //euphoria
390             typeOfEmotion = GlobalVariables.EMOTION_EUPHORIA;
            }
392     }
        else if (y < 0.5 && x >= 0.5)
394     {
            if (y>x){
396             //calmness
                typeOfEmotion = GlobalVariables.EMOTION_CALMNESS;
398         }
            else {
400             //euphoria
                typeOfEmotion = GlobalVariables.EMOTION_EUPHORIA;
402         }
        }
404     else if (y >= 0.5 && x < 0.5)
        {
406         if (y>x){
                //anxiety
408             typeOfEmotion = GlobalVariables.EMOTION_ANXIETY;
            }
410         else {
                //boredom
412             typeOfEmotion = GlobalVariables.EMOTION_BOREDOM;
            }
414     }
        else
416     {
            if (y>x){
```

```
418                 //boredom
                    typeOfEmotion = GlobalVariables.EMOTION_BOREDOM;
420             }
                else {
422                 //calmness
                    typeOfEmotion = GlobalVariables.EMOTION_CALMNESS;
424             }
            }
426         return typeOfEmotion;
        }
428
        /**
430      * checks if an emotion is "negative"
         * @param typeOfEmotion an int representing the type of emotion
432      * @return true if emotion is negative / false if it isnt
         */
434     private boolean emotionIsNegative(int typeOfEmotion) {
            boolean emotionIsNegative = false;
436         for (int i = 0; i<GlobalVariables.NEGATIVE_EMOTIONS.length; i++) {
                if (typeOfEmotion == GlobalVariables.NEGATIVE_EMOTIONS[i]) {
438                 emotionIsNegative = true;
                    break;
440             }
            }
442         return emotionIsNegative;
        }
444
        /**
446      * This method calculates the euclidean distance between the user
        feedback and the neural network output
         */
448     private double computeEuclideanDistance (double[] output, double[]
        idealOutput) {
            double euclideanDistance = 0;
450         for (int i = 0; i<output.length; i++) {
                euclideanDistance += Math.pow((output[i]-idealOutput[i]),2);
452         }
            euclideanDistance = Math.sqrt(euclideanDistance);
454         return euclideanDistance;
        }
456
        /**
458      * Sends output emotion to the server
         * @param outputs - the emotion output to be sent
460      */
        private void sendEmotionToServer(double[] outputs) {
462         LatLng currentPos = hWServ.getHwLocationListener().
        getActualposition();
            hWServiceHandler.post(new TaskSendEmotion(
464                 GlobalVariables.UUID,
                    outputs[GlobalVariables.
        NN_OUTPUT_ARRAY_INDEX_EUPHORIC_BORED],
466                 outputs[GlobalVariables.NN_OUTPUT_ARRAY_INDEX_ANXIOUS_CALM
        ],
                    currentPos.latitude,
468                 currentPos.longitude));
        }
470
        class EmotionRecognitionTask implements Runnable{
```

```
472         private double[] outputs;
            private double[] inputs;
474
           @Override
476         public void run() {
                //Only run if we have location, since we need it for the
        neural net!
478             if (hWServ.getHwLocationListener().getActualposition() != null
        ) {
                    try {
480                     /* Check if it is time to request user feedback */
                        if((lastEmoFeedbackReq + baseTimeToNextEmoFdbckReq) <
        System.currentTimeMillis()) {
482                         fetchInputsAndCompute();
                            //only fire notification if the service is still
        running!
484                         if (hWServ.isRunning()) {
                                showEmotionFeedbackNotification(inputs,
        outputs);
486                             postNotificationRemovalTask();
                            }
488                     } else {
                            /* first, check if our neural network has been
        behaving well enough for us to consider its output */
490                         float maxEuclideanDistance = (float) Math.sqrt(
        GlobalVariables.NN_OUTPUTS);
                            if (wMeanEuclideanDistance < GlobalVariables.
        MARGIN_PERCNT_MAX_EUCLD_DIST_ACCPT * maxEuclideanDistance) {
492                             fetchInputsAndCompute();
                                //fire a regular emotion notification, in case
         emotion is negative
494                             int typeOfEmotion = getTypeOfEmotion(outputs);
                                //only fire notification if the service is
         still running!
496                             if (hWServ.isRunning()) {
                                    if (emotionIsNegative(typeOfEmotion)) {
498                                     showNormalEmotionNotification(
        typeOfEmotion);
                                        postNotificationRemovalTask();
500                                 }
                                    sendEmotionToServer(outputs);
502                             }
                            }
504                     }
                    } catch (Exception e) {
506                     e.printStackTrace();
                    }
508             }
                //post a new emotion inference
510             scheduleEmotionRecog(System.currentTimeMillis() +
         nextEmotionExecutionMillis());
                /* Since we have posted our nex emotion inference, we
512             no longer need to keep the device's CPU awake. */
                wakeLock.release();
514         }

516         /**
             * Schedules the next emotion recognition through the
518          * AlarmManager class
```

```
         * @param triggerAtMillis - Time when the alarm should trigger
520      */
        protected void scheduleEmotionRecog(long triggerAtMillis) {
522         // Construct an intent that will execute the AlarmReceiver
            Intent intent = new Intent(hWServ.getApplicationContext(),
    EmotionWakefulReceiver.class);
524         // Create a PendingIntent to be triggered when the alarm goes
    off
            final PendingIntent pIntent = PendingIntent.getBroadcast(
    hWServ, EmotionWakefulReceiver.REQUEST_CODE, intent, PendingIntent.
    FLAG_UPDATE_CURRENT);
526         AlarmManager alarm = (AlarmManager) hWServ.getSystemService(
    hWServ.ALARM_SERVICE);
            // First parameter uses the wall clock time in UTC
528         // Interval is calculated based on nextEmotionExecutionMillis
            alarm.set(AlarmManager.RTC_WAKEUP, triggerAtMillis, pIntent);
530     }

532     protected void cancelEmotionRecog() {
            Intent intent = new Intent(hWServ.getApplicationContext(),
    EmotionWakefulReceiver.class);
534         final PendingIntent pIntent = PendingIntent.getBroadcast(
    hWServ, EmotionWakefulReceiver.REQUEST_CODE, intent, PendingIntent.
    FLAG_UPDATE_CURRENT);
            AlarmManager alarm = (AlarmManager) hWServ.getSystemService(
    hWServ.ALARM_SERVICE);
536         alarm.cancel(pIntent);
        }
538
        private void postNotificationRemovalTask() {
540         /* post a notificationRemovalTask, which will revert the
    notification in case the user takes too long to provide input.
            It runs a little after the expected expiration time.
542         */
            currentNotifRemovTask = new NotificationRemovalTask();
544         hWServiceHandler.postDelayed(currentNotifRemovTask, (long) (
    GlobalVariables.EXPIRE_EMOTION_MILLIS *1.05));
        }
546
        private void fetchInputsAndCompute() throws NoCurrentPosition {
548         outputs = new double[GlobalVariables.NN_OUTPUTS];
            inputs = collectInputs();
550         //compute the emotion
            network.compute(inputs, outputs);
552     }
    }
554
    /** This runnable task reverts our notification to its default
556  * state in case the inferred emotion has already expired.
     */
558 class NotificationRemovalTask implements Runnable {
        @Override
560     public void run() {
            if (System.currentTimeMillis() - lastEmotionNotifMillis >
    GlobalVariables.EXPIRE_EMOTION_MILLIS) {
562             //revert the notification
                Log.d("EMOTION NOTIFICATION","lastEmotioNotifMillis shows
    that current notification has expired. Reverting...");
564             hWServ.showNotification(false);
```

```
            }
566     }
    }
568
    /**
570  * Trains the neural network based on the Resilient Propagation
    heuristic
     */
572 class NeuralNetworkTrainingTask implements Runnable {
        private double[] inputs;
574     private double[] idealOutput;

576     public NeuralNetworkTrainingTask(double[] inputs, double[]
    idealOutput) {
            this.inputs = inputs;
578         this.idealOutput = idealOutput;
        }
580
        @Override
582     public void run() {
            double[][] trainingInput = {inputs};
584         double[][] trainingIdealOutput = {idealOutput};
            BasicMLDataSet trainingSet = new BasicMLDataSet(trainingInput,
     trainingIdealOutput);
586         ResilientPropagation rProp = new ResilientPropagation(network,
    trainingSet);
            //train the network
588         do {
                rProp.iteration();
590         } while(rProp.getError() >= GlobalVariables.
    NN_MAX_TRAINING_ERROR);

592         //save the new weights
            saveNNWeightsToPreferences(network.dumpWeights());
594     }
    }
596
    /**
598  * This class is responsible for updating the emotion accuracy
    feedback times and previous euclidean distance values
     */
600 class UpdateEmotionAccuracyTask implements Runnable {
        private double[] output;
602     private double[] idealOutput;

604     public UpdateEmotionAccuracyTask(double[] output, double[]
    idealOutput) {
            if (output.length != idealOutput.length) {
606             throw new AssertionError("output and idealOutput are of
    different sizes");
            }
608         else {
                this.output = output;
610             this.idealOutput = idealOutput;
            }
612     }

614     @Override
        public void run() {
```

```
616            /*calculate the euclidean distance
               This gives us an estimate on how accurate our last inference
        was. */
618            double euclideanDistance = computeEuclideanDistance(output,
        idealOutput);
               /*
620            This weighted mean will be used to check if our neural network
               is performing well enough to trigger notifications and send
        information to the server.
622            */
               wMeanEuclideanDistance = (float) (euclideanDistance *
        GlobalVariables.WEIGHT_OF_NEW_EUCLIDEAN_DISTANCE +
624                    wMeanEuclideanDistance * (1-GlobalVariables.
        WEIGHT_OF_NEW_EUCLIDEAN_DISTANCE));

626            float maxEuclideanDistance = (float) Math.sqrt(GlobalVariables
        .NN_OUTPUTS);
               /*
628            we compute a new feedback time through a direct linear
        variation
               based on the weighted mean of the euclidean distance
630
               number of milliseconds in an hour = 3600000
632             */
               long newFeedbackTime = (long) ((GlobalVariables.
        RECOG_MAX_HOURS_WITHOUT_FEEDBACK -
634                    ((GlobalVariables.RECOG_MAX_HOURS_WITHOUT_FEEDBACK -
        GlobalVariables.RECOG_MIN_HOURS_WITHOUT_FEEDBACK)*
                               wMeanEuclideanDistance/maxEuclideanDistance))
        *3600000);
636
               /* update the feedback time through a weighted arithmetic mean
        , with a bit of randomization */
638            baseTimeToNextEmoFdbckReq = (long) (newFeedbackTime *
        GlobalVariables.WEIGHT_OF_NEW_EMOTION_FEEDBACK_TIME +
                       baseTimeToNextEmoFdbckReq * (1-GlobalVariables.
        WEIGHT_OF_NEW_EMOTION_FEEDBACK_TIME));
640            Random rand = new Random();
               long margin = (long) (baseTimeToNextEmoFdbckReq*
        GlobalVariables.MARGIN_PERCNT_RANDOM_EMO_FDBCK_TIME);
642            //the final value will oscillate between baseValue -/+ (margin
        /2)
               baseTimeToNextEmoFdbckReq = (baseTimeToNextEmoFdbckReq-(margin
        /2)) +
644                    ((long) (rand.nextDouble()*margin));

646            /* update the last emotion feedback timestamp */
               lastEmoFeedbackReq = System.currentTimeMillis();
648        }
       }
650 }
```

References

1 R. Want, K. P. Fishkin, A. Gujar, and B. L. Harrison, "Bridging physical and virtual worlds with electronic tags," 1999.

2 P. Schramm, E. Naroska, P. Resch, J. Platte, and H. Linde, "Integration of limited servers into pervasive computing environments using dynamic gateway services," 2007.

3 T. Kindberg, J. Barton, J. Morgan, G. Becker, D. Caswell, P. Debaty, G. Gopal, M. Frid, V. Krishnan, H. Morris, J. Schettino, B. Serra, and M. Spasojevic, "People, places, things: Web presence for the real world," *Mob. Netw. Appl.*, vol. 7, pp. 365–376, October 2002. [Online]. Available: http://dx.doi.org/10.1023/A:1016591616731.

4 B. Traversat, M. Abdelaziz, D. Doolin, M. Duigou, J.-C. Hugly, and E. Pouyoul, "Project JXTA-C: Enabling a web of things," in *Proceedings of the 36th Annual Hawaii International Conference on System Sciences (HICSS'03)*, Washington, DC, USA: IEEE Computer Society, 2003. [Online]. Available: http://portal.acm.org/citation.cfm?id=820756.821825.

5 D. Guinard, V. Trifa, and E. Wilde, "A resource oriented architecture for the web of things," in *Proceedings of IoT 2010 (International Conference on the Internet of Things)*, Tokyo, Japan, November 2010.

6 X. Jiang, S. Dawson-Haggerty, and D. Culler, "smap: simple monitoring and actuation profile," in *Proceedings of the 9th ACM/IEEE International Conference on Information Processing in Sensor Networks*. New York, USA: ACM, 2010, pp. 374–375. [Online]. Available: http://doi.acm.org/10.1145/1791212.1791261.

7 L. Luo, A. Kansal, S. Nath, and F. Zhao, "Sharing and exploring sensor streams over geocentric interfaces," in *Proceedings of the 16th ACM SIGSPATIAL International Conference on Advances in Geographic Information Systems*. New York, USA: ACM, 2008, pp. 3:1–3:10. [Online]. Available: http://doi.acm.org/10.1145/1463434.1463439.

8 F. Calabrese, K. Kloeckl, and C. Ratti, "Wikicity: Real-time location-sensitive tools for the city," *Handbook of research on Urban Informatics: The practice and promise of the real-time city*, pp. 390–413, 2008.

9 A. Wood, L. Selavo, and J. Stankovic, "Senq: An embedded query system for streaming data in heterogeneous interactive wireless sensor networks," in *Distributed Computing in Sensor Systems*, Lecture Notes in Computer Science,

A Practical Introduction to Human-in-the-Loop Cyber-Physical Systems, First Edition.
David Nunes, Jorge Sá Silva and Fernando Boavida.

Companion Website URL: www.wiley.com/go/nunesloop

S. Nikoletseas, B. Chlebus, D. Johnson, and B. Krishnamachari, (eds), Springer, Berlin/Heidelberg, 2008, vol. 5067, pp. 531–543.

10 M. Musolesi, E. Miluzzo, N. D. Lane, S. B. Eisenman, T. Choudhury, and A. T. Campbell, "The second life of a sensor: Integrating real-world experience in virtual worlds using mobile phones," in *In Proc. of HotEmNets 08*, 2008.

11 N. Lathia, V. Pejovic, K. K. Rachuri, C. Mascolo, M. Musolesi, and P. J. Rentfrow, "Smartphones for large-scale behavior change interventions," *IEEE Pervasive Computing*, vol. 12, no. 3, pp. 66–73, 2013.

12 K. K. Rachuri, C. Mascolo, M. Musolesi, and P. J. Rentfrow, "Sociablesense: Exploring the trade-offs of adaptive sampling and computation offloading for social sensing," in *Proceedings of the 17th Annual International Conference on Mobile Computing and Networking*, New York, USA: ACM, 2011, pp. 73–84. [Online]. Available: http://doi.acm.org/10.1145/2030613.2030623.

13 C.-J. M. Liang, J. Stankovic, and S. Lin, "Reducing energy waste for computers by human-in-the-loop control," *IEEE Transactions on Emerging Topics in Computing*, vol. 99, no. PrePrints, p. 1, 2013.

14 Y. Agarwal, B. Balaji, S. Dutta, R. K. Gupta, and T. Weng, "Duty-cycling buildings aggressively: The next frontier in hvac control," in *2011 10th International Conference on. Information Processing in Sensor Networks (IPSN)*, IEEE, 2011, pp. 246–257, 2011.

15 M. Boulos, A. Rocha, A. Martins, M. Vicente, A. Bolz, R. Feld, I. Tchoudovski, M. Braecklein, J. Nelson, G. O Laighin, C. Sdogati, F. Cesaroni, M. Antomarini, A. Jobes, and M. Kinirons, "CAALYX: A new generation of location-based services in healthcare," *International Journal of Health Geographics*, vol. 6, pp. 1–6, 2007.

16 G. Schirner, D. Erdogmus, K. Chowdhury, and T. Padir, "The future of human-in-the-loop cyber-physical systems," *Computer*, vol. 46, no. 1, pp. 36–45, 2013.

17 W.-H. Rho and S.-B. Cho, "Context-aware smartphone application category recommender system with modularized bayesian networks." in *10th International Conference on Natural Computation (ICNC 2014)*, pp. 775–779, 2014.

18 R. E. Guinness, "Beyond where to how: A machine learning approach for sensing mobility contexts using smartphone sensors," *Sensors*, vol. 15, no. 5, pp. 9962–9985, 2015.

19 "The 'only' coke machine on the Internet," https://www.cs.cmu.edu/coke/history_long.txt, 1982.

20 G. E. Moore *et al*, "Cramming more components onto integrated circuits," 1965.

21 P. Middleton, P. Kjeldsen, and J. Tully, "Forecast: The Internet of things, worldwide, 2013," Gartner, Tech. Rep., November 2013.

22 S. Dawson-Haggerty, X. Jiang, G. Tolle, J. Ortiz, and D. Culler, "smap: a simple measurement and actuation profile for physical information," in *Proceedings of the 8th ACM Conference on Embedded Networked Sensor Systems*, New York, USA: ACM, 2010, pp. 197–210. [Online]. Available: http://doi.acm.org/10.1145/1869983.1870003.

23 A. Santanche, S. Nath, J. Liu, B. Priyantha, and F. Zhao, "Senseweb: Browsing the physical world in real time," in *Demo Abstract*, Nashville, TN, April 2006.

24 G. Montenegro, N. Kushalnagar, J. Hui, and D. Culler, "Transmission of IPv6 packets over IEEE 802.15. 4 Networks," Internet Engineering Task Force (IETF), Request for Comments 4944 (Proposed Standard), 2007.

25 J. Bryzek, "Roadmap for the trillion sensor universe," http://www-bsac.eecs.berkeley.edu/scripts/show_pdf_publication.php?pdfID=1365520205, April 2013.

26 M. Conner, "Sensors empower the 'Internet of things'", *EDN*, vol. 55, pp. 32–38, 2010.

27 "World Robotics 2015," International Federation of Robotics, 2015.

28 H. Sundmaeker, P. Guillemin, P. Friess, and S. Woelfflé, *Vision and challenges for realising the Internet of Things*. European Commision, Information Society and Media, 2010.

29 O. Vermesan and P. Friess, *Internet of Things: From Research and Innovation to Market Deployment*. River Publishers, 2014.

30 A. Koubâa and B. Andersson, "A vision of cyber-physical internet," in *Proc. of the Workshop of Real-Time Networks (RTN 2009), Satellite Workshop to (ECRTS 2009)*, 2009.

31 R. Baheti and H. Gill, "Cyber-physical systems," *The impact of control technology*, pp. 161–166, 2011.

32 S. Jeschke, "Everything 4.0? drivers and challenges of cyber physical systems," http://www.ima-zlw-ifu.rwth-aachen.de/fileadmin/user_upload/INSTITUTSCLUSTER/Publikation_Medien/Vortraege/download//Forschungsdialog4Dez2013.pdf, December 2013.

33 A. D. Wood and J. A. Stankovic, "Human in the loop: Distributed data streams for immersive cyber-physical systems," *SIGBED Rev.*, vol. 5, no. 1, pp. 20:1–20:2, January 2008.

34 P. Rawat, K. Singh, H. Chaouchi, and J. Bonnin, "Wireless sensor networks: A survey on recent developments and potential synergies," *The Journal of Supercomputing*, vol. 66, no. 1, pp. 1–48, October 2013.

35 M. Weiser, "The computer for the twenty-first century," *Scientific American*, vol. 265, no. 3, pp. 94–104, 1991.

36 P. Makris, D. N. Skoutas, and C. Skianis, "A survey on context-aware mobile and wireless networking: On networking and computing environments' integration," *Communications Surveys & Tutorials, IEEE*, vol. 15, no. 1, pp. 362–386, 2013.

37 C. Perera, A. Zaslavsky, P. Christen, and D. Georgakopoulos, "Context aware computing for the internet of things: A survey," *Communications Surveys & Tutorials, IEEE*, vol. 16, no. 1, pp. 414–454, 2014.

38 G. D. Abowd, A. K. Dey, P. J. Brown, N. Davies, M. Smith, and P. Steggles, "Towards a better understanding of context and context-awareness," in *Handheld and Ubiquitous Computing*. Springer, 1999, pp. 304–307.

39 S. H. Sigg and K. S David, "An alignment approach for context prediction tasks in ubicomp environments," *Pervasive Computing, IEEE*, 2010.

40 S. Fernandes and A. Karmouch, "Vertical mobility management architectures in wireless networks: A comprehensive survey and future directions," *Communications Surveys & Tutorials, IEEE*, vol. 14, no. 1, pp. 45–63, 2012.

41 A. Rahmati and L. Zhong, "Context-based network estimation for energy-efficient ubiquitous wireless connectivity," *Mobile Computing, IEEE Transactions on*, vol. 10, no. 1, pp. 54–66, 2011.

42 P. Bellavista, A. Corradi, M. Fanelli, and L. Foschini, "A survey of context data distribution for mobile ubiquitous systems," *ACM Computing Surveys (CSUR)*, vol. 44, no. 4, p. 24, 2012.

43 D. Niewolny, "How the Internet of things is revolutionizing healthcare," http://cache.freescale.com/files/corporate/doc/white_paper/IOTREVHEALCARWP.pdf, 2013.

44 H. Gao, A. Yuce, and J.-P. Thiran, "Detecting emotional stress from facial expressions for driving safety," in *International Conference on Image Processing (ICIP) 2014*, 2014.

45 S. Munir, J. A. Stankovic, C.-J. M. Liang, and S. Lin, "Cyber physical system challenges for human-in-the-loop control," in *Presented as part of the 8th International Workshop on Feedback Computing*. Berkeley, CA: USENIX, 2013.

46 N. B. Priyantha, A. Kansal, M. Goraczko, and F. Zhao, "Tiny web services: Design and implementation of interoperable and evolvable sensor networks," in *Proceedings of the 6th ACM Conference on Embedded network sensor systems*, New York, USA: ACM, 2008, pp. 253–266. [Online]. Available: http://doi.acm.org/10.1145/1460412.1460438.

47 D. Guinard, "Towards the web of things: Web mashups for embedded devices," in *In MEM 2009 in Proceedings of WWW 2009. ACM*, 2009.

48 Z. Shelby, K. Hartke, and C. Bormann, "The constrained application protocol (coap)," Internet Engineering Task Force (IETF) , Request for Comments 7252, June 2014. [Online]. Available: http://www.rfc-editor.org/rfc/rfc6275.txt.

49 MTA, "http://web.mta.info/developers/," http://web.mta.info/developers/, July 2014.

50 "Opendatabcn," http://opendata.bcn.cat/opendata/en/, July 2014.

51 "Open data: Toronto," http://www1.toronto.ca/wps/portal/contentonly?vgnextoid=9e56e03bb8d1e310VgnVCM10000071d60f89RCRD, July 2014.

52 "Edmonton's open data catalogue," https://data.edmonton.ca/, July 2014.

53 "Ottawa statistics," http://ottawa.ca/en/city-hall/get-know-your-city/statistics, July 2014.

54 "Vancouver open data catalogue," http://vancouver.ca/your-government/open-data-catalogue.aspx, July 2014.

55 G. Boone, "Reality mining: Browsing reality with sensor neworks," *Sensors Magazine*, vol. 9, 2004.

56 M. Srivastava, M. Hansen, J. Burke, A. Parker, S. Reddy, G. Saurabh, M. Allman, V. Paxson, and D. Estrin, "Wireless urban sensing systems," Center for Embedded Networked Sensing Systems, University of California, Los Angeles, Tech. Rep., 2006.

57 D. Gelles, "Yp, a mobile search firm, buys sense networks," http://dealbook.nytimes.com/2014/01/06/yp-a-mobile-ad-firm-buys-a-rival-sense-networks/?_r=0, January 2014.

58 J. Lifton, "Dual reality: An emerging medium," Ph.D. dissertation, Massachusetts Institute of Technology. Dept. of Architecture. Program in Media Arts and Sciences, 2007.

59 J. Lifton and J. A. Paradiso, "Dual reality: Merging the real and virtual," in *Proceedings of the First International ICST Conference on Facets of Virtual Environments (FaVE)*, Berlin, Germany, July 2009, pp. 27–29.

60 J. Lifton, M. Laibowitz, D. Harry, N.-W. Gong, M. Mittal, and J. A. Paradiso, "Metaphor and manifestation: Cross-reality with ubiquitous sensor/actuator networks," *IEEE Pervasive Computing*, vol. 8, pp. 24–33, July 2009. [Online]. Available: http://portal.acm.org/citation.cfm?id=1591886.1592128.

61 T. D. Tran and J. Silva, "A framework for integrating WSNs and external environments," in *5th International Conference on Management and Control of Production and Logistics*, Coimbra, Portugal, 2010.

62 ITU, "The world in 2011: ICT facts and figures," http://www.itu.int/ITU-D/ict/facts/2011/material/ICTFactsFigures2011.pdf, November 2011.

63 R. Duncombe and R. Boateng, "Mobile phones and financial services in developing countries: A review of concepts, methods, issues, evidence and future research directions," *Third World Quarterly*, vol. 30, no. 7, pp. 1237–1258, 2009.

64 J. James and M. Versteeg, "Mobile phones in Africa: How much do we really know?" *Social Indicators Research*, vol. 84, no. 1, pp. 117–126, 2007.

65 I. Demsky, "Cell phones help under-developed countries manage diseases," http://ur.umich.edu/1011/May23_11/2374-cell-phones-help, May 2011.

66 K. Fox, "Africa's mobile economic revolution," http://www.guardian.co.uk/technology/2011/jul/24/mobile-phones-africa-microfinance-farming, July 2011.

67 E. Brynjolfsson and A. McAfee, *The second machine age: Work, progress, and prosperity in a time of brilliant technologies*. WW Norton & Company, 2014.

68 M. Reardon, "Smartphones to outsell feature phones in 2013 for first time," http://www.cnet.com/news/smartphones-to-outsell-feature-phones-in-2013-for-first-time/, March 2013.

69 (2011, May) People are changing their facebook profile photo more often every year. Pixable Team. [Online]. Available: http://blog.pixable.com/2011/05/27/people-are-changing-their-facebook-profile-photo-more-often-every-year/.

70 (2011, February) Facebook photo trends. Pixable Team. [Online]. Available: http://blog.pixable.com/2011/02/14/facebook-photo-trends-infographic/.

71 "Social networking statistics," http://www.statisticbrain.com/social-networking-statistics/, July 2014.

72 "Facebook statistics," http://www.statisticbrain.com/facebook-statistics/, 2016.

73 "Statistics and facts about social networks," http://www.statista.com/topics/1164/social-networks/, October 2016.

74 K. M. Tsui, D.-J. Kim, A. Behal, D. Kontak, and H. A. Yanco, "I want that: Human-in-the-loop control of a wheelchair-mounted robotic arm," *Journal of Applied Bionics and Biomechanics*, vol. 8, 2011.

75 T. F. Sapata, "Look4mysounds: A remote monitoring platform for auscultation," Master's thesis, University of Coimbra, 2010.

76 I. F. Akyildiz, J. M. Jornet, and M. Pierobon, "Nanonetworks: A new frontier in communications," *Communications of the ACM*, vol. 54, no. 11, pp. 84–89, November 2011.

77 S. Pan, N. Wang, Y. Qian, I. Velibeyoglu, H. Y. Noh, and P. Zhang, "Indoor person identification through footstep induced structural vibration," in *Proceedings of the 16th International Workshop on Mobile Computing Systems and Applications*. ACM, 2015, pp. 81–86.

78 D. Wang, T. Abdelzaher, and L. Kaplan, *Social Sensing: Building Reliable Systems on Unreliable Data*. Morgan Kaufmann, 2015.

79 S. J. Stolfo, M. B. Salem, and A. D. Keromytis, "Fog computing: Mitigating insider data theft attacks in the cloud," in *Security and Privacy Workshops (SPW), 2012 IEEE Symposium*, 2012, pp. 125–128.

80 E. J. Hui and P. Thubert, "Compression format for IPv6 datagrams over IEEE 802.15.4-based networks," Internet Engineering Task Force (IETF), Request for Comments 6282, p. 28, September 2011.

81 Y. Song, Q. Hao, K. Zhang, M. Wang, Y. Chu, and B. Kang, "The simulation method of the galvanic coupling intrabody communication with different signal transmission paths." *IEEE T. Instrumentation and Measurement*, vol. 60, no. 4, pp. 1257–1266, 2011.

82 M. S. Wegmueller, M. Oberle, N. Felber, N. Kuster, and W. Fichtner, "Signal transmission by galvanic coupling through the human body." *IEEE T. Instrumentation and Measurement*, vol. 59, no. 4, pp. 963–969, 2010.

83 S. B. Eisenman, N. D. Lane, E. Miluzzo, R. A. Peterson, G.-S. Ahn, and A. T. Campbell, "Metrosense project: People-centric sensing at scale," in *WSW 2006 at Sensys*, 2006.

84 A. T. Campbell, S. B. Eisenman, N. D. Lane, E. Miluzzo, R. A. Peterson, H. Lu, X. Zheng, M. Musolesi, K. Fodor, and G.-S. Ahn, "The rise of people-centric sensing," *IEEE Internet Computing*, vol. 12, pp. 12–21, July 2008.

85 H. Lu, J. Yang, Z. Liu, N. D. Lane, T. Choudhury, and A. T. Campbell, "The jigsaw continuous sensing engine for mobile phone applications," in *Proceedings of the 8th ACM Conference on Embedded Networked Sensor Systems*. New York, USA: ACM, pp. 71–84, 2010.

86 H. Lu, W. Pan, N. D. Lane, T. Choudhury, and A. T. Campbell, "Soundsense: Scalable sound sensing for people-centric applications on mobile phones," in *Proceedings of the 7th International Conference on Mobile Systems, Applications, and Services*, New York, USA: ACM, 2009, pp. 165–178.

87 J. Lester, T. Choudhury, N. Kern, G. Borriello, and B. Hannaford, "A hybrid discriminative/generative approach for modeling human activities," in *Proceedings of the International Joint Conference on Artificial Intelligence (IJCAI)*, pp. 766–772, 2005.

88 S. Wang, W. Pentney, A.-M. Popescu, T. Choudhury, and M. Philipose, "Common sense based joint training of human activity recognizers," in *Proceedings of the 20th International Joint Conference on Artificial Intelligence*, pp. 2237–2242, 2007.

89 P. Zappi, "Activity recognition from on-body sensors by classifier fusion: Sensor scalability and robustness," *3rd International Conference on Intelligent Sensors, Sensor Networks and Information, 2007.*, pp. 281–286, 2007.

90 A. Jehad Sarkar, S. Lee, and Y.-K. Lee, "A smoothed naive bayes-based classifier for activity recognition," *IETE Technical Review*, vol. 27, no. 2, pp. 107–119, 2010.

91 S. Das, L. Green, B. Perez, and M. Murphy, "Detecting user activities using the accelerometer on Android smartphones," 2010.

92 E. M. Tapia, S. S. Intille, W. Haskell, K. Larson, J. Wright, A. King, and R. Friedman, "Real-time recognition of physical activities and their intensities using wireless accelerometers and a heart rate monitor," in *Proceedings of the 2007 11th IEEE International Symposium on Wearable Computers*, Washington, DC, USA: IEEE Computer Society, pp. 1–4, 2007.

93 N. D. Lane, Y. Xu, H. Lu, S. Hu, T. Choudhury, A. T. Campbell, and F. Zhao, "Enabling large-scale human activity inference on smartphones using community similarity networks (CSN)," in *UbiComp'11*, pp. 355–364, 2011.

94 G. Thatte, M. Li, S. Lee, A. Emken, S. S. Narayanan, U. Mitra, D. Spruijt-Metz, and M. Annavaram, "Knowme: An energy-efficient, multimodal body area network for physical activity monitoring," *ACM Transactions on Embedded Computing Systems*, 2010.

95 M. Berchtold, M. Budde, D. Gordon, H. R. Schmidtke, and M. Beigl, "Actiserv: Activity recognition service for mobile phones," in *ISWC'10*, pp. 1–8, 2010.

96 R. Wagenaar, I. Sapir, Y. Zhang, S. Markovic, and L. Vaina, "Continuous monitoring of functional activities using wearable, wireless gyroscope and accelerometer technology," *Conf Proc IEEE Eng Med Biol Soc. 2011*, pp. 4844–4847, 2011.

97 G. Bieber, A. Luthardt, C. Peter, and B. Urban, "The hearing trousers pocket: Activity recognition by alternative sensors," in *Proceedings of the 4th International Conference on Pervasive Technologies Related to Assistive Environments*, New York, USA: ACM, pp. 44:1–44:6, 2011.

98 J. Doyle, "Utilising mobile phone RSSI metric for human activity detection," *Signals and Systems Conference (ISSC 2009)*, pp. 1–6, 2009.

99 P. Rani, N. Sarkar, C. A. Smith, and J. A. Adams, "Affective communication for implicit human-machine interaction," in *Systems, Man and Cybernetics, 2003. IEEE International Conference on*, vol. 5. New York, pp. 4896–4903, 2003.

100 P. Rani, N. Sarkar, and J. Adams, "Anxiety-based affective communication for implicit human-machine interaction," *Adv. Eng. Inform.*, vol. 21, no. 3, pp. 323–334, Jul. 2007. [Online]. Available: http://dx.doi.org/10.1016/j.aei.2006.11.009.

101 J. M. Hektner, J. A. Schmidt, and M. Csikszentmihalyi, *Experience sampling method: Measuring the quality of everyday life*. Sage, 2007.

102 N. Lathia, K. K. Rachuri, C. Mascolo, and P. J. Rentfrow, "Contextual dissonance: Design bias in sensor-based experience sampling methods," in *Proceedings of the 2013 ACM International Joint Conference on Pervasive and Ubiquitous Computing* New York, USA: ACM, 2013, pp. 183–192. [Online]. Available: http://doi.acm.org/10.1145/2493432.2493452.

103 M. Liberman, K. Davis, M. Grossman, N. Martey, and J. Bell, "Emotional prosody speech and transcripts," 2002.

104 H.-J. Kim and Y. S. Choi, "Exploring emotional preference for smartphone applications," in *CCNC*. IEEE, pp. 245–249, 2012.

105 R. Li Kam Wa, Y. Liu, N. D. Lane, and L. Zhong, "Moodscope: Building a mood sensor from smartphone usage patterns," in *Proceeding of the 11th Annual International Conference on Mobile Systems, Applications, and Services*, New York, USA: ACM, pp. 389–402, 2013.

106 M. Kay, E. K. Choe, J. Shepherd, B. Greenstein, N. Watson, S. Consolvo, and J. A. Kientz, "Lullaby: A capture & access system for understanding the sleep environment," in *Proceedings of the 2012 ACM Conference on Ubiquitous Computing*, New York, USA: ACM, pp. 226–234, 2012.

107 K. Kelly, "Better than human: Why robots will—and must—take our jobs," http://www.wired.com/2012/12/ff-robots-will-take-our-jobs/all/, December 2012.

108 A. Weiss, D. Wurhofer, M. Lankes, and M. Tscheligi, "Autonomous vs. tele-operated: How people perceive human-robot collaboration with hrp-2," in

Proceedings of the 4th ACM/IEEE International Conference on Human Robot Interaction (New York, NY, USA), pp. 257–258, 2009.

109 C. Heyer, "Human-robot interaction and future industrial robotics applications," in *Proc. of IEEE/RSJ International Conference on Intelligent Robots and Systems*, Taiwan, 2010.

110 G. Amato, M. Broxvallx, S. Chessa, M. Dragone, C. Gennaro, and C. Vairo, "When wireless sensor networks meet robots," in *ICSNC 2012, The Seventh International Conference on Systems and Networks Communications*, pp. 35–40, 2012.

111 G.-J. M. Kruijff, F. Colas, T. Svoboda, J. van Diggelen, P. Balmer, F. Pirri, and R. Worst, "Designing intelligent robots for human-robot teaming in urban search and rescue," in *AAAI Spring Symposium: Designing Intelligent Robots*, 2012.

112 S. Haddadin, A. Albu-Schäffer, and G. Hirzinger, "Requirements for safe robots: Measurements, analysis and new insights," *The International Journal of Robotics Research*, vol. 28, no. 11–12, pp. 1507–1527, 2009.

113 E. A. Sisbot and R. Alami, "A human-aware manipulation planner," *Robotics, IEEE Transactions on*, vol. 28, no. 5, pp. 1045–1057, 2012.

114 S. Lallée, U. Pattacini, S. Lemaignan, A. Lenz, C. Melhuish, L. Natale, S. Skachek, K. Hamann, J. Steinwender, E. A. Sisbot *et al.*, "Towards a platform-independent cooperative human robot interaction system: III an architecture for learning and executing actions and shared plans," *Autonomous Mental Development, IEEE Transactions on*, vol. 4, no. 3, pp. 239–253, 2012.

115 M. Dorigo, D. Floreano, L. M. Gambardella, F. Mondada, S. Nolfi, T. Baaboura, M. Birattari, M. Bonani, M. Brambilla, A. Brutschy *et al.*, "Swarmanoid," *IEEE Robotics & Automation Magazine*, vol. 1070, no. 9932/13, 2013.

116 F. Schlachter, E. Meister, S. Kernbach, and P. Levi, "Evolve-ability of the robot platform in the symbrion project," in *Second IEEE International Conference on. Self-Adaptive and Self-Organizing Systems Workshops, 2008. SASOW 2008*. IEEE, 2008, pp. 144–149, 2008.

117 N. Casiddu, F. Cavallo, A. Divano, I. Mannari, E. Micheli, C. Porfirione, M. Zallio, M. Aquilano, and P. Dario, "Robot interface design of domestic and condominium robot for ageing population," in *Ambient Assisted Living: Italian Forum 2013*, Springer, pp. 53–60, 2014.

118 R. Schoenen and H. Yanikomeroglu, "User-in-the-loop: Spatial and temporal demand shaping for sustainable wireless networks," *Communications Magazine, IEEE*, vol. 52, no. 2, pp. 196–203, February 2014.

119 D. Nunes, J. S. Silva, C. Herrera, and F. Boavida, "Human-in-the-loop connectivity management in smartphones," in *International Conference on Wired/Wireless Internet Communication*, Springer, pp. 159–170, 2016.

120 K. Yano, S. Lyubomirsky, and J. Chancellor. (2012, December) Can technology make you happy? [Online]. Available: http://spectrum.ieee.org/at-work/innovation/can-technology-make-you-happy.

121 A. Waibel, R. Stiefelhagen, R. Carlson, J. Casas, J. Kleindienst, L. Lamel, O. Lanz, D. Mostefa, M. Omologo, F. Pianesi, L. Polymenakos, G. Potamianos, J. Soldatos, G. Sutschet, and J. Terken, *Handbook of Ambient Intelligence and Smart Environments*, Springer, New York, US Computers in the Human Interaction Loop, pp. 1071–1116, 2010.

122 J. Lu, T. Sookoor, V. Srinivasan, G. Gao, B. Holben, J. Stankovic, E. Field, and K. Whitehouse, "The smart thermostat: Using occupancy sensors to save energy in homes," in *Proceedings of the 8th ACM Conference on Embedded Networked Sensor Systems*, ser. SenSys '10, New York, USA: ACM, pp. 211–224, 2010. [Online]. Available: http://doi.acm.org/10.1145/1869983.1870005.

123 M. Boulos and A. Anastasiou, "A complete ambient assisted living experiment (caalyx) in second life (r)," in *Proceedings of MedNet2008: The 13th World Congress on the Internet in Medicine*, Saint Petersburg, Russia, October 2008, pp. 4–5.

124 R. S. Desmond, M. F. Dickerman, and J. A. Fleming, "A human-in-the-loop cyber physical system: Modular designs for semi-autonomous wheelchair navigation," Master's thesis, Worcester Polytechnic Institute, 2013.

125 D.-J. Kim and A. Behal, "Human-in-the-loop control of an assistive robotic arm in unstructured environments for spinal cord injured users," in, *2009 4th ACM/IEEE International Conference on Human-Robot Interaction (HRI)*. IEEE, pp. 285–286, 2009.

126 D. Arney, M. Pajic, J. M. Goldman, I. Lee, R. Mangharam, and O. Sokolsky, "Toward patient safety in closed-loop medical device systems," in *Proceedings of the 1st ACM/IEEE International Conference on Cyber-Physical Systems*. ACM, pp. 139–148, 2010.

127 "Highlight," Math Camp, Inc., December 2015. [Online]. Available: http://highlig.ht/about.html.

128 R. Lawler, "Highlight app combines facebook and gps to make real-world connections," January 2012. [Online]. Available: http://gigaom.com/2012/01/24/highlight-app/.

129 K. Hill, "Using facial recognition technology to choose which bar to go to," SceneTap, LLC., September 2011. [Online]. Available: http://www.forbes.com/sites/kashmirhill/2011/06/28/using-facial-recognition-technology-to-choose-which-bar-to-go-to/.

130 M.-Y. Chen, M.-N. Wu, C.-C. Chen, Y.-L. Chen, and H.-E. Lin, "Recommendation-aware smartphone sensing system," *Journal of Applied Research and Technology*, vol. 12, no. 6, pp. 1040–1050, 2014.

131 Statista, "Number of apps available in leading app stores as of July 2015," http://www.statista.com/statistics/276623/number-of-apps-available-in-leading-app-stores/, July 2015.

132 B. Yan and G. Chen, "Appjoy: Personalized mobile application discovery," in *Proceedings of the 9th International Conference on Mobile Systems, Applications, and Services*. ACM, pp. 113–126, 2011.

133 M. G. Berman, E. Kross, K. M. Krpan, M. K. Askren, A. Burson, P. J. Deldin, S. Kaplan, L. Sherdell, I. H. Gotlib, and J. Jonides, "Interacting with nature improves cognition and affect for individuals with depression," *Journal of Affective Disorders*, vol. 140, no. 3, pp. 300–305, 2012.

134 M. G. Berman, J. Jonides, and S. Kaplan, "The cognitive benefits of interacting with nature," *Psychological Science*, vol. 19, no. 12, pp. 1207–1212, 2008.

135 N. Weinstein, A. K. Przybylski, and R. M. Ryan, "Can nature make us more caring? Effects of immersion in nature on intrinsic aspirations and generosity," *Personality and Social Psychology Bulletin*, vol. 35, no. 10, pp. 1315–1329, 2009.

136 S. Haykin, *Neural Networks: A Comprehensive Foundation*, 2nd ed. Upper Saddle River, NJ, USA: Prentice Hall PTR, 1998.

137 H. Gunes and M. Pantic, "Automatic, dimensional and continuous emotion recognition," *International Journal of Synthetic Emotions*, vol. 1, no. 1, pp. 68–99, January 2010.

138 B. Osgood, "The Fourier transform and its applications," *Lecture Notes for EE*, vol. 261, p. 20, 2009.

139 J. W. Cooley, P. A. Lewis, and P. D. Welch, "The fast Fourier transform and its applications," *IEEE Transactions on Education*, vol. 12, no. 1, pp. 27–34, 1969.

140 J. Heaton, "Encog: Library of interchangeable machine learning models for java and c#," *Journal of Machine Learning Research*, vol. 16, pp. 1243–1247, 2015. [Online]. Available: http://jmlr.org/papers/v16/heaton15a.html.

141 K. G. Sheela and S. Deepa, "Review on methods to fix number of hidden neurons in neural networks," *Mathematical Problems in Engineering*, vol. 2013, 2013.

142 N. Lathia, K. Rachuri, C. Mascolo, and G. Roussos, "Open source smartphone libraries for computational social science," in *Proceedings of the 2013 ACM Conference on Pervasive and Ubiquitous Computing Adjunct Publication*, New York, USA: ACM, pp. 911–920, 2013. [Online]. Available: http://doi.acm.org/10.1145/2494091.2497345.

143 F. J. Oppermann, C. A. Boano, M. A. Zúniga, and K. Römer, "Automatic protocol configuration for dependable internet of things applications," in *Proceedings of the 10th IEEE International Workshop on Practical Issues in Building Sensor Network Applications (SenseApp)*, Clearwater Beach, FL, USA, October 2015.

144 S. W. Smith, "Humans in the loop: Human-computer interaction and security," *Security & Privacy, IEEE*, vol. 1, no. 3, pp. 75–79, 2003.

145 V. Kostakos and E. O'Neill, "Human-in-the-loop: Rethinking security in mobile and pervasive systems," in *CHI'08 Extended Abstracts on Human Factors in Computing Systems*. ACM, pp. 3075–3080, 2008.

146 L. F. Cranor, "A framework for reasoning about the human in the loop." *UPSEC*, vol. 8, pp. 1–15, 2008.

147 W. Qadeer, T. S. Rosing, J. Ankcorn, V. Krishnan, and G. D. Micheli, "Heterogeneous wireless network management," in *In PACS (2003)*, Springer, pp. 86–100, 2003.

148 (2015, May) Cisco visual networking index: Forecast and methodology, 2014–2019.

149 A. Balasubramanian, R. Mahajan, and A. Venkataramani, "Augmenting mobile 3G using wifi," in *Proceedings of the 8th international conference on Mobile systems, applications, and services*, New York, USA: ACM, pp. 209–222, 2010.

150 T. Pering, Y. Agarwal, R. Gupta, and R. Want, "Coolspots: Reducing the power consumption of wireless mobile devices with multiple radio interfaces," in *Proceedings of the 4th International Conference on Mobile Systems, Applications and Services*, New York, USA: ACM, pp. 220–232, 2006.

151 A. Rahmati and L. Zhong, "Context-for-wireless: Context-sensitive energy-efficient wireless data transfer," in *Proceedings of the 5th International Conference on Mobile Systems, Applications and Services*, New York, USA: ACM, pp. 165–178, 2007.

152 *IEEE Standard for Local and Metropolitan Area Networks: Part 21 Media Independent Handover Services*, IEEE Std 802.21-2008, Computer Society Ltd., January 2009.

153 E. Piri and K. Pentikousis, "IEEE 802.21," *The Internet Protocol Journal*, vol. 12, no. 2, pp. 7–27, June 2009.
154 A. Fladenmuller and R. De Silva, "The effect of mobile IP handoffs on the performance of TCP," *Mobile Networks and Applications*, vol. 4, no. 2, pp. 131–135, May 1999.
155 A. Rahmati, C. Shepard, C. Tossell, A. Nicoara, L. Zhong, P. T. Kortum, and J. P. Singh, "Seamless flow migration on smartphones without network support," *CoRR*, vol. abs/1012.3071, 2010.
156 C. Perkins, "Mobile IP," *IEEE Wireless Communications Magazine*, vol. 35, no. 5, pp. 84–99, 1997.
157 D. B. Johnson, C. E. Perkins, and J. Arkko, "Mobility support in IPv6," Internet Engineering Task Force (IETF), Request for Comments 6275, July 2011. [Online]. Available: http://www.rfc-editor.org/rfc/rfc6275.txt.
158 R. Koodli, "Mobile IPv6 fast handovers," Internet Engineering Task Force (IETF) , Request for Comments 5568, p. 28, July 2009.
159 S. Zaki and S. Razak, "Mitigating packet loss in mobile IPv6 using two-tier buffer scheme," *International Journal of Computer Science Letters*, vol. 3, no. 2, pp. 1–10, June 2011.
160 K. AI-Farabi and M. Kabir, "Reducing packet loss in mobile IPv6," in *14th International Conference on Computer and Information Technology*, pp. 38–43, 2011.
161 H. Soliman, C. Castelluccia, K. ElMalki, and L. Bellier, "Hierarchical mobile IPv6 (hmIPv6) mobility management," Internet Engineering Task Force (IETF), Request for Comments 5380, p. 28, October 2008.
162 C. Bernardos, "Proxy mobile IPv6 extensions to support flow mobility," NETEXT Working Group, Internet-Draft, October 2013.
163 S. Nirjon, A. Nicoara, C.-H. Hsu, J. Singh, and J. Stankovic, "Multinets: Policy oriented real-time switching of wireless interfaces on mobile devices," *Real-Time and Embedded Technology and Applications Symposium, IEEE*, vol. 0, pp. 251–260, 2012.
164 A. Ford, C. Raiciu, M. Handley, S. Barre, and J. Iyengar, "Architectural guidelines for multipath TCP development," Internet Engineering Task Force (IETF), Request for Comments 6182, p. 28, March 2011.
165 R. Stewart, "Stream control transmission protocol," Internet Engineering Task Force (IETF), Request for Comments 4960 (Proposed Standard), Internet Engineering Task Force, September 2007. [Online]. Available: http://www.ietf.org/rfc/rfc4960.txt.
166 A. A. E. Al, T. Saadawi, and M. Lee, "LS-SCTYP: A bandwidth aggregation technique for stream control transmission protocol," *Computer Communications*, vol. 27, no. 10, pp. 1012–1024, 2004.
167 J. Liao, J. Wang, and X. Zhu, "A multi-path mechanism for reliable voip transmission over wireless networks," *Computer Networks*, vol. 52, no. 13, pp. 2450–2460, 2008.
168 P. D. Amer, M. Becke, T. Dreibholz, N. Ekiz, J. R. Iyengar, P. Natarajan, R. R. Stewart, and M. Tuexen, "Load sharing for the stream control transmission protocol (SCTP)," IETF, Network Working Group, Internet Draft Version 07, Oct. 2013, draft-tuexen-tsvwg-sctp-multipath-07.txt, work in progress. [Online]. Available: http://tools.ietf.org/id/draft-tuexen-tsvwg-sctp-multipath-07.txt.

169 I. van Beijnum, "Multipath TCP," *IETF Journal*, vol. 5, no. 2, pp. 1, 8–10, September 2009. [Online]. Available: http://www.internetsociety.org/articles/multipath-tcp.
170 P. Hui, J. Crowcroft, and E. Yoneki, "Bubble rap: Social-based forwarding in delay-tolerant networks," *Mobile Computing, IEEE Transactions on*, vol. 10, no. 11, pp. 1576–1589, 2011.
171 J. Rodriguez, *Fundamentals of 5G Mobile Networks*. John Wiley & Sons, Ltd, 2015.
172 A. El-Rabbany, *Introduction to GPS: The Global Positioning System*. Artech House, 2002.
173 J. Liu, B. Priyantha, T. Hart, H. S. Ramos, A. A. F. Loureiro, and Q. Wang, "Energy efficient GPS sensing with cloud offloading," in *Proceedings of the 10th ACM Conference on Embedded Network Sensor Systems*, New York, USA: ACM, pp. 85–98, 2012. [Online]. Available: http://doi.acm.org/10.1145/2426656.2426666.
174 N. B. Priyantha, A. Chakraborty, and H. Balakrishnan, "The cricket location-support system," in *Proceedings of the 6th Annual International Conference on Mobile Computing and Networking*, New York, USA: ACM, pp. 32–43, 2000. [Online]. Available: http://doi.acm.org/10.1145/345910.345917.
175 Y.-C. Cheng, Y. Chawathe, A. LaMarca, and J. Krumm, "Accuracy characterization for metropolitan-scale wi-fi localization," in *Proceedings of the 3rd International Conference on Mobile Systems, Applications, and Services*, New York, USA: ACM, pp. 233–245, 2005. [Online]. Available: http://doi.acm.org/10.1145/1067170.1067195.
176 M. A. Caceres, F. Sottile, and M. A. Spirito, "Adaptive location tracking by kalman filter in wireless sensor networks," in *Proceedings of the 2009 IEEE International Conference on Wireless and Mobile Computing, Networking and Communications*, Washington, DC, USA: IEEE Computer Society, pp. 123–128, 2009. [Online]. Available: http://dx.doi.org/10.1109/WiMob.2009.30.
177 A. Harter, A. Hopper, P. Steggles, A. Ward, and P. Webster, "The anatomy of a context-aware application," in *Proceedings of the 5th Annual ACM/IEEE International Conference on Mobile Computing and Networking*, New York, USA: ACM, pp. 59–68, 1999. [Online]. Available: http://doi.acm.org/10.1145/313451.313476.
178 C.-L. Wu, L.-C. Fu, and F.-L. Lian, "WLAN location determination in e-home via support vector classification," in *2004 IEEE international conference on Networking, Sensing and Control*, vol. 2, pp. 1026–1031, 2004.
179 A. Shareef, Y. Zhu, and M. Musavi, "Localization using neural networks in wireless sensor networks," in *Proceedings of the 1st International Conference on MOBILe Wireless MiddleWARE, Operating Systems, and Applications*, ICST, Brussels, Belgium, Belgium: ICST (Institute for Computer Sciences, Social-Informatics and Telecommunications Engineering), pp. 4:1–4:7, 2007. [Online]. Available: http://dl.acm.org/citation.cfm?id=1361492.1361497.
180 B. Longstaff, S. Reddy, and D. Estrin, "Improving activity classification for health applications on mobile devices using active and semi-supervised learning," in *PervasiveHealth'10*, pp. 1–7, 2010.
181 A. Ortony, *The cognitive structure of emotions*. Cambridge University Press, 1990.
182 P. Ekman and W. V. Friesen, "Facial action coding system," http://face-and-emotion.com/dataface/facs/new_version.jsp, 2002.

183 J. A. Russell, "A circumplex model of affect," *Journal of Personality and Social Psychology*, vol. 39, no. 6, p. 1161, 1980.

184 R. S. Lazarus, *Emotion and adaptation.* Oxford University Press, 1991.

185 "Emotiv epoc," http://emotiv.com/, April 2014.

186 R. A. Brooks, "A robust layered control system for a mobile robot," *IEEE Journal of Robotics and Automation*, vol. 2, no. 1, pp. 14–23, 1986.

187 M. Musolesi, M. Piraccini, K. Fodor, A. Corradi, and A. T. Campbell, "Supporting energy-efficient uploading strategies for continuous sensing applications on mobile phones," in *Proceedings of the 8th International Conference on Pervasive Computing*, pp. 355–372, May 2010.

188 CNBC.com, "AP-CNBC Facebook IPO poll: Complete results & analysis," http://www.cnbc.com/id/47391504, May 2012.

189 E. C. Baig. (2012, March) Highlight app may tell others too much about you. [Online]. Available: http://www.usatoday.com/tech/columnist/edwardbaig/story/2012-03-20/highlight-app/53673820/1.

190 A. Robertson, "Crowd-detection app SceneTap tries to allay privacy fears after rocky San Francisco launch," May 2012. [Online]. Available: http://www.theverge.com/2012/5/18/3029229/scenetap-san-francisco-launch-backlash.

191 L. Anderson, "A night on the town with SceneTap," May 2012. [Online]. Available: http://www.theverge.com/2012/5/29/3043790/scene-tap-professional-pick-up-artist-smooth.

192 S. Das, L. Green, B. Perez, M. Murphy, and A. Perring, "Detecting user activities using the accelerometer on Android smartphones," *The Team for Research in Ubiquitous Secure Technology, TRUST-REU Carnefie Mellon University*, 2010.

193 G. Kasparov, "The chess master and the computer," *The New York Review of Books*, vol. 57, no. 2, pp. 16–19, 2010.

194 D. Rasskin-Gutman, *Chess metaphors: Artificial intelligence and the human mind.* MIT Press, 2009.

195 "Baxter," Rethink Robotics, December 2015. [Online]. Available: http://www.rethinkrobotics.com/baxter/.

196 "Pepper robot," Aldebaran Robotics, 2015. [Online]. Available: https://www.aldebaran.com/en/a-robots/who-is-pepper.

197 I. J. Good, "The mystery of Go," *New Scientist*, vol. 427, pp. 172–174, 1965.

198 D. Silver, A. Huang, C. J. Maddison, A. Guez, L. Sifre, G. van den Driessche, J. Schrittwieser, I. Antonoglou, V. Panneershelvam, M. Lanctot, S. Dieleman, D. Grewe, J. Nham, N. Kalchbrenner, I. Sutskever, T. Lillicrap, M. Leach, K. Kavukcuoglu, T. Graepel, and D. Hassabis, "Mastering the game of go with deep neural networks and tree search," *Nature*, vol. 529, no. 7587, pp. 484–489, 2016. [Online]. Available: http://dx.doi.org/10.1038/nature16961.

199 "Deepmind Alphago vs Lee Sedol," https://gogameguru.com/tag/deepmind-alphago-lee-sedol/, January 2017.

200 "Google's self-driving car project," https://www.google.com/selfdrivingcar/, September 2016.

201 J.-F. Bonnefon, A. Shariff, and I. Rahwan, "The social dilemma of autonomous vehicles," *Science*, vol. 352, no. 6293, pp. 1573–1576, 2016.

Index

A Practical Introduction to Human-in-the-Loop Cyber-Physical Systems, First Edition.
David Nunes, Jorge Sá Silva and Fernando Boavida.

Companion Website URL: www.wiley.com/go/nunesloop